Lipids and Biomembranes of Eukaryotic Microorganisms

CELL BIOLOGY: A Series of Monographs

EDITORS

D. E. BUETOW

Department of Physiology
and Biophysics
University of Illinois
Urbana, Illinois

I. L. CAMERON

Department of Anatomy
University of Texas
Medical School at San Antonio
San Antonio, Texas

G. M. PADILLA

Department of Physiology and Pharmacology
Duke University Medical Center
Durham, North Carolina

G. M. Padilla, G. L. Whitson, and I. L. Cameron (editors). THE CELL CYCLE: Gene-Enzyme Interactions, 1969

A. M. Zimmerman (editor). HIGH PRESSURE EFFECTS ON CELLULAR PROCESSES, 1970

I. L. Cameron and J. D. Thrasher (editors). CELLULAR AND MOLECULAR RENEWAL IN THE MAMMALIAN BODY, 1971

I. L. Cameron, G. M. Padilla, and A. M. Zimmerman (editors). DEVELOPMENTAL ASPECTS OF THE CELL CYCLE, 1971

P. F. Smith. THE BIOLOGY OF MYCOPLASMAS, 1971

Gary L. Whitson (editor). CONCEPTS IN RADIATION CELL BIOLOGY, 1972

Donald L. Hill. THE BIOCHEMISTRY AND PHYSIOLOGY OF *TETRA-HYMENA*, 1972

Kwang W. Jeon (editor). THE BIOLOGY OF AMOEBA, 1973

Dean F. Martin and George M. Padilla (editors). MARINE PHARMACOGNOSY: Action of Marine Biotoxins at the Cellular Level, 1973

Joseph A. Erwin (editor). LIPIDS AND BIOMEMBRANES OF EUKARYOTIC MICROORGANISMS, 1973

In preparation

A. M. Zimmerman, G. M. Padilla, and I. L. Cameron (editors). DRUGS AND THE CELL CYCLE

Stuart Coward (editor). CELL DIFFERENTIATION

LIPIDS AND BIOMEMBRANES OF EUKARYOTIC MICROORGANISMS

Edited by JOSEPH A. ERWIN

Department of Biology
Illinois Institute of Technology
Chicago, Illinois

 1973

ACADEMIC PRESS   New York and London

ACADEMIC PRESS, INC.
111 Fifth Avenue, New York, New York 10003

United Kingdom Edition published by
ACADEMIC PRESS, INC. (LONDON) LTD.
24/28 Oval Road, London NW1

LIBRARY OF CONGRESS CATALOG CARD NUMBER: 72-12211

PRINTED IN THE UNITED STATES OF AMERICA

Contents

List of Contributors

Numbers in parentheses indicate the pages on which the authors' contributions begin.

JOSEPH A. ERWIN (41), Department of Biology, Illinois Institute of Technology, Chicago, Illinois

G. S. GETZ (145), Departments of Pathology and Biochemistry, University of Chicago, Chicago, Illinois

T. W. GOODWIN (1), Department of Biochemistry, The University of Liverpool, Liverpool, England

THOMAS H. HAINES (197), Department of Chemistry, The City College of the City University of New York, New York, New York

S. HENRY* (259), Department of Genetics, University of California, Berkeley, California

A. D. KEITH† (259), Department of Genetics, University of California, Berkeley, California

D. MANGNALL (145), Departments of Pathology and Biochemistry, University of Chicago, Chicago, Illinois

ABRAHAM ROSENBERG (233), Department of Biological Chemistry, The Milton S. Hershey Medical Center, The Pennsylvania State University, Hershey, Pennsylvania

* Present address: Rosenstiel Basic Medical Sciences Research Center, Brandeis University, Waltham, Massachusetts

† Present address: Department of Biophysics, The Pennsylvania State University, University Park, Pennsylvania

J. C. WILLIAMS* (259), Department of Genetics, University of California, Berkeley, California

B. J. WISNIESKI† (259), Department of Genetics, University of California, Berkeley, California

    * Present address: Department of Biological Chemistry, Washington University School of Medicine, St. Louis, Missouri
    † Present address: Department of Bacteriology, University of California, Los Angeles, California

Preface

One of the most active fields of current research involves an attempt to construct a molecular biology of membranes. In this endeavor microbial systems are emerging as useful research tools, which they have been in other areas of biochemistry and cell biology. One universal fact has emerged from these studies: Membranes are composed of two types of molecules, proteins and lipids. While little has been learned in the area of membrane proteins in microorganisms, a great deal is now known about their lipids. However, the bulk of the literature on microbial lipids is both scattered and of relatively recent vintage; hence a book that synthesizes our knowledge of microbial lipids and relates this knowledge to microbial membranes is both timely and useful. This volume attempts such a synthesis for eukaryotic microorganisms.

Why does this study restrict itself to eukaryotic microorganisms? In part, practical considerations demanded a focus that was more contained than the entire microbial world. The recent literature on bacterial lipids and their possible function in membranes is so extensive that it would demand a separate volume.

While the upper limits of the microbial world cannot be easily defined (mushrooms and seaweeds, for example, could be classified as "microbial" only out of courtesy), microbial organisms fall into two very distinct categories; the prokaryotic, consisting of the bacteria and the blue-green algae, and the eukaryotic, consisting of all other algae plus yeasts, fungi, and protozoa (for a discussion of the two different cellular types, see Stanier *et al.*, 1963 and Stanier, 1970).* This distinction appears to be one of the most fundamental to be found in the biological realm, and it is to a large degree based upon the subject matter of this book, membranes and lipids. Eukaryotic microorganisms, in common with the cells of higher plants and animals, characteristically contain a variety of intracellular membranous organelles,

---

* Stanier, R. Y. (1970). *Symp. Soc. Gen. Microbiol.* **20**, 1–38.
  Stanier, R. Y., Doudoroff, M., and Adelberg, E. A. (1963). "The Microbial World." Prentice-Hall, Englewood Cliffs, New Jersey.

many of which are self-reproducing. Such organelles are lacking in prokaryotes; the simple intracellular vesicles of photosynthetic bacteria and the mesosomes common to many bacteria are derivatives of the bacterial limiting or plasma membrane (Stanier, 1970). Even the plasma membrane of prokaryotes differs from that of eukaryotes. Endocytosis, for example (a term that embrances related phenomena such as pinocytosis, phagocytosis, and reversed or secretory pinocytosis), is a fundamental property of the plasma membranes of eukaryotes but is never found among prokaryotes (Stanier, 1970).

The chemical composition of eukaryotic and prokaryotic organisms is also strikingly different. Sterols and polyunsaturated fatty acids are ubiquitous components of all eukaryotic cells—plant, animal, and microbial (Chapters 1 and 2). In bacteria, however, polyunsaturated fatty acids are completely absent, and sterols, if present at all, are found only in trace amounts. [Blue-green algae are an intermediate group in that some species contain sterols and polyunsaturated fatty acids but others do not (Chapters 1 and 2).] Galactolipids and sulfolipids are the universal and principal components of the photosynthetic membranes of all eukaryotes but are absent in photosynthetic bacteria (see Chapter 5). Again, the glycerophospholipids—phosphatidyl choline, phosphatidyl serine, and phosphatidyl inositol—commonly found among eukaryotes (see Chapter 3) are rarely found among prokaryotes. In contrast, phosphatidyl glycerol, a major and common lipid component of prokaryotes, is either a minor component of eukaryotic lipids or absent altogether (Chapter 3). Similarly, lipids containing long-chain bases in lieu of glycerol (the ceramides and sphingolipids) are widespread among eukaryotes and are virtually never found in prokaryotes.

Thus different groups of eukaryotic microorganisms may display variations in membrane morphology and in lipid composition, but these are simply variations on a common theme characteristic of the entire eukaryotic world—plant, animal, and microbial. Hence the results of studies on the role of lipids in the formation and function of membranes in eukaryotic microorganisms are very likely to be applicable to the bulk of the biological world. We would have to employ considerably greater caution if we attempted to extrapolate the results of studies on the membrane systems of bacteria to the eukaryotic world, even though the apparent simplicity of the membrane systems of bacteria make them attractive as research tools.

In Chapters 1–4 each of the major classes of lipids—sterols, fatty acids, phospholipids, and sulfolipids—is considered separately. In each case an attempt has been made to provide a comprehensive summary and to evaluate critically the literature on the occurrence and biosynthesis of these compounds in yeasts, fungi, algae, and protozoa. Physiological functions of these lipids, particularly their role in the membranes of the organisms, are discussed. In some cases attention has been called to the possible usefulness of lipids as taxonomic criteria.

A reading of these chapters reveals that much has been learned about eukaryotic microbial lipids and their biosynthesis during the last decade or so, but obvious vacuums of information exist. Perhaps the most serious of these are the paucity of data on the lipid composition of specific types of membranes in various eukaryotic microorganisms and our lack of understanding of the regulatory mechanisms governing the control of different lipids in these organisms (see Chapter 3).

In Chapters 5 and 6 experimental systems for studying the relation between the structure of lipids and their function in biomembranes are discussed. These systems include the photosynthetic membranes in organisms such as *Euglena, Chlorella,* and *Chlamydomonas* in which the formation of the chloroplasts is susceptible to experimental control (Chapter 5) and fatty acid auxotrophic mutants of yeasts and *Neurospora* in which the fatty acid composition of the membrane lipids can be altered by the experimenter (Chapter 6).

It is hoped that this volume will prove useful to lipid biochemists, microbial physiologists, taxonomists, and cell biologists who are interested in the molecular aspects of biomembranes.

JOSEPH A. ERWIN

A Note on Taxonomy and Biological Nomenclature

The members of the classic major groups of eukaryotic microorganisms (algae, fungi, yeasts, and protozoa) are highly heterogeneous and their boundaries are indistinct, which has led to jurisdictional disputes among taxonomists; for example, a number of groups of unicellular and colonial flagellated microorganisms such as euglenids, chrysomonads, dinoflagellates, and cryptomonads contain both photosynthetic and nonphotosynthetic forms. All of these are often termed phytoflagellates and are assigned by zoologists to the phylum protozoa (Honigberg et al., 1964).* Botanists, on the other hand, usually classify these same organisms as algae (Smith, 1950).† Similarly, the slime molds are considered to be protozoa by many zoologists, but botanists classify them as fungi (Cronquist, 1971).** The fungi and the yeasts also present a taxonomic problem; it is not always possible to distinguish a true yeast from fungi that have yeastlike stages in their life cycle. It should be clear then that the designation of some groups of eukaryotic microorganisms as algae or protozoa or as yeasts or fungi is often arbitrary and largely a matter of personal taste. As editor of this volume I have made no attempt to impose my own tastes in this matter on the other authors. The nonbiologist should not be unduly upset to find the same organism called a protozoan in one chapter and an alga or fungus in the following chapter.

Microorganisms can also be catagorized on the basis of their mode of nutrition (Lwoff, 1951).‡ This has led to widespread use of a terminology that may sometimes be confusing to the nonbiologist. Organisms that

---

* Honigberg, B. M., Balmuth, N., Bovee, E. C., Corliss, J. O., Gojdics, A., Hall, R. P., Kudo, R. R., Levine, N. D., Loeblich, A. R., Jr., Weiser, J., and Wenrich, D. A. (1964). *J. Protozool.* **11,** 7.

† Smith, G. M. (1950). "The Fresh Water Algae of The United States." McGraw-Hill, New York.

** Cronquist, A. (1971). "Introductory Botany," pp. 125–222. Harper, New York.

‡ Lwoff, A. (1951). *In* "Biochemistry and Physiology of Protozoa" (A. Lwoff, ed.), Vol. 1, pp. 1–28. Academic Press, New York.

grow on mineral media in the light and that derive all their organic carbon compounds from the fixation of carbon dioxide via photosynthesis are designated as photoautotrophs. If such organisms require supplementation of their mineral media with vitamins, they are termed photoauxotrophs. If growth supported by photosynthesis is supplemented by utilization of substrate levels of organic compounds supplied in the medium, such growth is referred to as photoheterotrophic. Microorganisms that grow in the absence of light utilizing exogenously supplied organic compounds as sources of carbon and energy are termed heterotrophs; those that utilize only dissolved nutrients are osmotrophs, while those that can take up particulate matter (including smaller microorganisms) are phagotrophs.

Many microscopic algae and fungi, most yeasts, and some protozoa can be cultured in the laboratory in pure culture (free of all other living organisms). Such cultures are often termed axenic and are usually clonal (they are derived from a single cell via multiplication by asexual means and hence are essentially genetically homogeneous). Cultures that contain more than one type of living organism are termed nonaxenic. Cultures containing a single type of algae (usually clonal) but contaminated with bacteria are designated by the botanists as unialgal.

JOSEPH A. ERWIN

CHAPTER 1

Comparative Biochemistry of Sterols in Eukaryotic Microorganisms

T. W. Goodwin

I. Distribution

A. ALGAE

1. *Introduction*

The major sterols in leafy material of higher plants are frequently sitosterol (I) and stigmasterol (II), which are accompanied by a complex

1

mixture of minor sterols that can vary considerably from species to species. Compared with cholesterol (III), which until recently was considered a typical animal sterol but which is now known to occur widely in traces in higher plants, these two sterols exhibit structural features characteristic and widespread in phytosterols: (1) an additional side chain, which can be methyl or methylene (1-C) or ethyl or ethylidene (2-C), at C-24, and (2) a double bond at C-22. The major sterols present in the different algal classes vary much more considerably, and this much greater variety is also reflected in the carotenoid distribution in algae compared with higher plants.

Recently, analytical techniques for detecting sterols have improved greatly and have become much more sensitive. In particular, mass spectrometry combined with gas-liquid chromatography has provided the means of identifying very small amounts of sterols with a reliability and confidence that was unknown 5 years ago. Some earlier identifications must therefore be considered today with some caution; they are summarized by Miller (1962).

2. Chlorophyta

a. Sterols. The main microalgae examined in this class belong to the order Chlorococcales and are Chlorella sp. They tend to fall into three main categories in which the major sterols are ergosterol ($\Delta^{5,7}$) (IV), chondrillasterol (Δ^7) (V), and poriferasterol (Δ^5) (VI). The additional minor sterols are indicated in Table I. Scenedesmus obliquus (two strains) falls into the third group in accumulating chondrillasterol, ergost-7-enol (VII), and chondrillast-7-enol (VIII) in about the same relative amounts as do the Chlorella sp. (Patterson, 1967). Earlier investigations had indicated a different distribution in S. obliquus (Bergmann and Feeney, 1950). The absolute configuration at C-24 is β in all sterols in Chlorella sp. This contrasts sharply with the C-24 alkylated sterols in higher plants, which always have the α configuration at C-24; compare, for example, poriferasterol (VI) with stigmasterol (II). One member of the Chlorococcales, Hydrodictyon reticulatum, is said to synthesize spinasterol (XI), which has the 24-α configuration (Hunek, 1969), but the identification is not unequivocal.

The colonial Chlorophyta so far examined are limited, but results indicate a sharp differentiation from the Chlorococcales. In the Uvales, Ulva pertusa contains 0.12% dry weight of sterol, of which 74% is cholesterol, 2.5% 24-methylene cholesterol (XII), and 14.5% fucosterol (XIII) (Ikekawa et al., 1968), while the major component of U. lactuca was 28-isofucosterol (XIV) (Δ^5-avenasterol) (Knights, 1965; Gibbons et al.,

1968). Earlier, sitosterol had been reported as the main sterol of *U. lactuca* (Heilbron *et al.*, 1935), but sitosterol and 28-isofucosterol were not separable by the techniques then available. Similarly, the major sterol of *Entero-morpha intestinalis* (Gibbons *et al.*, 1968) and *E. linza* (K. Tsuda and Sakai, 1960) is 28-isofucosterol, and this is probably true for *E. compressa*,

(I)

(II)

(III)

(IV)

(V)

(VI)

(VII)

(VIII)

which was reported in 1935 to contain sitosterol. *Monostroma nitidum* is said to contain haliclonasterol (XV), a C-20 isomer of campesterol (K. Tsuda and Sakai, 1960). In the only member of the Cladophorales so far examined, *Chaetomorpha crassa*, the major sterol is sitosterol (I), and it is accompanied by smaller amounts of cholesterol (III), 24-methylene cholesterol (XII), campesterol (XVI), and brassicasterol (XVII) (Ikekawa *et al.*, 1968). A sterol glycoside is reported in *Oedogonium* sp. (Heilbron *et al.*, 1935).

b. *4-Methyl and 4,4-Dimethyl Sterols.* Cycloartenol (XVIII) and 24-methylene cycloartanol (XIX) but not lanosterol (XX) have been detected in small amounts in *Enteromorpha linza* and *Ulva lactuca* (Gibbons, 1968), and 24-methylene lophenol (XXI) and 24-ethylidene lophenol (XXII) have been identified in *E. intestinalis* (Gibbons *et al.*, 1968).

TABLE I

STEROLS OF *Chlorella* SP. (% OF TOTAL STEROL)[a,b]

Species	$\triangle^{5,7}$-Sterols		$\triangle^{7}$-Sterols			$\triangle^{5}$-Sterols		
	1	2	3	4	5	6	7	8
C. vannielii	76	24						
C. sorokiniana	75	25						
C. nocturna	66	34						
C. simplex	70	30						
C. candida	76	24						
C. protothecoides var. *nannophila*	35	65						
C. protothecoides var. *communis*	16	84						
C. ellipsoidea						28	56	16
C. saccharophila						30	60	7
C. vulgaris[c]			25	65	10			
C. glucotropha			15	75	10			
C. fusca			27	59	12			
C. emersonii			28	62	9			
C. miniata			16	59	9			

[a] From Patterson (1971).

[b] Key to numbers: 1, ergosterol (IV); 2, unidentified $\triangle^{5,7}$-sterols; 3, ergost-7-enol (VII); 4, chondrillasterol (V); 5, ergost-5-enol (IX); 6, chondrillast-7-enol (VII); 7, poriferasterol (VI); 8, clionasterol (X).

[c] See also Tomita *et al.* (1970).

3. Rhodophyta

a. Sterols. Very few unicellular red algae exist, and they are members of the order Porphyridiales. Only one, *Porphyridium cruentum,* has been examined in detail; it was said to contain no sterols when cultured on a

(IX)

(X)

(XI)

(XII)

(XIII)

(XIV)

(XV)

(XVI)

chemically defined medium (Aaronson and Baker, 1961), but a reinvestigation has shown that sterols are present and that 22-dehydrocholesterol is the major component. It is accompanied by smaller amounts of ergosterol and cholesterole (Beastall *et al.*, 1971a). Of the colonial forms, members of six orders have been examined, and in almost all cases cholesterol (III), or

(XVII)

(XVIII)

(XIX)

(XX)

(XXI)

(XXII)

(XXIII)

(XXIV)

TABLE II
STEROLS OF RED ALGAE (RHODOPHYTA)

Order	Species	Major sterols[a]	References
Porphyridiales	*Porphyridium cruentum*	1, 2	Beastall *et al.* (1972)
Bangiales	*Porphyra purpurea*	3	Gibbons *et al.* (1967)
Gelidiales	*Acanthopeltis japonica*	4	K. Tsuda *et al.* (1958a)
	Gelidium amansii	4	K. Tsuda *et al.* (1958a)
	G. japonicum	4	K. Tsuda *et al.* (1957)
	G. subcostatum	4	K. Tsuda *et al.* (1957)
	Pterocladia tenuis	4	K. Tsuda *et al.* (1957)
Cryptonemiales	*Dilsea earnosa*	4	Gibbons *et al.* (1967)
	Corallina officinalis	4	Gibbons *et al.* (1967)
	Gloiopeltis fureata	4	K. Tsuda *et al.* (1958b)
	Tichocarpus crinitus	4	K. Tsuda *et al.* (1958b)
	Grateloup elliptica	4	K. Tsuda *et al.* (1958b)
	Cyrtymenia sparsa	4	K. Tsuda *et al.* (1958b)
	Polyides caprinus	4	Gibbons *et al.* (1967
	P. rotundus	4	Idler *et al.* (1968)
Gigartinales	*Gracilaria verrucosa*	4	Patterson (1971)
	Plocamium vulgare	4	Gibbons *et al.* (1967)
	Furcellaria fastigiata	4	Gibbons *et al.* (1967)
	Hypnea japonica	1	K. Tsuda *et al.* (1960)
	Ahnfeltia stellata	4	Gibbons *et al.* (1967)
	Chondrus crispus	4	Alcaide *et al.* (1968); Saito and Idler (1966)
	C. giganteus	4	K. Tsuda *et al.* (1958b)
	C. ocellatus	4	K. Tsuda *et al.* (1958b)
	Gigartina stellata	4	Gibbons *et al.* (1967)
	Iridophycus cornucopiae	4	K. Tsuda *et al.* (1958b)
	Rhodoglossum pulchrum	4	K. Tsuda *et al.* (1957)
Rhodymeniales	*Halosaccion ramentaceum*	3, 4	Idler *et al.* (1968)
	Rhodymenia palmata	3, 4	Gibbons *et al.* (1967); Idler *et al.* (1968); Alcaide *et al.* (1968)
	Coeloseira pacifica	4	K. Tsuda *et al.* (1958b)
Ceramiales	*Ceramium rubrum*	4	Patterson (1971)
	Chondria dasyphylla	4	Patterson (1971)
	Laurencia pinnatifida	4	Gibbons *et al.* (1967)
	Polysiphonia nigrescens	4	Gibbons *et al.* (1967)
	P. lanosa (fastigata)	4	Gibbons *et al.* (1967)
	P. subtillissima	4	Patterson (1971)
	Rhodomela conferoides	4	Idler *et al.* (1968); Alcaide *et al.* (1968)
	R. larix	4	K. Tsuda *et al.* (1958b)
	Dasya pedicellata	4	Patterson (1971)
	Grinnellia americana	3, 4	Patterson (1971)
	Rytiphlea tinctoria	5 or 6	Alcaide *et al.* (1969)

[a] Key to numbers: 1, 22-dehydrocholesterol; 2, ergosterol; 3, desmosterol; 4, cholesterol; 5, campesterol; 6, ergost-5-enol.

its immediate precursor desmosterol (XXIII), is the major sterol present
and frequently the only one reported (Table II). However, the relative
amounts can vary with the season; for example, desmosterol ranges from
30.6 to 97.2% of the total in different samples of *Rhodymenia palmata*
(Idler and Wiseman, 1970). In *Hypnea japonica*, 22-dehydrocholesterol
(XXIV) is the major sterol (K. Tsuda *et al.*, 1960) and it is also present in
traces in other Rhodophyta (Idler *et al.*, 1968; Gibbons *et al.*, 1967). Traces
of 24-alkyl sterols were noted in *Rhodomela conferoides, Chondrus crispus*,
and *Rhodymenia palmata* (Alcaide *et al.*, 1968), although they were not
detected by Saito and Idler (1966) and Gibbons *et al.* (1967). This problem
has probably been resolved by the observation of Idler and colleagues that
the sterol levels vary with the time of year and with the environment.
However, in one member of the Ceramiales, *Rytiphlea tinctoria*, a 24-ethyl
sterol, with one double bond at C-5 and another in the side chain, was the
main sterol present (Alcaide *et al.*, 1969).

b. 4-Methyl and 4,4-Dimethyl Sterols. Cycloartenol (XVIII) but not
lanosterol (XX) has been detected in *Rhodomela conferoides, Chondrus
crispus*, and *Rhodymenia palmata* (Alcaide *et al.*, 1968).

4. *Phaeophyta*

a. Sterols. Following the pioneer work of Carter *et al.* (1939), it has
frequently been confirmed that fucosterol (XIII) is the major sterol of all
Phaeophyta examined. An earlier claim that sargasterol (the C-20 isomer
of fucosterol) replaces fucosterol in *Sargassum ringgoldianum* (K. Tsuda
et al., 1958:) has not been substantiated by later experiments with more
refined techniques (Ikekawa *et al.*, 1966, 1968). A new sterol, saringosterol
(XXV), noted in a number of Phaeophyta (Ikekewa *et al.*, 1966, 1968;
Patterson, 1968), may be an oxidative artifact arising from fucosterol
(Knights, 1970:). Traces of other sterols are often present in addition to
fucosterol [for example, 24-methylene cholesterol and desmosterol in
Laminaria faeroensis and *L. digitata* (Patterson, 1968)], and these traces
tend to increase in old or milled samples (Knights, 1970b). 24-Oxocholesterol,
recently reported in *Pelvetia canaliculata*, may also be an artifact
(Motzfeldt, 1970). Brown algae examined for sterols are listed in Table III.

b. 4-Methyl and 4,4-Dimethyl Sterols. Cycloartenol (XVIII) and 24-
methylene cycloartanol (XIX), but not lanosterol (XX), have been
detected in *Fucus spiralis* (Goad and Goodwin, 1969).

5. *Euglenophyta*

Sterols. Fairly recent work indicated that ergosterol (IV) was the major
sterol of *Euglena gracilis* strain Z (Stern *et al.*, 1960; Avivi *et al.*, 1967),
but a thorough reinvestigation has shown that the major free sterols in

TABLE III
STEROLS OF BROWN ALGAE (PHAEOPHYTA)

Order	Species	Major sterols[a]	References
Ectocarpales	*Pylaiella littoralis*	1	Heilbron (1942)
	Spongonema tomentosum (*Ectocarpus tomentosus*)	1	Heilbron (1942)
Sphacelariales	*Cladostephus spongisus*	1	Heilbron (1942)
	Sphacelaria pennata (*cirrosa*)	1	Heilbron (1942)
	Stypocaulon scaparium	1	Heilbron (1942)
Dictyotales	*Dictyopteris divaricata*	1	Ikekawa *et al.* (1966, 1968)
	Dictyota dichotoma	1	Heilbron (1942)
	Padina arborescens	1	Ito *et al.* (1959)
Chordariales	*Heterochordaria abietina*	1	Ito *et al.* (1959)
Dictysiphonales	*Myelophycus caespitosus*	1	K. Tsuda *et al.* (1958a)
Laminariales	*Alaria crassifolia*	1	K. Tsuda *et al.* (1958a); Ikekawa *et al.*, (1968)
	Chorda filum	1	Heilbron (1942)
	Costaria costata	1, 2	K. Tsuda *et al.* (1958b)
	Eisenta bicyclis	1	Patterson (1971)
	Laminaria angustata	1	Ito *et al.* (1959)
	L. digitata	1, 2	Heilbron (1942); Patterson (1968)
	L. faeroensis	1, 2	Patterson (1968)
	L. hyperborea (*cloustonii*)	1	Black and Cornhill (1951)
	L. japonica	1	Ito *et al.* (1959)
	L. saccharina	1	Black and Cornhill (1951)
Fucales	*Ascophyllum nodosum*	1	Heilbron (1942); Patterson (1968); Black and Cornhill (1951); Knights (1970b)
	Cystophyllum hakodatense	1	K. Tsuda *et al.* (1958a,b)
	Fucus gardneri	1	Reiner *et al.* (1962)
	F. evanescens	1, 2	K. Tsuda *et al.* (1958a,b); Ikekawa *et al.* (1968)
	F. diviarcarpus	1, 2	Ciereszko *et al.* (1968)
	F. ceranoides	1	Heilbron (1942)
	F. serratus	1	Black and Cornhill (1951)
	F. spiralis	1	Black and Cornhill (1951)
	F. vesiculosus	1	Black and Cornhill (1951)
	Halidrys siliquosa	1	Heilbron (1942)
	Pelvetia wrightii	1	K. Tsuda *et al.* (1958b)
	P. canaliculata	1	Black and Cornhill (1951)
	Sargassum muticum	1	Reiner *et al.* (1962)
	S. confusum	1	Ikekawa *et al.* (1968)
	S. thunbergii	1, 2	Ikekawa *et al.* (1968)
	S. ringgoldianum	1, 2	Ikekawa *et al.* (1968)

[a] Key to numbers: 1, fucosterol; 2, 24-methylene cholesterol.

TABLE IV

Nature of the Sterols and Their Relative Amounts (%) of Total Fraction in
Free and Bound Fractions from Green and White *Euglena gracilis* Z[a]

			Bound					
	Free		Esters		Water-soluble, acid-treated		Water-soluble pyrogallol	
Sterols	Green	White	Green	White	Green	White	Green	White
Cholest-7-enol (XXVI)	Trace	Trace						
Episterol (XXVII)		4						
Ergost-7-enol (VII)	33	30						
Chondrillast-7-enol (VIII)	12	7						
Ergosta-5,22-dienol (XXVIII)	18	6						
Chondrillasterol (V)	35	—						
Cholesta-5,7-dienol (XXV)	Trace	Trace						
Chondrillasta-5,7-dienol (XXIX)		5		8	2	6	2	3
Cholesterol (III)	1	12	Trace	16	52	42	54[b]	53
Chalinasterol (XXX)				6				
22-Dihydrobrassicasterol (XXXI)		Trace		3				1
Clionasterol (X)		16		50	28	39	31	31
Poriferasterol (VI)		3		6	4		5	3

[a] From Brandt *et al.* (1970).
[b] Up to 85% in a second experiment where no other sterols were detected.

E. gracilis grown autotrophically, that is, photosynthetically, are Δ^7-sterols and not $\Delta^{5,7}$-sterols such as ergosterol. Only a trace of cholesta-5,7-dienol was detected (Brandt *et al.*, 1970). The sterols and their relative distribution are indicated in Table IV. *Euglena gracilis* can also be grown heterotrophically in the dark, when it is colorless. Under these conditions, chondrillasterol (V), the major sterol of green cells, is absent, but additional sterols, cholesterol (III), clionasterol (X), and poriferasterol (VI) appear, together with 5% of chondrillasta-5,7-dienol (Table IV).

Although only a trace of esterified cholesterol is present in green cells, far more esterified sterols are present in white cells, but they are confined to the Δ^5- and $\Delta^{5,7}$-sterols; no esterified Δ^7-sterols were observed. Similarly, water-soluble sterols are confined to the same group as the esterified sterols. The nature of the solubilizing agent for the water-soluble sterols is not yet known. Sterols represent 0.005–0.07% (dry weight), and free sterols predominate in green *Euglena*, while the esterified forms predominate in white cells (Brandt *et al.*, 1969; 1970).

(XXV)

(XXVI)

(XXVII)

(XXVIII)

(XXIX)

(XXX)

(XXXI)

(XXXII)

6. *Chrysophyta*

Sterols. Early work on *Apistonema carteri, Thallochrysis litoralis,* and *Gloechrysis maritima* indicated that fucosterol (XIII) was the main sterol (Heilbron, 1942).

The phytoflagellate *Ochromonas malhamensis* contains poriferasterol (VI) as the major sterol (Williams *et al.,* 1966; Gershengorn *et al.,* 1968), that is, the 24-*S* isomer and not stigmasterol (II) (24-*R*), as previously suggested (Bazzano, 1965; Avivi *et al.,* 1967). Ergosterol was previously reported in this organism (Aaronson and Baker, 1961) and this was confirmed in the later investigations. The related *O. dancia* contains, in addition to poriferasterol (VI), which is the main component, brassicasterol (XVI), 22-dihydrobrassicasterol (XXXI), clionasterol (X), and probably 7-dehydroporiferasterol (XXXII), all 24-*S* sterols (Gershengorn *et al.,* 1968). The presence of ergosterol in *O. danica* was first reported by Stern *et al.* (1960), Aaronson and Baker (1961), and Halevy *et al.* (1966), but the last-named authors claimed that the main sterol was stigmasterol (II). They also reported stigmasterol as the major sterol in *O. sociabilis;* this needs to be checked, as does the report that *Synura petersenii* contains cholesterol (III) and sitosterol (I) (Collins and Kalnins, 1969).

7. *Charophyta*

The early report of sitosterol (I) and fucosterol (XIII) in *Nitella opaca* (Heilbron, 1942) may have to be reassessed, but it is interesting that this is one of the two algae so far reported to contain sterol glycosides (Heilbron *et al.,* 1935). Such compounds are relatively common in higher plants.

8. *Bacillophyta*

The only relatively recent report is that *Navicula pelliculosa* contains chondrillasterol (V) (Low, 1955).

9. *Blue-Green Algae and Bacteria*

Although prokaryotes, we must consider briefly the blue-green algae and bacteria for completeness. It was originally thought that the blue-green algae, like the other prokaryotes, the bacteria, did not synthesize sterols (Levin and Bloch, 1964). However, cholesterol and 24-ethyl derivatives have recently been detected in *Anacystis nidulans, Fremyella diplosiphon* (Reitz and Hamilton, 1968), and *Phormidium luridum* var. *olivaceae* (de Souza and Nes, 1968). In *P. luridum,* 24-ethylcholest-7-enol was the major compound, and it was accompanied by Δ^{7}-, $\Delta^{5,7}$-, and Δ^{22}-ethyl derivatives of cholesterol. The stereochemistry at C-24 has not yet been determined. It is the general impression that sterols are not present in bacteria, but traces have been reported in *Azotobacter chroococcum* (0.01% dry matter), *Escherichia coli* (0.0004%) (Schubert *et al.*, 1964), and *Streptomyces olivaceous* (0.0035%) (Schubert *et al.*, 1967, 1968, and references therein). The sterols detected in *A. chroococcum* were ergost-7-enol, ergosta-7,22-dienol, and ergosterol; cholesterol was not detected, but it is the major sterol in *Escherichia coli* and *Streptomyces olivaceous*. The unique position of *Azotobacter chroococcum*, with a relatively high level of sterols, might be related to the fact that it has a much more complex membrane structure than other bacteria.

No sterols were detected (i.e., levels were lower than 0.0001% dry weight) in *Aerobacter cloacae*, *Bacillus subtilis*, *Lactobacillus bulgaricus*, *Corynebacterium pseudodiphtheriticum*, *Mycobacterium smegmatis*, and *Mycobacterium tuberculosis* var. *bov.* (Schubert *et al.*, 1968) or in *Staphylococcus* sp. (Suzue *et al.*, 1968). Recently, sterols have been reported in relatively large amounts in *Methylococcus capsulatus* (Bird *et al.*, 1971).

B. PROTOZOA

Williams *et al.* (1966) made a survey of sterols in a number of protozoa (Table V). It is interesting that the major sterol in most organisms is

TABLE V

STEROL DISTRIBUTION IN SOME PROTOZOA[a]

Species	Sterol[b]
Astasia ocellata	1, 2
Chilomonas paramoecium	1, 3
Crithidia oncopelti	1, 2
Haematococcus pluvialis[c]	1, 2
Hartmannella rhysodes	Absent
Peranema tricophorum	1, 3, 4, 5
Polytoma uvella	1, 2
Polytomella coeca	Probably absent
Prototheca zopfii	1
Tetrahymena pyriformis	Absent[d]
Trypanosoma mega	1
Trypanosoma rhodesiense	
Blood form	4
Culture form	1

[a] From Williams *et al.* (1966).

[b] Key to numbers: 1, ergosterol; 2, spinasterol or chondrillasterol; 3, poriferasterol or stigmasterol; 4, cholesterol; 5, unknown.

[c] Also classed in the Chlorophyta.

[d] Contains tetrahymanol (XXXIII).

ergosterol (IV), which is accompanied in a number of cases by a sterol (A), which could not be distinguished at that time from either spinasterol (XI) or chondrillasterol (V); the latter is more likely to be present by analogy with sterol distribution with other microorganisms. Cholesterol (III) accompanies ergosterol and another unknown sterol (B) in *Peranema trichophorum* and most interestingly is present in the blood form of *Trypanosoma rhodesiense*, although ergosterol is the major sterol of the culture form.

It is known that protozoa can take up sterols from growth media (Britt and Bloch, 1961; Conner and Unger, 1964), but in the study summarized in Table V all except *Peranema tricophorum* and *Trypanosoma rhodesiense* were cultured on sterol-free media. As *P. tricophorum* was cultured on a medium containing cow's milk, the cholesterol found could have arisen from this source. The cholesterol found in the blood form of *T. rhodesiense* could also have arisen from the host, but no ergosterol was present in the medium on which the culture form was grown.

No sterol was noted in *Polytomella coeca*, *Tetrahymena pyriformis*, or *Hartmannella rhysodes*. The major triterpene in *T. pyriformis* is tetrahymanol (XXXIII) (Mallory *et al.*, 1963; Y. Tsuda *et al.*, 1965).

C. FUNGI

1. *Introduction*

It has been generally accepted for many years that the major sterol of most fungi is ergosterol (IV), so that the characteristic structural features of a fungal sterol would be (1) a $\Delta^{5,7}$-diene system, (2) a *trans* Δ^{22} double bond, and (3) a 24-β methyl group. However, recent detailed reinvestigations of some fungi reveal the presence of significant amounts of other sterols and often the complete absence of ergosterol.

2. *Phycomycetes*

a. Sterols. A recent survey of members of three orders of Phycomycetes (Table VI) reveals that only the Mucorales fit into the previously accepted pattern in synthesizing ergosterol, and even in some *Mucor* sp., for example, *Mucor hiemalis*, it is not the major component, its place being taken by 22,23-dihydroergosterol (XXXIV) (McCorkindale *et al.*, 1969). Cholesterol has also been noted in a number of Mucorales. Ergosterol has been obtained from *Blakeslea trispora* (Goad *et al.*, 1966) and *Phycomyces blakesleeanus* (Goulston and Mercer, 1969); however, in *P. blakesleeanus*, episterol (XXVII) represents about 20% of the total sterol present (Goulston, 1969), and ergosta-5,7,24-trien-3β-ol (XXXV) has also been isolated from the same sources (Goulston and Mercer, 1969).

TABLE VI

COMPOSITION OF STEROL MIXTURES FROM SOME PHYCOMYCETES
(RELATIVE AMOUNTS AS % OF TOTAL)[a,b]

Order	Species	\multicolumn{6}{c}{Sterol}
		1	2	3	4	5	6
Saprolegniales							
	Saprolegnia fera	4	13		68		15
	S. megasperma	23	1		47		29
	Leptolegnia caudata	2	14		70		14
	Aplanopsis terrestris	34			6		60
	Achlya caroliniana	72	27		1		
	Pythiopsis cymosa	23			73		4
Leptomitales							
	Apodachlya minima	85	0.5		6		1
	A. brachynema	68	2		8		22
	Apodachlyella completa	3			41		56
Mucorales							
	Mucor hiemalis (+)	1		7		85	
	M. hiemalis (−)	15		6		52	
	M. dispersus			80		10	
	Rhizopus stolonifer (−)			96		2	
	Phycomyces blakesleeanus (+)			60		20	
	P. blakesleeanus (−)			80		10	
	Absidia glauca (+)			12		83	
	A. glauca (−)	1		3		93	

[a] From McCorkindale *et al.* (1969).

[b] Key to numbers: 1, cholesterol; 2, desmosterol; 3, ergosterol; 4, 24-methylene cholesterol; 5, 22-dihydroergosterol; 6, fucosterol.

Two of the three members of the order Leptomitales examined by McCorkindale *et al.* (1969) synthesize mainly cholesterol (III) and fucosterol (XIII). The same general type of distribution is also observed in six Saprolegniales (Table VI). Perhaps the most unexpected finding is the presence of fucosterol (XIII) in all but one of the fungi examined in these two orders; previously it had been considered unique to the brown algae.

One member of the Saprolegniales, *Achlya bisexualis*, produces in its female form antheridiol (XXXVI), a sterol that is excreted into the medium and that stimulates growth of antheridiae in the male forms and thus sexual reproduction in the mold (Arsenault *et al.*, 1968; Edwards *et al.*, 1969).

Perhaps equally unexpectedly, the Peronosporales, *Phytophthora cactorum*

(Elliott *et al.*, 1964), *Phytophthora infestans, Pythium ultimum,* and *Pythium debaryarum* (McCorkindale *et al.*, 1969), do not synthesize detectable amounts of sterols.

Early work indicated that fungisterol (ergost-7-en-3β-ol) (VII) is present in *Rhizopus japonicus* (Wieland and Coutelle, 1941).

(XXXIII)

(XXXIV)

(XXXV)

(XXXVI)

(XXXVII)

(XXXVIII)

(XXXIX)

(XL)

The resting spores of the parasitic *Plasmodiophora brassicae* contain sterols similar to those found in the plant that it parasitizes, which suggests that they originated in the plant (Knights, 1970a).

b. 4,4-Dimethyl and 4α-Methyl Sterols. Lanosterol (XX), 24-methylene-24,25-dihydrolanosterol (XXXVII), and 4,4-dimethyl-5α-ergosta-8,24(28)-dien-3β-ol (XXXVIII) are present in *Phycomyces blakesleeanus*, and 4α-methyl-5α-ergosta-8,24(28)-dien-3β-ol (XXXIX) has been detected in the same organism (Goulston *et al.*, 1967, 1972; Goad and Goodwin, 1967).

3. *Ascomycetes*

Most of the work with Ascomycetes has been done on yeast, from which, in addition to ergosterol (IV), the following have been isolated: lanosterol (XX) (Wieland *et al.*, 1937), ergosta-7,22-dienol (VII) (Wieland *et al.*, 1941), ascosterol (XL) (Wieland *et al.*, 1941; Fürst, 1966), episterol (XXVII) (Wieland *et al.*, 1941; Barton, 1945), fecosterol (XLI) (Wieland *et al.*, 1941; Barton, 1945), and ergosta-5,7,22,24(28)-tetraenol [24-dehydroergosterol (XLII)] (Breivick *et al.*, 1954; Petzoldt *et al.*, 1967). Ergosterol has been detected in a number of marine Ascomycetes (Table VII), although it is absent from others (Kirk and Catalfomo, 1970).

Ergosterol is the main sterol of *Aspergillus flavus* (Vacheron and Michel, 1968), *A. fumigatus* (Fiore, 1948), *A. niger* (Bimpson, 1970), and certain *Penicillium* sp., but, somewhat unexpectedly, cholesterol (III) has been found in *Penicillium funiculosum* (Chen and Haskins, 1963). Ergosta-5,7,14,22-tetraen-3β-ol (XLIII) has been obtained from *A. niger* (Barton

TABLE VII

DISTRIBUTION OF ERGOSTEROL IN MARINE ASCOMYCETES[a]

Organism	Isolate	Ergosterol[b]
Corollospora trifurcata	R-1	+
	R-3	−
	R-562	−
Corollospora maritima	R-19	+
	R-563	+
Cerioporopsis calyptrata	R-612	−
Cerioporopsis halima	R-546	−
	R-552	+
Halosphaeria appendiculata	R-558	−

[a] From Kirk and Catalfomo (1970).
[b] +, present; −, absent.

and Bruun, 1951), but ergosterol peroxide (XLIV) (Wieland and Prelog, 1947; Bauslaugh *et al.*, 1964; Tanahashi and Takahashi, 1966; Clarke and McKenzie, 1967; Hamilton and Castrejon, 1966) and cerevisterol (XLV) (Wieland and Coutelle, 1941; Alt and Barton, 1954) are possibly artifacts.

(XLI)

(XLII)

(XLIII)

(XLIV)

(XLV)

(XLVI)

(XLVII)

(XLVIII)

4. *Basidiomycetes*

a. Sterols. Ergosterol (IV) has been reported in a number of Basidiomycetes (Milazzo, 1965), and 22-dihydroergosterol (XXXIV) was obtained from *Polyporus pargamenus* (Singh and Rangaswami, 1966). Ergosterol and 5α-ergosta-7,24(28)-dien-3β-ol have been isolated from *Agaricus campestris* (Goulston *et al.*, 1972).

b. 4,4-Dimethyl and 4α-Methyl Sterols. Lanosterol (XX), 24-methylene-24,25-dihydrolanosterol, 4,4-dimethyl-5α-ergosta-8,24(28)-dien-3β-ol (XXXVIII), and 4α-methyl-5α-ergosta-8,24(28)-dien-3β-ol (XXXIX) have been isolated from *Agaricus campestris* (Goulston *et al.*, 1972).

A characteristic of *Polyporus* sp. is the elaboration of a number of lanosterol derivatives in which the methyl substituent at C-20 has been oxidized to either CH_2OH or $COOH$. Typical examples are pinicolic acid A (XLVI) (Guider *et al.*, 1954), lanosta-4,9(11),24-trien-3β,21-diol (XLVII) (Halsall and Sayer, 1959), and lanosta-8,24-dien-3α-ol-21-oic acid (XLVIII) (Beereboom *et al.*, 1957). It is interesting that the last-mentioned compound has the 3α configuration.

The uredospores of the flax rust *Melampsora lini* contain stigmast-7-enol (XLIX) and, in traces, stigmasta-5,7-dienol (L) (Jackson and Frear, 1968). Although the nature of the sterols of the host plant was not investigated, it would appear that the rust sterols were derived from the flax sterols.

5. *Fungi Imperfecti*

Ergosterol (IV) has been found in two Fungi Imperfecti, *Zalerion maritima* and *Pyrenochaeta* sp., but was apparently absent from *Culcitalna achraspora*, *Flagellospora* sp., and *Clavatospora stellatacula* (Kirk and Catalforno, 1970). Ergosterol (IV), 22-dihydroergosterol (XXXIV), and 7-dehydroclionasterol (LI) were identified by gas-liquid chromatography in *Fusarium roseum* (Tillman and Bean, 1970). The ergosterol peroxide (XLIV) and cerevisterol (XLV) found in *Fusarium oxysporum* (Sharratt and Madhosingh, 1967) were possibly artifacts.

The so-called protosterols are produced by Fungi Imperfecti; for example, helvolic acid (LII) is found in *Cephalosporium caerulens* (Okuda *et al.*, 1968) and fusidic acid (LIII) is found in *Fusidium coccineun* (Godtfredson *et al.*, 1965).

D. MYXOMYCETES

Only one of the Myxomycetes (true slime molds) has been thoroughly examined for sterols. *Physarum polycephalum* contains stigmasterol (II),

(XLIX)

(L)

(LI)

(LII)

(LIII)

(LIV)

(LV)

(LVI)

sitosterol (I), stigmastanol (LIV), campesterol (XVI), campestanol (LV), and cholesterol (III) (Lenfant et al., 1970). These sterols are characteristic of higher plants and algae and not of fungi; most interestingly, however, the characteristic cyclic precursors found in yeasts, lanosterol (XX) and 24-methylene-24,25-dihydrolanosterol (XXXVII), were detected. Those associated with sterol synthesis in higher plants and algae are cycloartenol (XVIII) and 24-methylene cycloartanol (XIX).

E. ACRASIALES

The only species of Acrasiales (cellular slime molds) examined is *Dictyostelium discoideum*, which contains stigmast-22-en-3β-ol (LVI) as its major component (Heftmann et al., 1960).

II. Biosynthesis

A. ALGAE

1. *Introduction*

The main details of sterol biosynthesis have been worked out in animal (liver) systems in which the final product is cholesterol. The first specific precursor is mevalonic acid (MVA), and the overall pathway from MVA to squalene and the first cyclized product, lanosterol, is indicated in Fig. 1. Without considering the exact sequence of events the changes involved in the transformation of lanosterol (XX) into cholesterol (III) are (1) oxidative removal of the methyl groups at C-4 and C-14, (2) conversion of Δ^8 to Δ^5, and (3) saturation at C-24. Additional steps required in producing phytosterols are alkylation at C-24 and desaturation at C-22. Before these steps are discussed, the role of lanosterol as an intermediate must be considered.

2. *First Cyclic Precursor*

Lanosterol (XX) has rarely been observed in photosynthetic tissues (see Goad, 1970), and its place appears to have been taken by cycloartenol (XVIII); the presence of the latter in algae has already been noted. Investigations that led to the conclusion that cycloartenol and not lanosterol is the first stable cyclic product have involved the use of stereospecifically labeled species of MVA. Experiments with [2-¹⁴C-4R-³H₁]MVA* (LVII)† and [2-¹⁴C-4S-³H₁]MVA* (LVIII)† showed that in the formation of one

---

* R and S represent the absolute configuration around an asymmetric carbon atom according to the rules of Cahn and Ingold (1951).

† See page 30 for structures (LVII) and (LVIII).

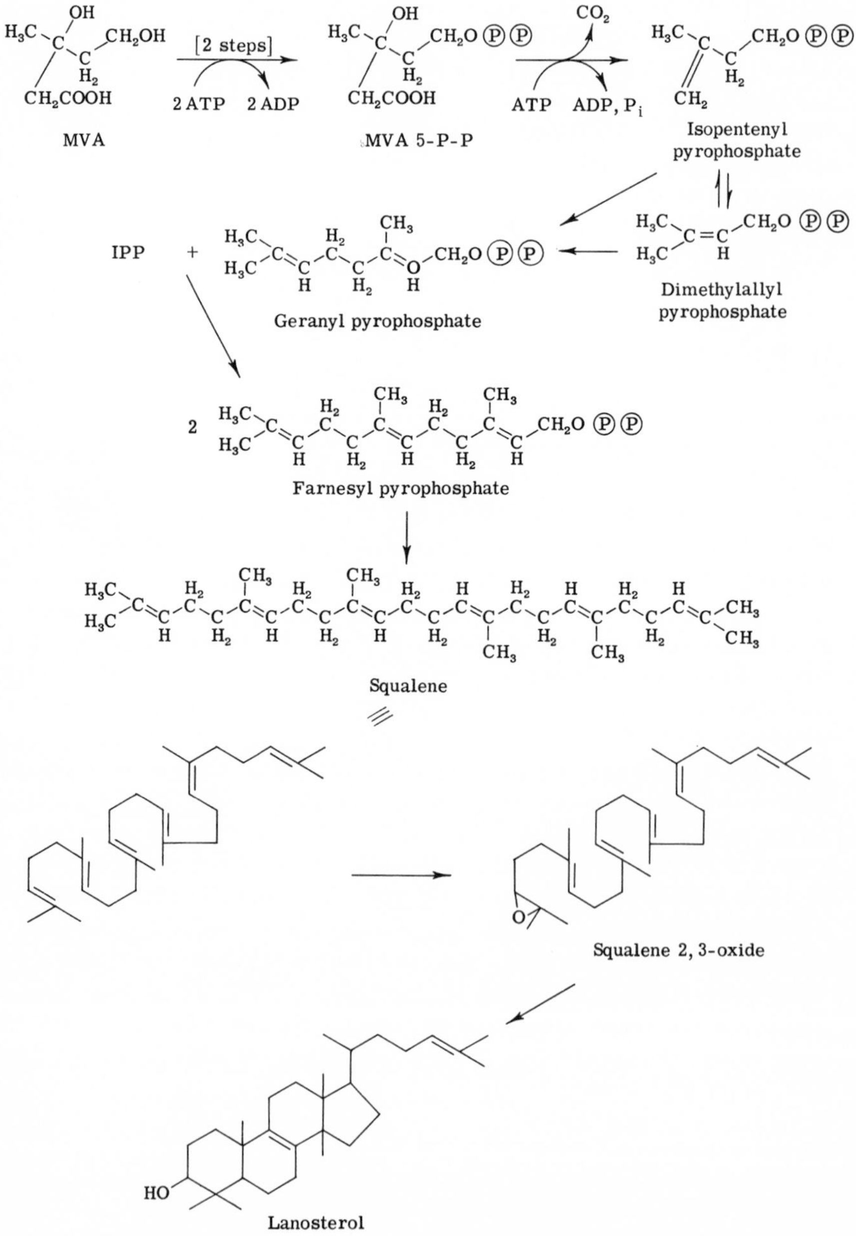

Fig. 1. Biosynthesis of lanosterol from mevalonic acid.

molecule of squalene from six molecules of MVA, a 4-pro-R hydrogen* is retained. The squalene molecule synthesized from [2-^{14}C-4R-^{3}H$_1$]MVA will have the distribution of labeling indicated in (LIX)† (^{14}C indicated by ●) (Popják and Cornforth, 1966).

In the cyclization process squalene is converted first into squalene 2,3-epoxide, which then cyclizes by proton attack to give an unstable carbonium ion (enzyme-substrate complex), which, by a backward rearrangement stabilized by expulsion of a proton from C-9, yields lanosterol (Fig. 2). The labeling in lanosterol formed in liver preparations when [2-^{14}C-4R-4-^{3}H$_1$]MVA is the substrate is also indicated in Fig. 2, from which it can be seen that compared with squalene it has lost one tritium atom and its ^{14}C-^{3}H atomic ratio is 6:5 compared with 6:6 for squalene. If cycloartenol were formed by the isomerization of lanosterol, its atomic ratio would also be 6:5. On the other hand, if cycloartenol were formed directly from squalene 2,3-oxide, a hydrogen would not be expelled from C-9 but would move to C-8, as a hydrogen is lost from the methyl group at C-10 as the cyclopropane ring is formed. In this case a tritium would not be lost from the molecule and the ^{14}C-^{3}H ratio for cycloartenol would be the same as that for squalene. This has been demonstrated experimentally in potato leaves (Rees et al., 1968) and in Fucus spiralis (Goad and Goodwin, 1969). Furthermore, in the experiment with potato leaves the tritium was located at C-8. Cycloartenol is therefore not formed in higher plants and algae from lanosterol but by direct cyclization of squalene 2,3-oxide. Furthermore, under anaerobic conditions cell-free preparations from Ochromonas malhamensis will convert squalene 2,3-oxide into cycloartenol but not into lanosterol (Rees et al., 1969); in liver preparations lanosterol accumulates under these conditions, because the next step, demethylation, is an aerobic process. In the presence of oxygen, cycloartenol is very effectively converted into phytosterols in Ochromonas malhamensis (Hall et al., 1969; Lenton et al., 1971). These and other experiments on higher plants (see, e.g., Goad, 1970; Goodwin, 1972) suggest that cycloartenol and not lanosterol is the precursor of sterols in photosynthetic organisms. However, labeled lanosterol is effectively converted into phytosterols in Ochromonas malhamensis (Hall et al., 1969; Lenton et al., 1971), Euphorbia peplus (Baisted et al., 1968), and tissue cultures of Nicotiana tabacum (Hewlins et al., 1969). This means that the enzymes concerned with the metabolism of 4,4-dimethyl sterols in plants are not completely specific and that

---

* The nomenclature of Cahn and Ingold (1951) as modified to name paired ligands, g, g at a tetrahedron atom X$ggij$ (Hanson, 1966). In the present context a pro-R hydrogen is that hydrogen that if replaced by deuterium or tritium would confer the R configuration at the carbon atom involved.

† See page 30 for structure (LIX).

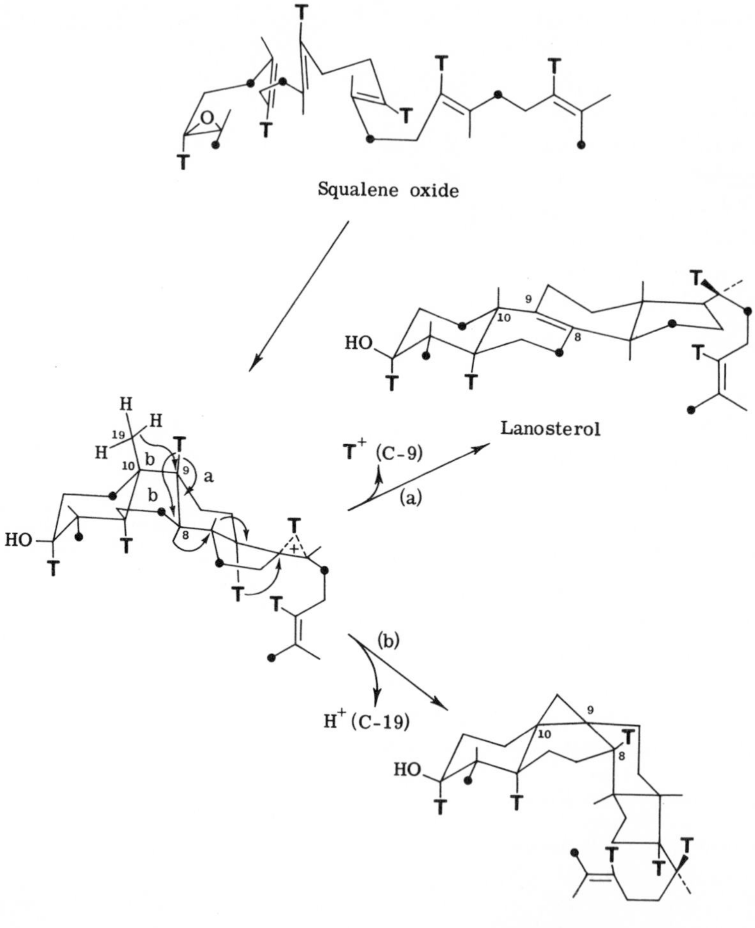

Fig. 2. Conversion of squalene oxide into lanosterol and cycloartenol. ●, Carbons from C-2 of MVA; **T**, tritiums from $[4R - {}^3H_1]MVA$.

lanosterol, although not a naturally occurring intermediate, can be utilized in the synthesis of sterols. Recently, the 2,3-oxidosqualene cycloartenol cyclase from *Ochromonas malhamensis* has been purified some 25-fold; it produces no lanosterol, only cycloartenol (Beastall *et al.*, 1971b).

3. *Possible Pathway of Synthesis*

The conversion of cycloartenol into phytosterols involves (1) ring opening of the cyclopropane ring; (2) demethylation at C-4 and C-14;

576.11 En94p
C.1

(3) conversion of a Δ^8-sterol into a Δ^7-, $\Delta^{5,7}$, or Δ^5-sterol; (4) alkylation at C-24; and (5) insertion of a side-chain double bond at C-22.

Before considering the details of these metabolic transformations, let us indicate the probable sequence of reactions involved in sterol formation in photosynthetic organisms (Fig. 3). Evidence for this pathway is based to some extent upon the observation that all the compounds listed have been found to occur naturally. Insofar as algae are concerned, 24-methylene cycloartanol (XIX) has been found in a number of species (see Section II,A,2), and [2-^{3}H$_2$]-24-methylene cycloartanol is converted into phytosterols in Ochromonas malhamensis. Furthermore, 24-methylene lophenol (XXI) and 24-ethylidene lophenol (XXII) are present in Enteromorpha intestinalis (Gibbons et al., 1968).

4. Demethylation at C-14 and C-4

Experiments with liver preparations have shown that demethylation of lanosterol at C-14 involves the formation of a double bond at C-14–C-15; and a possible mechanism is indicated in Fig. 4 (see also Goad, 1970; Goodwin, 1972). It is also now clear that in demethylation at C-4, in photosynthetic organisms the 4α-methyl group is first removed and a 3-oxo intermediate is involved. Reduction of this group to 3β-hydroxy results in the remaining methyl group, originally the 4β-methyl group, taking up the α configuration. This is then eventually removed presumably by a mechanism similar to that involved in removing the first methyl group (Ghisalberti et al., 1969).

5. Transformations in Ring B

It is now generally accepted that the intramolecular movement of the double bond at Δ^8 to Δ^5 in ring B involves the following transformations: $\Delta^8 \rightarrow \Delta^7 \rightarrow \Delta^{5,7} \rightarrow \Delta^5$. Details of the stereochemistry of the reactions have been worked out for cholesterol (see Goad, 1970); all that is known of the details in algae is that in the $\Delta^8 \rightarrow \Delta^7$ step it is the β-hydrogen that is eliminated from C-7 in Ochromonas malhamensis (A. R. H. Smith et al., 1968a) and the α-hydrogen that is removed from Fucus spiralis (Bimpson, 1970). These experiments were made using [2R-2-^{3}H$_1$]MVA (LX)* and [2S-2^{3}H$_1$]MVA (LXI)*; these substrates give rise to cycloartenol with tritium at C-7 in the α and β position, respectively. Poriferasterol biosynthesized by Ochromonas malhamensis contained tritium at C-7 when [2R-2-^{3}H$_1$]MVA was the substrate but only little when [2S-2-^{3}H$_1$] was the substrate (Fig. 5). It should be noted that in the eventual reduction of this double bond there is an inversion, and the 7α-hydrogen takes up the 7β position as a hydride ion from the B face of NADPH enters at the

---

* See page 30 for structures (LX) and (LXI).

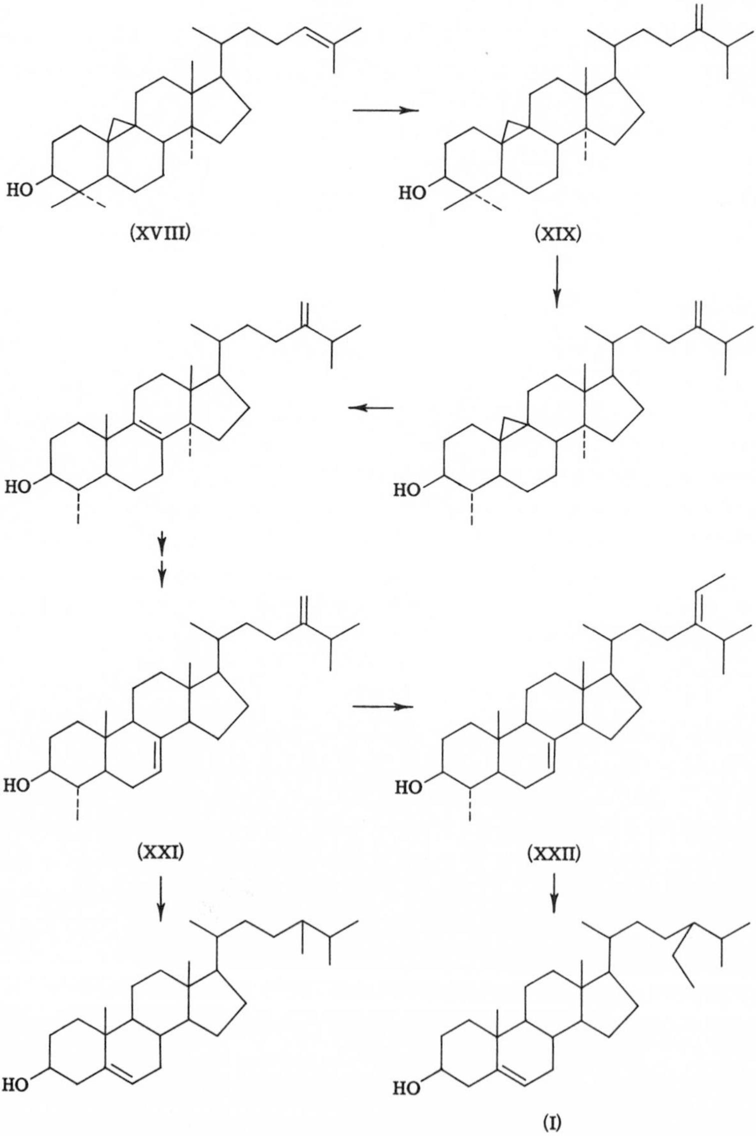

Fig. 3. Possible pathway of biosynthesis of phytosterols from cycloartenol (Goad, 1970).

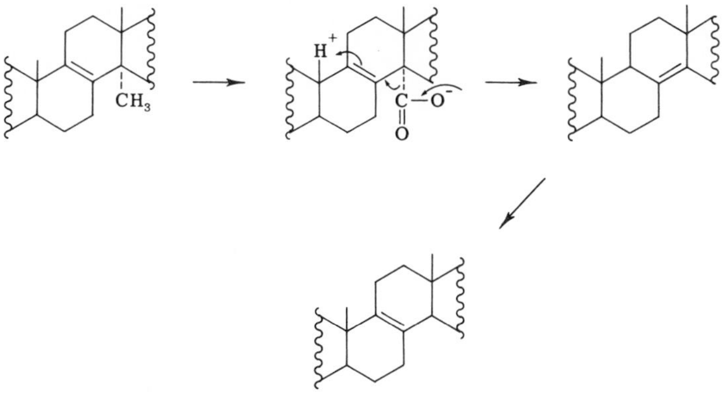

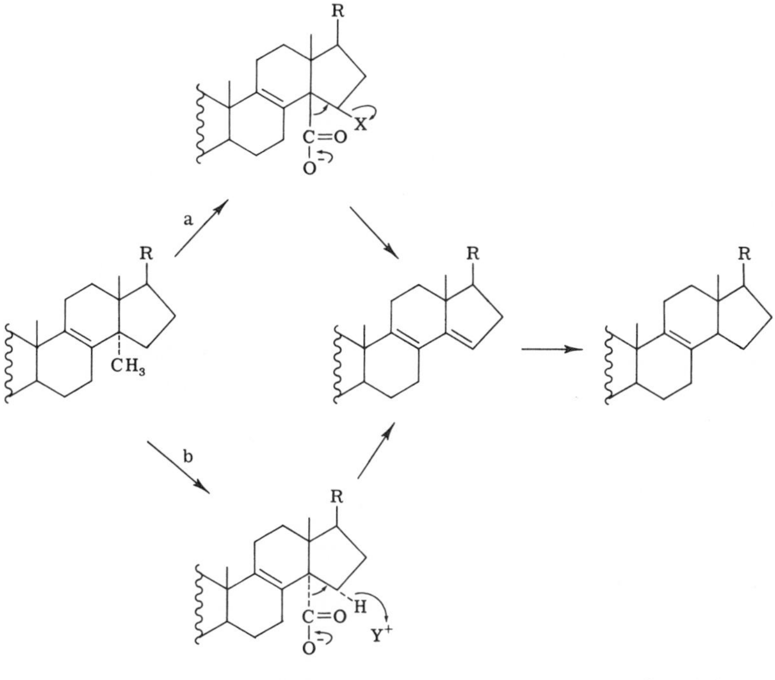

Fig. 4. Mechanisms of demethylation of lanosterol at C-24 (Goad, 1970).

α position. Demonstration of the addition of a proton at 8β in liver preparation (which exhibits the same stereochemistry at C-7 as *O. malhamensis*) fully defines the stereochemistry of the reduction (Fig. 5) (see Goad, 1970). In eliminating the 7α-hydrogen from cycloartenol, *Fucus spiralis* aligns itself with the fungi (see Section II,B,2).

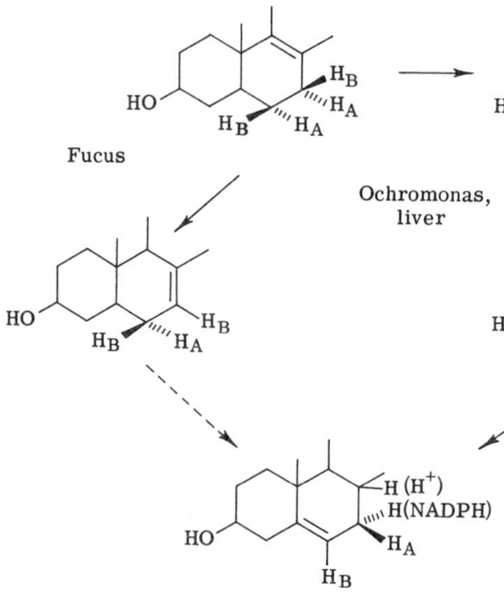

Fig. 5. Stereochemistry of formation of a $\triangle^5$ bond from a $\triangle^8$ bond.

The formation of the $\triangle^5$ double bond has been studied in *O. malhamensis*, and experiments with [5R-5-³H₁]MVA have demonstrated that the 6α-hydrogen is eliminated (A. R. H. Smith *et al.*, 1968b). Experiments on cholesterol biosynthesis have further shown that the 5α-hydrogen as well as the 6α-hydrogen is removed in forming the $\triangle^5$ bond (see Goad, 1970), and there is no reason to doubt that the same mechanism is involved in poriferasterol biosynthesis.

The insertion of the double bond at C-5 into ring B is an aerobic process, and it has been postulated that a hydroxylated intermediate is first involved and that this is the oxygen-requiring step. Topham and Gaylor (1970) have now demonstrated the anaerobic dehydration of ergosta-7,22-diene-3β,5α-diol (LXII)* to yield ergosterol by a purified enzyme system from bakers yeast; this gives strong support to the view that a 5α-hydroxy sterol is an intermediate in the reaction. Other hydroxy compounds previously investigated were 6α and 6β derivatives.

6. *Alkylation at C-24*

It is well established that methylation at C-24 arises by transmethylations from *S*-adenosyl methionine (Jauréguiberry *et al.*, 1965) and that the ethyl group arises by a sequential double transmethylation (see Lederer, 1969). More than one mechanism appears to be involved in sterol alkylation

* See page 30 for structure (LXII).

in algae (Fig. 6). Route (a)–(c), involving the intermediation of an ethylidene sterol, appears to function in the case of *Ochromonas malhamensis*. This route demands that the proton originally at C-24 shifts to C-25 and that only four hydrogens from two methyl groups of methionine are incorporated into the sterol molecule. With [4R-³H₁]MVA as substrate, a tritium will appear at C-24 in the unalkylated precursor (cycloartenol); in fucosterol (XIII) synthesized by *Fucus spiralis* it was found that the C-24 tritium was still present but had migrated to C-25 during alkylation (Goad and Goodwin, 1965, 1969); this was also the case with poriferasterol synthesized by *O. malhamensis* (A. R. H. Smith, 1969). Furthermore, mass spectra studies demonstrated that *O. malhamensis* retained only four deuterium atoms in its ethyl side chain when [CD₃]methionine was incorporated into the molecule (A. R. H. Smith *et al.*, 1967). Thus it does seem that route (a)–(c) (Fig. 6) is operating in *O. malhamensis*. It still remains to be shown whether an ethylidene sterol is a true intermediate. At the moment 28-isofucosterol appears to be the most likely intermediate for it is widely distributed in trace amounts, while fucosterol is found only in the Phaeophyta and certain fungi. Furthermore, tritiated fucosterol was not incorporated into poriferasterol (VI) (24β-ethyl) in *Chlorella ellipsodea*; it was, however, incorporated into clionasterol (X) (24α-ethyl) (Patterson

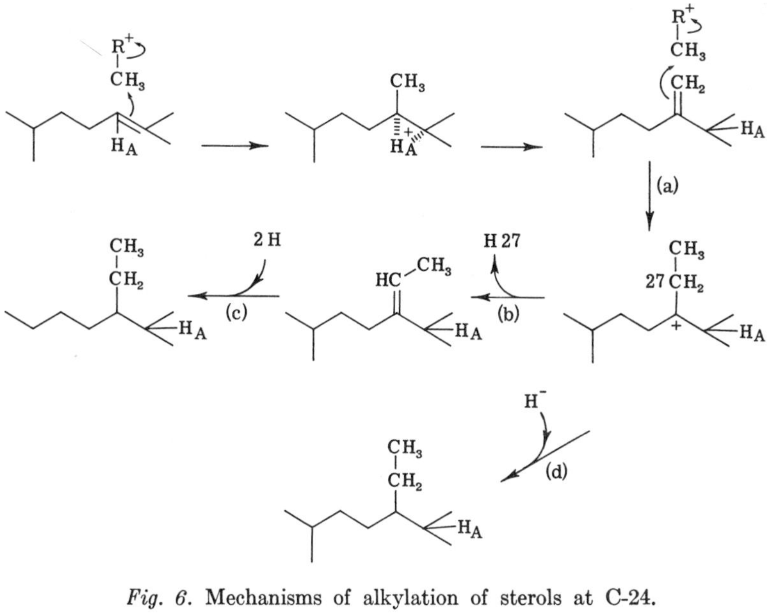

Fig. 6. Mechanisms of alkylation of sterols at C-24.

and Karlander, 1967), which leaves one with the intriguing thought that maybe fucosterol leads to the 24α series and isofucosterol to the 24β series. Experiments in the author's laboratory are in hand in an attempt to settle the problem.

However, route (a)–(c) does not function in *Chlorella vulgaris*, where all five hydrogens from the two methyl residues of methionine are retained in stigmasterol (Tomita *et al.*, 1970); thus pathway (a)–(d) (Fig. 6) is most likely in this organism.

(LVII)

(LVIII)

(LIX)

(LX)

(LXI)

(LXII)

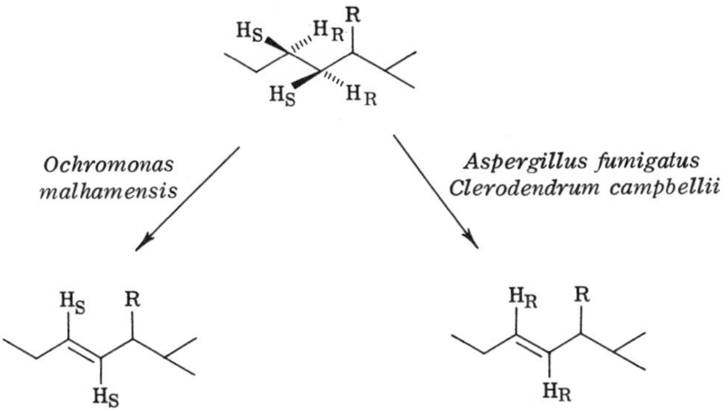

Fig. 7. Stereochemistry of hydrogen elimination in forming the Δ^{22} double bond in an alga (*Ochromonas malhamensis*), a higher plant (*Clerodendrum campbellii*), and a fungus (*Aspergillus fumigatus*).

7. Formation of the Δ^{22}-trans Double Bond

The formation of the Δ^{22}-*trans* double bond, a very characteristic reaction of plants, does not appear to be directly related to the alkylation reaction, although the formation of a Δ^{25} double bond in certain higher-plant sterols is so related (Bolger *et al.*, 1970). Furthermore, the enzyme concerned has not been isolated, and the immediate precursor has not been identified; however, the stereochemistry of the desaturation has been determined. In a sterol molecule, C-22 and C-23 originate from C-2 and C-5 of MVA, respectively, so that experiments with [$2R_1$-^{3}H$_1$]MVA and [$2S$-^{3}H$_1$]MVA and with [$5R$-^{3}H$_1$]MVA (LXIII) and [$5S$-^{3}H$_1$]MVA (LXIV) should indicate whether there is stereospecific removal of the hydrogens during desaturation and, if so, which hydrogens are involved. Appropriate degradations of poriferasterol synthesized by *Ochromonas malhamensis* in the presence of [$2R$-^{3}H$_1$]MVA, [$2S$-^{3}H$_1$]MVA, and [$5R$-^{3}H$_1$]MVA ([$5S$-^{3}H$_1$]MVA was not available at the time) demonstrated that the desaturation was stereospecific and that the 22-pro-R and 22-pro-S hydrogens were eliminated (A. R. H. Smith *et al.*, 1968a) (Fig. 7). A similar stereochemistry was observed at C-23 in ergosterol (IV) synthesized by *Ochromonas danica* (A. R. H. Smith, 1969). It will be seen in Section II,B,4 that exactly the opposite stereochemistry exists in ergosterol formation in fungi.

(LXIII)   (LXIV)

B. Fungi

1. *First Cyclic Precursor*

There is no doubt that lanosterol is present in fungi and that, as in animals but not photosynthetic plants, it is a precursor of sterols. Recently, it has been shown that squalene 2,3-oxide is converted into lanosterol in cell-free systems from yeast (Barton *et al.*, 1966; 1969) and *Phycomyces blakesleeanus* (Mercer and Johnson, 1969).

2. *Transformations in Ring B*

In the formation of the Δ^7 double bond in ring B it is the 7α-hydrogen that is eliminated in yeast (Caspi and Ramm, 1969) and in *Aspergillus niger* (Bimpson, 1970). The same stereochemistry was found in *Fucus spiralis*, but in *Ochromonas malhamensis* the opposite hydrogen (7β) is lost.

3. *Alkylation at C-24*

Work on *Neurospora* demonstrated that only two hydrogens of the methyl group of methionine appeared in the methyl side chain of ergosterol (see Lederer, 1969); thus a 24-methylene sterol was likely to be an intermediate and the pathway involved route (a)–(c) in Fig. 6. This view is supported by the demonstration that the hydrogen originally at C-24 in the unalkylated precursor migrates to C-25 in ergosterol (Akhtar *et al.*, 1967; Stone and Hemming, 1968) and the fact that many 24-methylene derivatives have been observed in fungi. However, in the slime mold *Dictyostelium discoideum* five methionine methyl hydrogens are retained in stigmast-22-en-3β-ol (LVI) (Lenfant *et al.*, 1967). This means that although a 24-methylene compound can be intermediate, a 24-ethylidene derivative cannot.

There seems to be some confusion as to the exact site in the biosynthetic sequence at which alkylation occurs, or it might be that different pathways are followed in different fungi. In yeast itself alkylation appears to take place after complete demethylation of lanosterol, because in the presence of [^{14}CH$_3$]methionine two metabolites were obtained, one of which was probably ergosta-5,7,22,24(28)-tetraen-3β-ol (LXV), but no methylated derivatives of lanosterol were observed (Katsuki and Bloch, 1967). Furthermore, (LXV) was converted into ergosterol by intact yeast cells but not by cell-free extracts (Turner and Parks, 1965; Katsuki and Bloch, 1967), and zymosterol (LXVI) is converted into (LXV) by intact cells (Katsuki and Bloch, 1967). The conclusion that methylation is a late stage in the biosynthetic sequence is supported by studies on the purified enzyme

S-adenosyl methionine:Δ^{24}-sterol methylase; the most effective substrate for this enzyme is zymosterol, while various 4-methyl sterols were very poor substrates (Moore and Gaylor, 1970). In contrast to these findings there are reports that 24-methylene-24,25-dihydrolanosterol is converted into ergosterol by yeast (Akhtar et al., 1966; Barton et al., 1966) and that lanosterol but not zymosterol is a precursor (Schwenk and Alexander, 1958).

In other fungi, methylation can occur at a much earlier point because 24-methylene-24,25-dihydrolanosterol can be isolated from *Phycomyces blakesleeanus* and methylated lanosterol derivatives are widespread in the wood-rotting fungi.

4. *Insertion of a* Δ^{22} *Double Bond*

Although the exact point in the biosynthetic sequence to ergosterol when the Δ^{22} double bond is inserted is not yet known, the stereochemistry of the desaturation has been established in *Aspergillus fumigatus* by the technique described for poriferasterol synthesis in *Ochromonas malhamensis* (Section II,A,7). Exactly the opposite stereochemistry is involved in *A. fumigatus*; the 22-pro-S and 23-pro-S hydrogens are eliminated (Bimpson et al., 1969a). A similar result for C-23 has been obtained for ergosterol synthesized by *Blakeslea trispora* (Bimpson, 1970).

5. *Biosynthesis of Protosterols*

In the formation of sterols, squalene 2,3-oxide cyclizes to form the cation (enzyme-substrate complex) (LXVII); stabilization of this by loss of hydrogen without any rearrangement would yield protosterol (LXVIII), found in *Fusidium coccineum*. This, by various secondary transformations, could yield fusidic acid (LIII) with tritiums in positions indicated when [4R-^{3}H$_1$]MVA is the substrate, and this has been demonstrated experimentally (Caspi and Mulheirn, 1970). It also contains radioactivity from [2-^{14}C]MVA at the carbons in (LIII), indicated by ● (D. Arigoni, quoted by Caspi and Mulheirn, 1970). As (LXVIII), found in *F. coccineum*, is probably a precursor of fusidic acid, it follows that the 4α-carbon of fusidic acid arises from the original 4α-carbon of the protosterol, which itself arises from C-2 of MVA (Stone et al., 1969, Moss and Nicolaidis, 1969). This is exactly opposite to what happens in liver (Sharpless et al., 1969) and higher plants (Ghisalberti et al., 1969), where the 4α-methyl is first removed and the 4β-methyl group takes up the 4α position before being removed itself. However, the situation in helvolic acid (LII), synthesized by *Cephalosporium caerulens*, is the same as in fusidic acid; that is, the 4β-methyl group is the first to be removed (Okuda et al., 1968).

C. Protozoa

The pentacyclic triterpene tetrahymanol (XXXIII) is of considerable biosynthetic interest because it appears that squalene is converted directly, that squalene 2,3-oxide is not an intermediate, and that the oxygen in the tetrahymanol molecule does not arise from atmospheric oxygen. Thus cyclization is initiated by a direct proton attack on squalene (Caspi *et al.*, 1968a,b).

(LXV)

(LXVI)

(LXVII)

(LXVIII)

(LXIX)

III. Metabolism

Although *Tetrahymena pyriformis* cannot synthesize sterols, it does metabolize cholesterol to cholesta-5,7,22-trien-3β-ol (LXIX) (Mallory *et al.*, 1968). The stereochemistry of the formation of the Δ^{22} double bond is the same as that observed in *Ochromonas malhamensis*; that is, the 22-pro-*R* and 23-pro-*S* hydrogens* are lost (Bimpson *et al.*, 1969b). The formation of the Δ^7 double bond involves the loss of the β-hydrogens from C-7 and C-8 (Bimpson *et al.*, 1969b). A similar *cis* elimination has been observed in the cockroach (Clayton and Edwards, 1963).

Pythium periplocum, which cannot synthesize sterols, will metabolize cholesterol, but the reaction is very slight and the metabolites have not yet been obtained in sufficient amounts for complete identification (Hendrix *et al.*, 1970).

IV. Function

A. INTRODUCTION

Little is known of the function of sterols in algae, but in fungi a number of functions seem to be well established. These will be summarized here, but for a full discussion a recent review by Hendrix (1970) should be consulted.

B. GROWTH

The growth in liquid culture of a number of fungi is stimulated by sterols, for example, *Aspergillus niger* (Jefferson and Sisco, 1961), *Pythium* sp. (Lenny and Klemmer, 1966), and *Phytophthora* sp. (Elliott *et al.*, 1966). The response in *Aspergillus niger* (Jefferson and Sisco, 1961) and *Phytophthora parasitica* var. *nicotianae* (Hendrix *et al.*, 1969) is significant only in aerated cultures.

C. REPRODUCTION

A number of workers reported almost simultaneously that exogenous sterols are necessary for oospore formation in the pythiaceous fungi (Elliott *et al.*, 1964; Haskins *et al.*, 1964; Hendrix, 1964; Leal *et al.*, 1964), and Hendrix (1964) also showed that they are necessary for zoosporangium

---

* Because of the *R/S* rules, the 23-pro-*S* hydrogen of cholesterol is equivalent to the 23-pro-*R* hydrogen of poriferasterol and ergosterol.

production. Most of the *Phytophthora* sp. that respond in this way are homothallic; only three out of eight pairs of heterothallic species examined produced oospores when cultured on a defined medium containing sterols (Hunter *et al.*, 1965). The great difficulty in obtaining pure sterols and preventing oxidative change during experiments has made a number of careful investigations on the structural requirements for activity in oosporogenesis of doubtful value, although it is clear that a 3β-hydroxyl, a Δ⁵ double bond, and a side chain of eight carbon atoms are essential (Elliott *et al.*, 1966). Hendrix (1970) quotes concentrations of about 10^{-6} M for minimum activity and 10^{-5} to 10^{-4} M for maximum activity.

As already pointed out, the female forms of *Achlya bisexualis* produce antheridiol (XXXVI), which stimulates growth of antheridiae in male forms and thus allows reproductive activity (Barksdale, 1969; Edwards *et al.*, 1969).

D. Membrane Phenomena

There is not a great deal of compelling evidence for the intracellular localization of sterols in sterol-synthesizing fungi or in sterol-free organisms that can bind exogenous sterols, but certain functional activities are indicated and discussed fully by Hendrix (1970); they are summarized here. In Mycoplasma, sterols are involved as carriers for uptake of carbon sources (P. F. Smith, 1969). The antifungal polyene antibiotics appear to function by interacting with cellular sterols and thus disrupting cellular permeability, and it may be significant that the non-sterol-synthesizing Pythiaceous fungi are relatively resistant to these antibiotics.

Yeast grown anaerobically produces no ergosterol and only promitochondria, which are much less efficient than mitochondria from aerobically grown yeast, which contains its full complement of ergosterol. It is claimed that yeast grown anaerobically in the presence of ergosterol has higher cytochrome levels (Kellerman *et al.*, 1969) and, when aerated, synthesizes mitochondrial enzymes faster than yeast grown under the same conditions in the absence of ergosterol (Hebb and Slebodnik, 1958). However, the non-sterol-containing fungi contain mitochondria, as revealed by electron microscopy (see, e.g., Chapman and Vujicic, 1965), and mitochondria that appear structurally normal have been observed in *Pythium ultimum* grown in the absence of exogenous sterol (Marchant and Smith, 1968).

References

Aaronson, S., and Baker, H. (1961). *J. Protozool.* **8**, 274.

Adam, H. K., Campbell, I. M., and McCorkindale, N. J. (1967). *Nature (London)* **216**, 397.

Akhtar, M., Hunt, P. F., and Parvez, M. A. (1966). *Chem. Commun.* p. 565.
Akhtar, M., Hunt, P. F., and Parvez, M. A. (1967). *Biochem. J.* **109,** 877.
Alcaide, A., Devys, M., and Barbier, M. (1968). *Phytochemistry* **7,** 329.
Alcaide, A., Barbier, M., Potier, P., Magueur, A. M., and Teste, J. (1969). *Phytochemistry* **8,** 2301.
Alt, G. H., and Barton, D. H. R. (1954). *J. Chem. Soc., London* p. 1356.
Arsenault, G. P., Bremann, K., Barksdale, A. W., and McMorris, T. C. (1968). *J. Amer. Chem. Soc.* **90,** 5635.
Avivi, L., Iaron, O., and Halevy, S. (1967). *Comp. Biochem. Physiol.* **21,** 321.
Baisted, D. J., Gardner, R. L., and McReynolds, L. A. (1968). *Phytochemistry* **7,** 945.
Barksdale, A. W. (1969). *Science* **166,** 831.
Barton, D. H. R. (1945). *J. Chem. Soc., London* p. 813.
Barton, D. H. R., and Brunn, T. (1951). *J. Chem. Soc.,* 2728.
Barton, D. H. R., Harrison, D. M., and Moss, G. P. (1966). *Chem. Commun.* p. 595.
Barton, D. H. R., Gosden A. F., Mellows, G., and Widdowson, D. A. (1968). *Chem. Commun.* p. 1067.
Bauslauigh, G., Just, G., and Blank, F. (1964). *Nature (London)* **202,** 1218.
Bazzano, G. (1965). University Microfilms 65-4100.
Beastall, G. H., Rees, H. H., and Goodwin, T. W. (1971a). *Tetrahedron Lett.* No. **52,** 4935.
Beastall, G. H., Rees, H. H., and Goodwin, T. W. (1971b). *Febs. Lett.* **18,** 175.
Bergmann, W., and Feeney, R. J. (1950). *J. Org. Chem.* **15,** 812.
Beereboom, J. J., Fazakerley, H., and Halsall, T. G. (1957). *J. Chem. Soc.* 3437.
Bimpson, T. (1970). Ph.D. Thesis, University of Liverpool.
Bimpson, T., Goad, L. J., and Goodwin, T. W. (1969a). *Chem. Commun.* p. 297.
Bimpson, T., Goad, L. J., and Goodwin, T. W. (1969b). *Biochem. J.* **115,** 857.
Bird, C. W., Lynch, J. M., Pirt, F. J., Reid, W. W., Brooks, C. J. W., and Middleditch, B. S. (1971). *Nature (London)* **230,** 473.
Black, W. A. P., and Cornhill, W. J. (1951). *J. Sci. Food Agr.* **2,** 387.
Bolger, L. M., Rees, H. H., Ghisalberti, E. L., Goad, L. J., and Goodwin, T. W. (1970). *Biochem. J.* **118,** 197.
Brandt, R. D., Ourisson, G., and Pryce, R. J. (1969). *Biochem. Biophys. Res. Commun.* **37,** 399.
Brandt, R. D., Pryce, R. J., Anding, C., and Ourisson, G. (1970). *Eur. J. Biochem.* **17,** 344.
Breivick, O. N., Owades, J. L., and Light, R. F. (1954). *J. Org. Chem.* **19,** 1734.
Britt, J. J., and Bloch, K. (1961). *Comp. Biochem. Physiol.* **2,** 202.
Bryce, T. A., Campbell, I. M., and McCorkindale, N. J. (1967). *Tetrahedron* **23,** 3427.
Cahn, R. S., and Ingold, C. K. (1951). *J. Chem. Soc., London.* p. 612.
Carter, P. W., Heilbron, I. M., and Lythgoe, B. (1939). *Proc. Roy. Soc. B.* **128,** 82.
Caspi, E., and Mulheirn, L. J. (1970). *J. Amer. Chem. Soc.* **92,** 404.
Caspi, E., and Ramm, P. J. (1969). *Tetrahedron Lett.* No. 3, p. 181.
Caspi, E., Zander, J. M., Greig, J. B., Mallory, F. M., Conner, R. L., and Landrey, J. R. (1968a). *J. Amer. Chem. Soc.* **90,** 3563.
Caspi, E., Greig, J. B., and Zander, J. M. (1968b). *Biochem. J.,* **109,** 931.
Chapman, J. A., and Vujicic, R. (1965). *J. Gen. Microbiol.* **41,** 275.
Chen, Y. S., and Haskins, R. H. (1963). *Can. J. Chem.* **41,** 1647.
Ciereszko, L. S., Johnson, M. A., Schmidt, R. W., and Koons, C. B. (1968). *Comp. Biochem. Physiol.* **24,** 899.
Clarke, S. M., and McKenzie, M. (1967). *Nature (London)* **213,** 504.
Clayton, R. B., and Edwards, A. M., (1963). *J. Biol. Chem.* **238,** 1966.

Collins, R. P., and Kalnins, K. (1969). *Comp. Biochem. Physiol.* **30**, 779.

Conner, R. L., and Ungar, F. (1964). *Exp. Cell Res.* **36**, 134.

de Souza, N. J., and Nes, W. R. (1968). *Science* **162**, 363.

Edwards, J. A., Mills, J. S., Sundeen, J., and Fried, J. H. (1969). *J. Amer. Chem. Soc.* **91**, 1248.

Elliott, C. G., Hendrie, M. R., Knights, B. A., and Parker, W. (1964). *Nature (London)* **203**, 427.

Elliott, C. G., Hendrie, M. R., and Knights, B. A. (1966). *J. Gen. Microbiol.* **42**, 425.

Fiore, J. V. (1948). *Arch. Biochem.* **16**, 161.

Fürst, W. (1966). *Justus Liebigs Ann. Chem.* **699**, 206.

Gershengorn, M. C., Smith, A. R. H., Goulston, G., Goad, L. J., Goodwin, T. W., and Haines, T. H. (1968). *Biochemistry* **7**, 1698.

Ghisalberti, E. L., de Souza, N. J., Ress, H. H., Goad, L. J., and Goodwin, T. W. (1969). *Chem. Commun.* p. 1403.

Gibbons, G. F. (1968). Ph.D. Thesis, University of Liverpool.

Gibbons, G. F., Goad, L. J., and Goodwin, T. W. (1967). *Phytochemistry* **6**, 677.

Gibbons, G. F., Goad, L. J., and Goodwin, T. W. (1968). *Phytochemistry* **6**, 677.

Goad, L. J. (1970). *Biochem. Soc. Symp.* **29**, 45.

Goad, L. J., and Goodwin T. W. (1965). *Biochem. J.* **96**, 79P.

Goad, L. J., and Goodwin, T. W. (1967). *Proc. Jap. Conf. Biochem. Lipids. 9th*, 1967, p. 37.

Goad, L. J., and Goodwin T. W. (1969). *Eur. J. Biochem.* **7**, 502.

Goad, L. J., Hamman, A. S. A., Dennis, A., and Goodwin, T. W. (1966). *Nature (London)* **210**, 1322.

Godtfredsen, W. O., von Daehne, W., Vangedal, S., Marquet, A., Arigoni, D., and Melera, A. (1965). *Tetrahedron* **21**, 3563.

Godtfredsen, W. O., Horck, H., Van Tamelen, E. E., Willett, J. D., and Clayton R. B. (1968). *J. Amer. Chem. Soc.* **90**, 208.

Goodwin, T. W. (1971). *Biochem. J.* **123**, 293.

Goulston, G. (1969). Ph.D. Thesis, University College of Wales, Aberystwyth.

Goulston, G., and Mercer, E. I. (1969). *Phytochemistry* **8**, 1945.

Goulston, G., Goad, L. J., and Goodwin, T. W. (1967). *Biochem. J.*, **102**, 15C.

Goulston, G., Mercer, E. I., and Goad, L. J. (1972). In press.

Guider, J. M., Halsall, T. G., and Jones, E. R. H. (1954). *J. Chem. Soc.* 4471.

Halevy, S., Avivi, L., and Katan, H. (1966). *J. Protozool.* **13**, 480.

Hall, J., Smith, A. R. H., Goad, L. J., and Goodwin, T. W. (1969). *Biochem. J.* **112**, 129.

Halsall, T. G., and Sayer, G. C. (1959). *J. Chem. Soc.* 2031.

Hamilton, J. G., and Castrejon, R. N. (1966). *Fed. Proc., Fed. Amer. Soc. Exp. Biol.* **25**, 221.

Hanson, K. R. (1966). *J. Amer. Chem. Soc.* **88**, 2731.

Haskins, R. H., Tulloch, A. P., and Micetich, R. G. (1964). *Can. J. Microbiol.* **10**, 187.

Hebb, C. R., and Slebodnik, J. (1958). *Exp. Cell Res.* **14**, 286.

Heftmann, E., Wright, B. E., and Liddel, C. V. (1960). *Arch. Biochem. Biophys.* **91**, 266.

Heilbron, I. M. (1942). *J. Chem. Soc., London* p. 79.

Heilbron, I. M., Parry, E. G., and Phipers, R. F. (1935). *Biochem. J.* **29**, 1376.

Hendrix, J. W. (1964). *Science* **144**, 1028.

Hendrix, J. W. (1970). *Annu. Rev. Plant Pathol.* **8**, 111.

Hendrix, J. W., Guttman, S. M., and Wightman, D. L. (1969). *Phytopathology* **59**, 1620.

Hendrix, J. W., Bennett, R. D., and Heftmann, E. (1970). *Microbios* **5**, 11.

Hewlins, M. J. E., Ehardt, J. D., Hirth, L., and Ourisson, G. (1969). *Eur. J. Biochem.* **8**, 184.

Hunek, S. (1969). *Phytochemistry* **8**, 1313.

Hunter, J. H., Berg, L. A., Harnish, W. H., and Merz, W. G. (1965). *Proc. W. Va. Acad. Sci.* **37**, 75.

Idler, D. R., and Wiseman, P. (1970). *Comp. Biochem. Physiol.* **35**, 679.

Idler, D. R., Saito, A., and Wiseman, P. (1968). *Steroids* **11**, 465.

Ikekawa, N., Tsuda, K., and Morisaki, N. (1966) *Chem. Ind. (London)* p. 1179.

Ikekawa, N., Morisaki, N., Tsuda, K., and Ycshida, T. (1968). *Steroids* **12**, 41.

Ito, S., Tamura, T., and Matsumoto, T. (1959). *Nippon Daigaku Kogaku Kenkyusho Iho* **13**, 99.

Jackson, L. L., and Frear, D. S. (1968). *Phytochemistry* **7**, 651.

Jauréguiberry, G., Law, J. H., McCloskey, J. A., and Lederer, E. (1965). *Biochemistry* **4**, 347.

Jefferson, W. E., Jr., and Sisco, G. (1961). *J. Gen. Physiol.* **44**, 1029.

Katsuki, H., and Bloch, K. (1967). *J. Biol. Chem.* **242**, 222.

Kellerman, G. M., Briggs, D. R., and Linnane, A. W. (1969). *J. Cell Biol.* **42**, 378.

Kirk, P. W., Jr., and Catalfomo, P. (1970). *Phytochemistry* **9**, 595.

Knights, B. A. (1965). *Phytochemistry* **4**, 857.

Knights, B. A. (1970a). *Phytochemistry* **9**, 903.

Knights, B. A. (1970b). *Phytochemistry* **9**, 701.

Leal, J. A., Friend, J., and Holliday, P. (1964). *Nature (London)* **203**, 545.

Lederer, E. (1969). *Quart. Rev., Chem. Soc.* **23**, 453.

Lenfant, M., Zissman, E., and Lederer, E. (1967). *Tetrahedron Lett.* No. 12, p. 1049.

Lenfant, M., Lecompte, M. F., and Farrugia, G. (1970). *Phytochemistry* **9**, 2529.

Lenny, J. F., and Klemmer, H. W. (1966). *Nature (London)* **209**, 1365.

Lenton, J. R., Hall, J., Smith, A. R. H., Ghisalberti, E. L., Rees, H. H., Goad, L. J., and Goodwin, T. W. (1971). *Arch. Biochem. Biophy.* **143**, 664.

Levin, E. Y. and Bloch, K. (1964). *Nature, London* **202**, 90.

Low, E. M. (1955). *J. Mar. Res.* **14**, 199.

McCorkindale, N. J., Hutchinson, S. A., Pursey, B. A., Scott, W. T., and Wheeler, R. (1969). *Phytochemistry* **8**, 861.

Mallory, F. B., Gordon, J. T., and Conner, R. L. (1963). *J. Amer. Chem. Soc.* **85**, 1382.

Mallory, F. B., Conner, R. C., Landray, J. R., and Iyengar, C. L. (1968). *Tetrahedron Lett.* No. 58, p. 6103.

Marchant, R., and Smith, D. G. (1968). *J. Gen. Microbiol.* **50**, 391.

Mercer, E. I., and Johnson, M. W. (1969). *Phytochemistry* **8**, 2329.

Milazzo, F. H. (1965). *Can. J. Bot.* **43**, 1347.

Miller, J. D. A. (1962). *In* "Physiology and Biochemistry of Algae" (R. A. Lewin, ed.), p. 357. Academic Press, New York.

Moore, J. T., Jr. and Gaylor, J. L. (1970). *J. Biol. Chem.* **245**, 4684.

Moss, G. P., and Nicolaidis, S. A. (1969). *Chem. Commun.* p. 1072.

Motzfeldt, A.-M. (1970). *Acta Chem. Scand.* **24**, 1814.

Okuda, T., Sato, Y., Hattori, T., and Igarashi, H. (1968). *Tetrahedron Lett.* No. 46, p. 4769.

Patterson, G. W. (1967). *Plant Physiol.* **42**, 1457.

Patterson, G. W. (1968). *Comp. Biochem. Physiol.* **24**, 501.

Patterson, G. W. (1971). *Lipids* **6**, 120.

Patterson, G. W., and Karlander, E. P. (1967). *Plant Physiol.* **42**, 1651.

Petzoldt, K., Kuhne, M., Blanke, E., Kieslich, K., and Kasper, E. (1967). *Justus Liebigs Ann. Chem.* **709**, 203.

Popják, G., and Cornforth, J. W. (1966). *Biochem. J.* **101**, 553.

Rees, H. H., Goad, L. J., and Goodwin, T. W. (1968). *Biochem. J.* **107**, 417.

Rees, H. H., Goad, L. J., and Goodwin, T. W. (1969). *Biochim. Biophys. Acta.* **176**, 892.

Reiner, E., Topliff, J., and Wood, J. D. (1962). *Can. J. Biochem. Physiol.* **40**, 1401.

Reitz, R. C., and Hamilton, J. G. (1968). *Comp. Biochem. Physiol.* **25**, 401.

Saito, A., and Idler, D. R. (1966). *Can. J. Biochem.* **44**, 1195.

Schubert, K., Rose, G., Tümmler, R., and Ikekawa, N. (1964). *Hoppe-Seyler's Z. Physiol. Chem.* **339**, 293.

Schubert, K., Rose, G., and Hörhold, C. (1967). *Biochim. Biophys. Acta* **137**, 168.

Schubert, K., Rose, G., Wachtel, H., Hörhold, C., and Ikekawa, N. (1968). *Eur. J. Biochem.* **5**, 246.

Schwenk, E., and Alexander, G. J. (1958). *Arch. Biochem. Biophys.* **76**, 65.

Sharpless, K. B., Snyder, T. E., Spencer, T. A., Maheshwari, K. K., Nelson, J. A., and Clayton, R. B. (1969). *J. Amer. Chem. Soc.* **91**, 3394.

Singh, P., and Rangaswami, S. (1966). *Curr. Sci.* **35**, 515.

Smith, A. R. H. (1969). Ph.D. Thesis, University of Liverpool.

Smith, A. R. H., Goad, L. J., Goodwin, T. W., and Lederer, E. (1967). *Biochem. J.* **104**, 56C.

Smith, A. R. H., Goad, L. J., and Goodwin, T. W. (1968a). *Chem. Commun.* p. 926.

Smith, A. R. H., Goad, L. J., and Goodwin, T. W. (1968b). *Chem. Commun.* p. 1259.

Smith, P. F. (1969). *Lipids* **4**, 331.

Sharratt, A. N., and Madhosingh, C. (1967). *Can. J. Microbiol.* **13**, 1351.

Stern, A. I., Schiff, J. A., and Klein, H. P. (1960). *J. Protozool.* **7**, 52.

Stone, K. J., and Hemming, F. W. (1968). *Biochem. J.*, **109**, 877.

Stone, K. J., Roeske, W. R., Clayton, R. B., and Van Tamelen, E. E. (1969). *Chem. Commun.* p. 530.

Suzue, G., Tsukada, K., Nakai, C., and Tanaka, S. (1968). *Arch. Biochem. Biophys.* **123**, 644.

Tanahashi, Y., and Takahashi, T. (1966). *Bull. Chem. Soc. Jap.* **39**, 848.

Tillman, R. W., and Bean, G. A. (1970). *Mycologia* **62**, 428.

Tomita, Y., Uomori, A., and Minato, H. (1970). *Phytochemistry* **9**, 555.

Topham, R. W., and Gaylor, J. L. (1970). *J. Biol. Chem.* **245**, 2319.

Tsuda, K., and Sakai, K. (1960). *Chem. Pharm. Bull.* **8**, 554.

Tsuda, K., Akagi, S., and Kishida, Y. (1957). *Science* **126**, 927.

Tsuda, K., Akagi, S., and Kishida, Y. (1958a). *Pharm. Bull.* **6**, 101.

Tsuda, K., Akagi, S., Kishida, Y., Hayatsu, R., and Sakai, K. (1958b). *Pharm. Bull.* **6**, 724.

Tsuda, K., Hayatsu, R., Kishida, Y., and Akagi, S. (1958c). *J. Amer. Chem. Soc.* **80**, 921.

Tsuda, K., Sakai, K., Tanabe, K., and Kishida, Y. (1960). *J. Amer. Chem. Soc.* **82**, 1442,

Tsuda, Y., Morimoto, A., Sano, T., Inubishi, Y., Mallory, F. B., and Gordon, J. T. (1965). *Tetrahedron Lett.* No. 19, p. 1427.

Turner, J. R., and Parks, L. W. (1965). *Biochim. Biophys. Acta.* **98**, 394.

Vacheron, M. J., and Michel, G. (1968). *Phytochemistry* **7**, 1645.

Wieland, H., and Coutelle, G. (1941). *Justus Liebigs Ann. Chem.* **548**, 275.

Wieland, H., and Prelog, V. (1947). *Helv. Chim. Acta* **30**, 1028.

Wieland, H., Pasedach, H., and Ballauf, A. (1937). *Justus Liebigs Ann. Chem.* **529**, 68.

Wieland, H., Rath, F., and Hesse, H. (1941). *Justus Liebigs Ann. Chem.* **548**, 19 and 34.

Williams, B. L., Goodwin, T. W., and Ryley, J. F. (1966). *J. Protozool.* **13**, 227.

CHAPTER 2

Comparative Biochemistry of Fatty Acids in Eukaryotic Microorganisms

J. Erwin

I. Introduction*

Long-chain fatty acids are prominent components of both the polar lipids of biomembranes and the triglycerides of the storage fats of cells and microorganisms. The development of thin-layer chromatographic and gas-liquid chromatographic techniques has enabled investigators to rapidly separate and identify fatty acids obtained from small samples of biological materials. This, in turn, has fostered investigations of the fatty acids of the lipids of a wide variety of microorganisms. Much remains to be done, but a substantial amount of information is now available concerning the types of fatty acid structures found in different organisms and their distribution among different classes of complex lipids.

While environmental factors greatly influence the fatty acid composition of an organism (see Section IV,C), the capacities of an organism for synthesizing various fatty acids is under genetic control (see Chapter 6), and these capacities might be expected to vary from one group of organisms to the next. Present information demonstrates that this is indeed so, and several phyletic patterns of fatty acid composition and biosynthesis have been found among eukaryotic microorganisms, plants, and animals. These patterns, along with their possible relationship to cellular physiology, will be discussed in some detail in this chapter.

A. Notes on Taxonomic Classification

In this chapter I have followed the scheme of Cronquist (1971) for dividing the algae into major divisions and classes and have relied heavily on G. M. Smith (1950) and Silva (1962) for assignment to orders. I have treated the phytoflagellates as algae and employed the usual formal nomenclature. I have included the slime molds with the protozoa; indeed, except for the exclusion of the phytoflagellates, I have adhered to the semiofficial taxonomic classification of protozoa developed by the committee on taxonomy of The Society of Protozoologists (see Honigberg et al., 1964). In defining and classifying yeasts I have followed the rules outlined by Lodder and associates (see Lodder, 1970).

B. Notes on Fatty Acid Nomenclature

In addition to using systematic and trivial names of fatty acids, I have also employed a shorthand notation (see Schlenk and Sand, 1967). With

* Abbreviations: CoA, coenzyme A; ACP, acyl carrier protein; NADH, reduced nictotinamide adenine dinucleotide; NADPH, reduced nicotinamide adenine dinucleotide phosphate; ATP, adenosine triphosphate; GLC, gas-liquid chromatography; TLC, thin-layer chromatography; PC, phosphatidyl choline; MGDG, monogalactosyl diglyceride; NAC, N-acetylcysteamine.

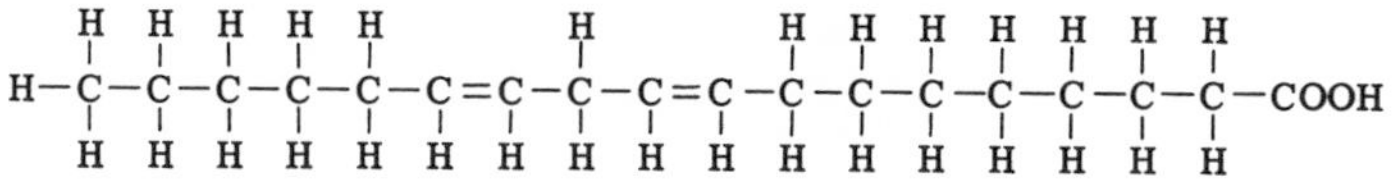

All *cis*-9, 12-octadecadienoic acid; 18:2ω6

Fig. 1. Linoleic acid.

this notation system, linoleic acid would be (Fig. 1) written as 18:2ω6, the number before the colon designating the number of carbon atoms in the molecule and the number following the colon indicating the number of methylene-interrupted double bonds in the molecule. The number following the omega (ω) symbol is the number of carbon atoms from the last double bond to the terminal methyl group. Since all the fatty acids for which this nomenclature is employed have the methylene-interrupted rhythm of unsaturation, the position of the ω terminal double bond indicates the positions of all other double bonds in the molecule. The configuration of unsaturated fatty acids is assumed to be all *cis* unless otherwise designated. (Thus oleic acid is written as 18:1ω9, while its *trans* isomer, elaidic acid, is written as *trans*-18:1ω9.) The prefix *a* indicates an *anteiso* branched-chain fatty acid, while the prefix *i* indicates an *iso* branched-chain fatty acid (see Fig. 3). The symbol Δ (see Law *et al.*, 1963) is used to designate a cyclopropane fatty acid (see Fig. 4).

II. Saturated Fatty Acids

A. TYPICAL SATURATED FATTY ACIDS OF MICROORGANISMS

Among the saturated fatty acids of the eukaryotic microorganisms, straight-chain compounds of 12 to 20 carbon atoms predominate, primarily palmitic and stearic acids (Fig. 2). Branched-chain fatty acids of the *iso* and *anteiso* type (Fig. 3), if present, are usually minor fatty acid components. Cyclopropane fatty acids (Fig. 4) are rarely encountered in eukaryotic microorganisms, having been reported in only a small group of flagellated protozoa (Table XII). This is in contrast to bacteria where *iso* and *anteiso* branched-chain fatty acids and cyclopropane fatty acids are not only widely distributed but often exceed in concentration the straight-chain fatty acids. The prokaryotic blue-green algae resemble the eukaryotic microorganisms in that they synthesize primarily palmitic and stearic acids and do not synthesize either cyclopropane or branched-chain fatty acids (Table III).

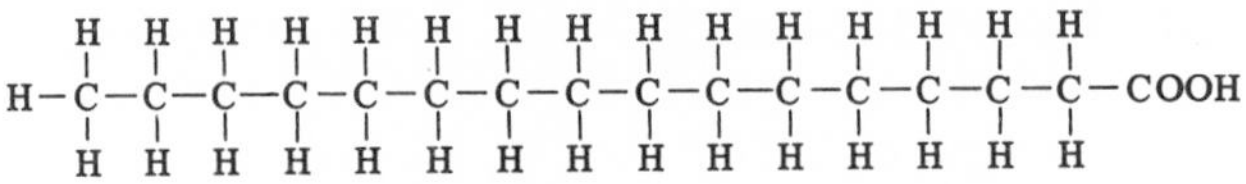

$$H-\underset{\underset{H}{|}}{\overset{\overset{H}{|}}{C}}-\underset{\underset{H}{|}}{\overset{\overset{H}{|}}{C}}-\underset{\underset{H}{|}}{\overset{\overset{H}{|}}{C}}-\underset{\underset{H}{|}}{\overset{\overset{H}{|}}{C}}-\underset{\underset{H}{|}}{\overset{\overset{H}{|}}{C}}-\underset{\underset{H}{|}}{\overset{\overset{H}{|}}{C}}-\underset{\underset{H}{|}}{\overset{\overset{H}{|}}{C}}-\underset{\underset{H}{|}}{\overset{\overset{H}{|}}{C}}-\underset{\underset{H}{|}}{\overset{\overset{H}{|}}{C}}-\underset{\underset{H}{|}}{\overset{\overset{H}{|}}{C}}-\underset{\underset{H}{|}}{\overset{\overset{H}{|}}{C}}-\underset{\underset{H}{|}}{\overset{\overset{H}{|}}{C}}-\underset{\underset{H}{|}}{\overset{\overset{H}{|}}{C}}-\underset{\underset{H}{|}}{\overset{\overset{H}{|}}{C}}-\underset{\underset{H}{|}}{\overset{\overset{H}{|}}{C}}-COOH$$

Palmitic acid (hexadecanoic acid; 16:0)

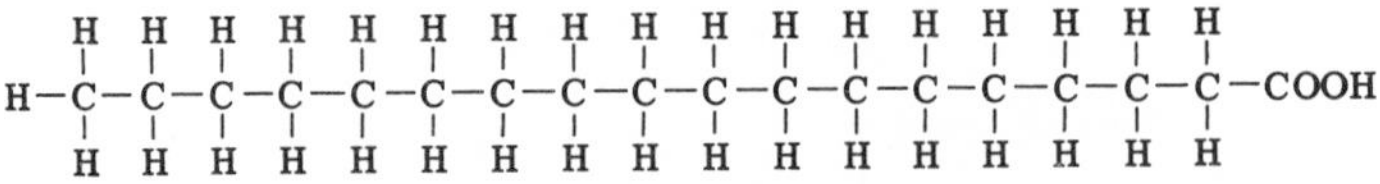

$$H-\underset{\underset{H}{|}}{\overset{\overset{H}{|}}{C}}-\underset{\underset{H}{|}}{\overset{\overset{H}{|}}{C}}-\underset{\underset{H}{|}}{\overset{\overset{H}{|}}{C}}-\underset{\underset{H}{|}}{\overset{\overset{H}{|}}{C}}-\underset{\underset{H}{|}}{\overset{\overset{H}{|}}{C}}-\underset{\underset{H}{|}}{\overset{\overset{H}{|}}{C}}-\underset{\underset{H}{|}}{\overset{\overset{H}{|}}{C}}-\underset{\underset{H}{|}}{\overset{\overset{H}{|}}{C}}-\underset{\underset{H}{|}}{\overset{\overset{H}{|}}{C}}-\underset{\underset{H}{|}}{\overset{\overset{H}{|}}{C}}-\underset{\underset{H}{|}}{\overset{\overset{H}{|}}{C}}-\underset{\underset{H}{|}}{\overset{\overset{H}{|}}{C}}-\underset{\underset{H}{|}}{\overset{\overset{H}{|}}{C}}-\underset{\underset{H}{|}}{\overset{\overset{H}{|}}{C}}-\underset{\underset{H}{|}}{\overset{\overset{H}{|}}{C}}-\underset{\underset{H}{|}}{\overset{\overset{H}{|}}{C}}-\underset{\underset{H}{|}}{\overset{\overset{H}{|}}{C}}-COOH$$

Stearic acid (octadecanoic acid; 18:0)

Fig. 2. Saturated straight-chain fatty acids.

Iso acids:

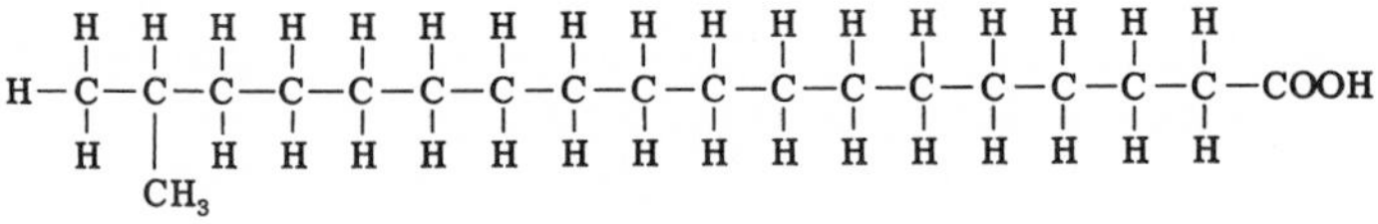

17-Methyloctadecanoic acid (i-19:0)

Anteiso acids:

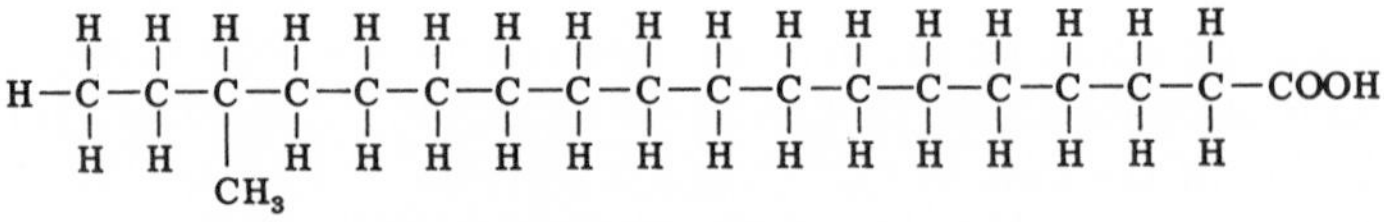

16-Methyloctadecanoic acid (a-19:0)

Fig. 3. Branched-chain fatty acids.

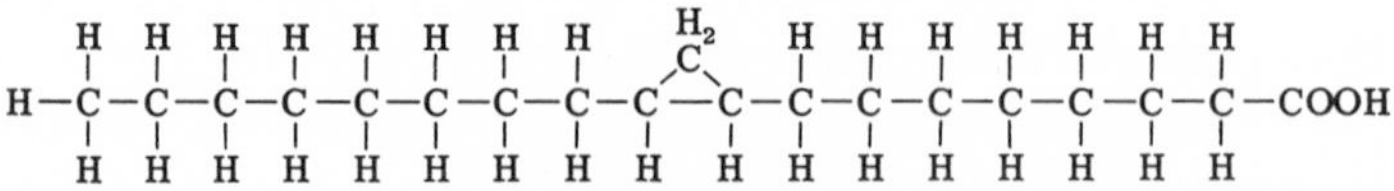

Lactobacillic acid
(9,10-methyleneoctadecanoic acid; 9-19:0Δ)

Fig 4. Cyclopropane fatty acids.

Little information exists concerning the occurrence of hydroxy acids among eukaryotic microorganisms. They have not been reported to exist in protozoa and most algae but are found in yeasts (see Fulco, 1967b), at least some fungi (see Table X), and particularly in the spores of plant rusts (see Tulloch, 1964; Tulloch and Ledingham, 1964).

B. BIOSYNTHESIS OF STRAIGHT-CHAIN FATTY ACIDS

1. *Fatty Acid Synthetase System*

At one time it was widely believed that long-chain fatty acids were synthesized biologically by a simple reversal of the β-oxidation pathway of fatty acid catabolism (Lynen, 1961). However, Wakil, Lynen, and others (Wakil, 1958; Bressler and Wakil, 1961; Lynen, 1961; Brady, 1958) demonstrated that in avian and mammalian liver and in the yeast *Saccharomyces cerevisiae* a fatty acid synthetase system exists that synthesizes long-chain fatty acids via the sequential condensation of malonyl-CoA units.

Malonate itself can be synthesized from acetate and carbon dioxide (Wakil and Gibson, 1960; Lynen, 1961) by the biotin-containing enzyme acetyl-CoA carboxylase. This enzyme is accessory to, but not a part of, the fatty acid synthetase system of animals and yeasts (Wakil and Gibson, 1960; Rasmussen and Klein, 1968a,b; Lynen, 1961; Lynen *et al.*, 1968; Phillips *et al.*, 1970). Fatty acid synthesis in enzyme preparations obtained from yeast and animals is stimulated by citrate and isocitrate; this effect is exerted not on the synthetase itself but on acetyl-CoA carboxylase, and it appears to be allosteric in nature (Rasmussen and Klein, 1968a,b,c).

Malonate supplies all the carbon atoms of the long-chain fatty acid with the exception of the two methyl terminal carbons. The latter are supplied by acetate (Fig. 5), and thus acetate may be regarded as a primer for fatty acid biosynthesis (Lynen, 1961). The use of acetate as the primer results exclusively in the formation of even-chain fatty acids (Fig. 5), usually palmitate (Bressler and Wakil, 1961; Lynen, 1961). Substitution of propionate for acetate as primer (Fig. 5) results in the formation of odd-chain fatty acids (Bressler and Wakil, 1961; Lynen, 1961).

The detailed mechanism of fatty acid biosynthesis from malonyl-CoA was finally resolved by studies with the individual isolated enzymes of the soluble fatty acid synthetase system of *Escherichia coli*. The essential features of this mechanism had, however, been originally postulated by Lynen on the basis of experiments with the particulate fatty acid synthetase system of yeast. This mechanism is outlined in Fig. 6. The substrates, intermediates, and products in fatty acid biosynthesis are acyl-ACP derivatives (Fig. 6); thus the initial reactions in fatty acid synthesis consist

Even-chain fatty acids:

$$CH_3C\overset{O}{\underset{}{\diagdown}}S-CoA \ + \ n \ HOOCCH_2C\overset{O}{\underset{}{\diagdown}}S-CoA \ + \ 2n \ NADPH \ + \ 2n \ H^+$$

$$\downarrow$$

$$CH_3(CH_2)n\text{-}COOH \ + \ 2n \ NADP^+ \ + \ n \ CO_2 \ + \ n\text{-}1\,H_2O \ + \ n\text{+}1 \ CoA$$

Odd-chain fatty acids:

$$CH_3CH_2C\overset{O}{\underset{}{\diagdown}}S-CoA \ + \ n \ HOOCCH_2C\overset{O}{\underset{}{\diagdown}}S-CoA \ + \ 2n \ NADPH \ + \ 2n \ H^+$$

$$\downarrow$$

$$CH_3CH_2(CH_2)n\text{-}COOH \ + \ 2n \ NADP^+ \ + \ n \ CO_2 \ + \ n\text{-}1\,H_2O \ + \ n\text{+}1 \ CoA$$

Fig. 5. Overall reactions of the biosynthesis of saturated fatty acids.

of the transfer of acetate and malonate from their CoA to ACP coenzyme carriers (Fig. 6). Acetyl-ACP and malonyl-ACP then condense to form acetoacetyl-ACP, which is accompanied by the release of carbon dioxide. The acetoacetyl-ACP is then reduced to butyryl-ACP (Fig. 6). Butyryl-ACP can then condense with a second molecule of malonyl-ACP, and the series of condensation and reduction reactions continues until a long-chain fatty acid results.

ACP was originally isolated as a heat-stable protein cofactor from the *E. coli* fatty acid synthetase system (Lennarz *et al.*, 1962a; Goldman *et al.*, 1963; Pugh *et al.*, 1964); the characteristics of the protein and its role in fatty acid synthesis were defined in the laboratories of Vagelos and Wakil (Majerus *et al.*, 1964; Alberts *et al.*, 1964, 1965, Wakil *et al.*, 1964; Pugh and Wakil, 1965). The ACP of the fatty acid synthetase of *E. coli* (Fig. 7) is a low-molecular-weight protein (about 9000); its prosthetic group is identical with that of CoA (4′-phosphopantetheine), and, as in CoA derivatives, the acyl groups are attached to the sulfhydryl moiety of the prosthetic group (Majerus *et al.*, 1964, 1965; Pugh and Wakil, 1965). In *E. coli* the 4′-phosphopantetheine moiety of ACP probably arises by transfer from CoA itself (Alberts and Vagelos, 1966; Prescott *et al.*, 1969).

Protein coenzymes containing 4′-phosphopantetheine and having properties generally similar to those of the *E. coli* ACP have been either isolated from or shown to be present in the fatty acid synthetase systems of other organisms. These include the bacteria *Clostridium butyricum* (Ailhaud *et al.*, 1967) and *Arthrobacter* (Simoni *et al.*, 1967; Matsumura and Stumpf,

Acetyl transacylase:

$$CH_3-\overset{O}{\overset{\|}{C}}-S\text{-}CoA \;+\; ACP-SH \;\rightleftharpoons\; CH_3-\overset{O}{\overset{\|}{C}}-S-ACP \;+\; CoA\text{-}SH$$

Malonyl transacylase:

$$HOOC-CH_2-\underset{O}{\overset{}{\underset{\|}{C}}}-S\text{-}CoA \;+\; ACP-SH \;\rightleftharpoons\; HOOC-CH_2-\overset{O}{\overset{\|}{C}}-S-ACP \;+\; CoA\text{-}SH$$

β-Ketoacyl-ACP synthetase:

$$HOOC-CH_2-\overset{O}{\overset{\|}{C}}-S-ACP \;+\; CH_3-\overset{O}{\overset{\|}{C}}-S-ACP \;\rightleftharpoons\;$$

$$CH_3-\overset{O}{\overset{\|}{C}}-CH_2-\overset{O}{\overset{\|}{C}}-S-ACP \;+\; CO_2 \;+\; ACP-SH$$

β-Ketoacyl-ACP reductase:

$$CH_3-\overset{O}{\overset{\|}{C}}-CH_2-\overset{O}{\overset{\|}{C}}-S-ACP \;+\; NADPH \;+\; H^+ \;\rightleftharpoons\;$$

$$CH_3-\overset{OH}{\overset{}{\underset{}{CH}}}-CH_2-\overset{O}{\overset{\|}{C}}-S-ACP \;+\; NADP^+$$

Enoyl-ACP hydrase:

$$CH_3-\overset{OH}{\overset{}{\underset{}{CH}}}-CH_2-\overset{O}{\overset{\|}{C}}-S-ACP \;\rightleftharpoons\; CH_3-CH=CH-\overset{O}{\overset{\|}{C}}-S-ACP \;+\; H_2O$$

Enoyl-ACP reductase:

$$CH_3-CH=CH-\overset{O}{\overset{\|}{C}}-S-ACP \;+\; NADPH \;+\; H^+ \;\rightleftharpoons\;$$

$$CH_3-CH_2-CH_2-\overset{O}{\overset{\|}{C}}-S-ACP \;+\; NADP^+$$

Fig. 6. Enzymes and reactions of the fatty acid synthetase system of *Escherichia coli.*

that increase the chain length of preexisiting long-chain fatty acids by addition of two carbon units. One of these, found in the mitochondria of animals (Wakil, 1961; Harlan and Wakil, 1964) and plants (Barron *et al.*, 1961), utilizes acetyl-CoA and not malonyl-CoA as the source of two carbon units; it has sometimes been mistaken for a malonate-independent system of *de novo* fatty acid biosynthesis. A second system found in the microsomal fraction of mammalian cells employs malonyl-CoA as the source of two carbon units and produces CoA-bound intermediates (Nugteren, 1965; Wakil, 1964).

While rarely studied, enzymes for chain elongation must exist in most eukaryotic microorganisms since it is common experience that radioactivity from exogenously supplied ^{14}C-palmitate is readily incorporated into longer-chain fatty acids by yeasts, algae, and protozoa without extensive break-down of palmitate and reutilization of the label (e.g., see Erwin *et al.*, 1964; Korn *et al.*, 1965; H. Meyer and Holz, 1966). *Euglena* extracts chain-elongate dodecanoyl- and myristoyl-ACP esters to give stearic and arachidic acids (Nagai and Bloch, 1967), but this probably represents simply the ability of the soluble fatty acid synthetase system to utilize exogenously supplied acyl-ACP compounds rather than a chain elongation system per se. The yeasts *Saccharomyces cerevisiae* and *Candida utilis* can chain-elongate arachidic acid (20:2) to docosahexanoic acid (26:0) and 2-hydroxyeico-sanoic acids, but the nature of the enzymes involved is unknown (Fulco, 1967b).

C. Biosynthesis of Branched-Chain Fatty Acids

Branched-chain fatty acids of both the *iso* and *anteiso* types apparently result from the substitution of appropriate short-chain, branched fatty acyl-CoA compounds for acetyl-CoA as the primer in the fatty acid syn-thetase reactions. Amino acids may serve as metabolic sources of the short-chain branched fatty acid primers (Fig. 8). Thus incorporation of α-methylbutyric acid, derived from isoleucine, into the methyl end of a long-chain fatty acid would produce an *anteiso* fatty acid (Fig. 8). Similarly, incorporation of isovaleric acid, derived from leucine, and isobutyric acid, derived from valine, would produce, respectively, odd-chain and even-chain *iso* fatty acids (Fig. 8). Evidence that this is indeed the mechanism for the synthesis of bacterial branched-chain fatty acids was provided by Lennarz (1961), who showed that ^{14}C-labeled isoleucine, leucine, or valine or their short-chain branched fatty acid derivatives was incorporated exclusively into the methyl terminal end of the appropriate *iso* and *anteiso* long-chain fatty acids of *Micrococcus lysodeikticus*. Similar results were obtained with several species of *Bacilli* (see Kaneda, 1966). Furthermore, it has been

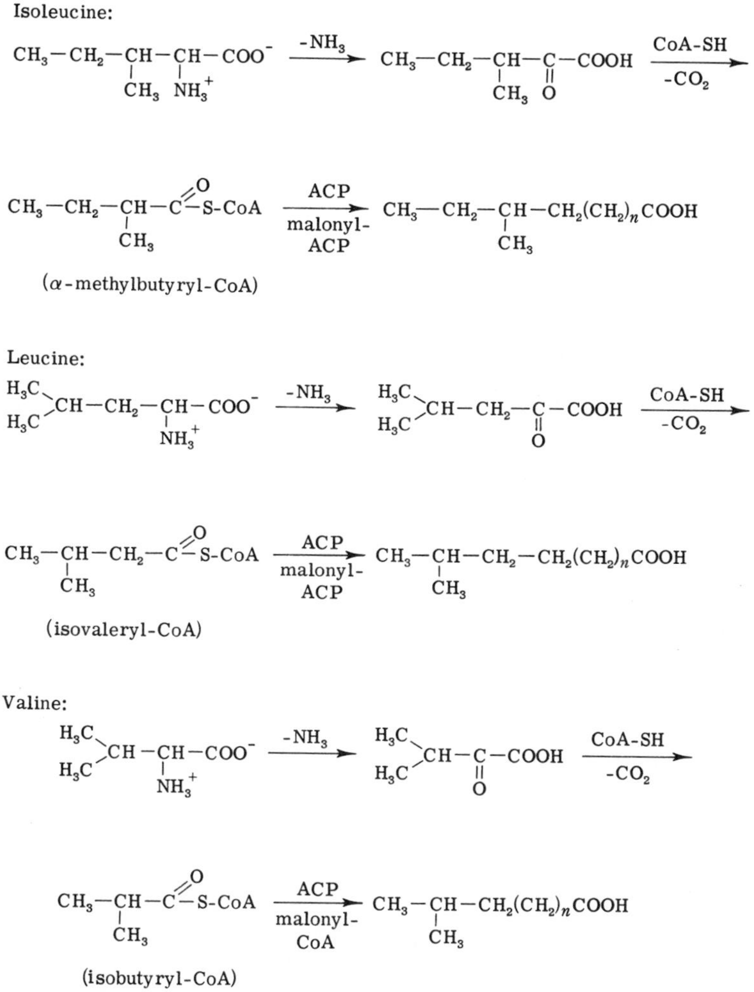

Isoleucine:

$$CH_3-CH_2-\underset{\underset{CH_3}{|}}{CH}-\underset{\underset{NH_3^+}{|}}{CH}-COO^- \xrightarrow{-NH_3} CH_3-CH_2-\underset{\underset{CH_3}{|}}{CH}-\underset{\underset{O}{\|}}{C}-COOH \xrightarrow[-CO_2]{CoA-SH}$$

$$CH_3-CH_2-\underset{\underset{CH_3}{|}}{CH}-\overset{\overset{O}{\diagup}}{C}-S\text{-CoA} \xrightarrow[\substack{malonyl-\\ACP}]{ACP} CH_3-CH_2-\underset{\underset{CH_3}{|}}{CH}-CH_2(CH_2)_n COOH$$

(α-methylbutyryl-CoA)

Leucine:

$$\underset{H_3C}{\overset{H_3C}{\diagdown}}CH-CH_2-\underset{\underset{NH_3^+}{|}}{CH}-COO^- \xrightarrow{-NH_3} \underset{H_3C}{\overset{H_3C}{\diagdown}}CH-CH_2-\underset{\underset{O}{\|}}{C}-COOH \xrightarrow[-CO_2]{CoA-SH}$$

$$CH_3-\underset{\underset{CH_3}{|}}{CH}-CH_2-\overset{\overset{O}{\diagup}}{C}-S\text{-CoA} \xrightarrow[\substack{malonyl-\\ACP}]{ACP} CH_3-\underset{\underset{CH_3}{|}}{CH}-CH_2-CH_2(CH_2)_n COOH$$

(isovaleryl-CoA)

Valine:

$$\underset{H_3C}{\overset{H_3C}{\diagdown}}CH-\underset{\underset{NH_3^+}{|}}{CH}-COO^- \xrightarrow{-NH_3} \underset{H_3C}{\overset{H_3C}{\diagdown}}CH-\underset{\underset{O}{\|}}{C}-COOH \xrightarrow[-CO_2]{CoA-SH}$$

$$CH_3-\underset{\underset{CH_3}{|}}{CH}-\overset{\overset{O}{\diagup}}{C}-S\text{-CoA} \xrightarrow[\substack{malonyl-\\CoA}]{ACP} CH_3-\underset{\underset{CH_3}{|}}{CH}-CH_2(CH_2)_n COOH$$

(isobutyryl-CoA)

Fig. 8. Amino acids as the source of primers for *iso* and *anteiso* branched long-chain fatty acids. (*n* is an odd number.)

demonstrated that cell-free fatty acid synthetase systems from rat adipose tissue (Horning *et al.*, 1961) and *Bacillus subtilis* (Butterworth and Bloch, 1970) incorporate isobutyrate into the methyl terminal portion of long-chain fatty acids when isobutyryl-CoA is supplied as a primer.

Protozoa appear to synthesize branched-chain fatty acids of the *iso* and *anteiso* type by the same mechanism that has been demonstrated in the

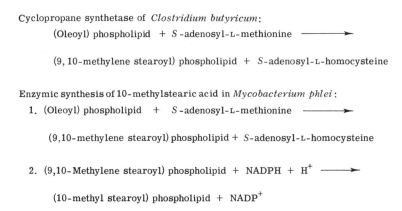

Cyclopropane synthetase of *Clostridium butyricum*:

  (Oleoyl) phospholipid + S-adenosyl-L-methionine ⟶

  (9,10-methylene stearoyl) phospholipid + S-adenosyl-L-homocysteine

Enzymic synthesis of 10-methylstearic acid in *Mycobacterium phlei*:

  1. (Oleoyl) phospholipid + S-adenosyl-L-methionine ⟶

     (9,10-methylene stearoyl) phospholipid + S-adenosyl-L-homocysteine

  2. (9,10-Methylene stearoyl) phospholipid + NADPH + H^+ ⟶

     (10-methyl stearoyl) phospholipid + $NADP^+$

Fig. 9. Biosynthesis of cyclopropane fatty acids and 10-methylstearic acids in bacteria.

bacteria and in adipose tissue. Growth of the ciliated protozoan *Tetrahymena paravorax* on U-^{14}C-leucine results in the preferential incorporation of the radioactivity into i-15:0 and i-19:0 (Erwin and Bloch, 1963a). Similarly, growth of the zooflagellates *Crithidia fasciculata*, *Blastocrithidia culicis*, *Leishmania tarentolae*, and *Leptomonas leptoglossi* on either U-^{14}C-leucine or U-^{14}C-isovaleric acid (see Fig. 8) also results in preferential incorporation of the radioactivity into *iso* branched-chain fatty acids (H. Meyer and Holz, 1966).

The amounts of *iso* and *anteiso* acids produced by some eukaryotic microorganisms are greatly influenced by the presence or absence of appropriate precursors in the growth medium. For example, the ciliated protozoan *Tetrahymena pyriformis* normally contains very low concentrations of branched-chain fatty acids (Erwin and Bloch, 1963a; Shorb, 1963). However, growth on isobutyrate produces ciliates in which even-chain *iso* acids constitute 30% of the total long-chain fatty acids; growth on α-methylbutyrate produces comperable increases in the *anteiso* fatty acids of this ciliate (Shorb, 1963). Similarly, in the yeast *Schizosaccharomyces liquefaciens* addition of isobutyrate, leucine, valine, or isoleucine to the culture medium also produces comparable increases in the production of *iso* and *anteiso* fatty acids (Baraud *et al.*, 1970).

D. Biosynthesis of Cyclopropane Fatty Acids

Hofmann and others demonstrated that the cyclopropane fatty acids of bacteria arise through the methylation of monounsaturated fatty acids (for a review, see Hofmann, 1963). The enzyme for the reaction, cyclopropane synthetase, has been isolated from the bacterium *Clostridium butyricum* and has been shown to use *S*-adenosyl methionine as the donor for the methylene group (Zalkin *et al.*, 1963). The enzymic conversion of monounsaturated fatty acids to cyclopropane fatty acids in bacteria occurs at the level of an intact phospholipid (Fig. 9) rather than at the level of a free fatty acid (Thomas and Law, 1966). An analogous enzyme system prepared from the bacterium *Mycobacterium phlei* (Akamatsu and Law, 1970) converts oleic acid in the form of an acyl phospholipid to 10-methyl-stearic acid with *S*-adenosyl methionine serving as the methyl donor (Fig. 9).

Among the eukaryotic microorganisms, cyclopropane acids are quite rare. However, they are found in some zooflagellates, namely *Leptomonas leptoglossi*, *Leishmania tarentolae*, and five species of *Crithidia* (H. Meyer and Holz, 1966). These organisms contain substantial amounts of 9,10-methyleneoctadecanoic acid (4–13%) (Table XII). In *C. fasciculata* the compound is found exclusively in a single phospholipid, phosphatidyl ethanolamine, where it constitutes 23% of the total fatty acids of this lipid (H. Meyer and Holz, 1966). Other phospholipids and the neutral lipids of the protozoan are devoid of cyclopropane fatty acids (H. Meyer and Holz, 1966).

When either *Crithidia fasciculata* or several of the other zooflagellates were incubated with 1-^{14}C-oleic acid, the cyclopropane fatty acid 9-18:0Δ was the only saturated fatty acid to become labeled. Incubation of the protozoan with Me-^{14}C-methionine produced incorporation of the label exclusively into 9-18:0Δ (H. Meyer and Holz, 1966). Thus these zooflagellates appear to synthesize their cyclopropane fatty acids by a mechanism similar to that demonstrated in bacteria.

Monoenoic acids:

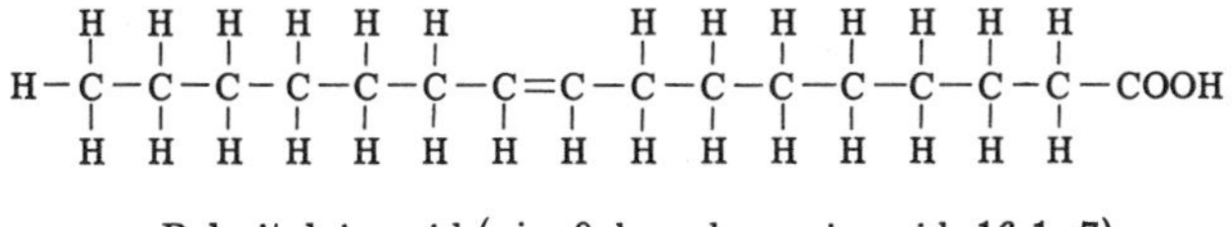

Palmitoleic acid (*cis*-9-hexadecenoic acid; 16:1ω7)

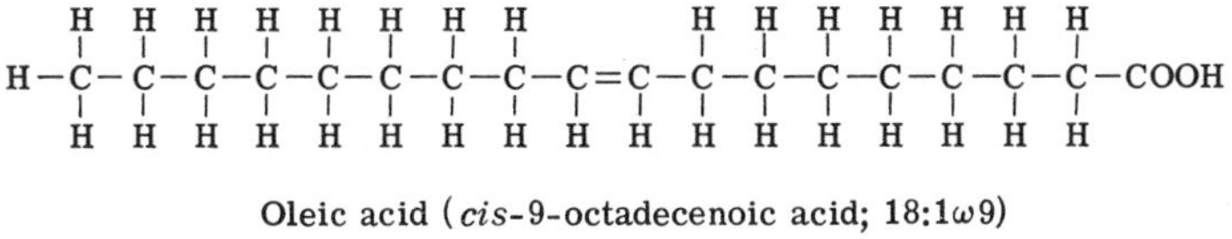

Oleic acid (*cis*-9-octadecenoic acid; 18:1ω9)

Fig. 10. Unsaturated fatty acids.

III. Monounsaturated Fatty Acids

A. TYPICAL MONOUNSATURATED FATTY ACIDS OF MICROORGANISMS

Fatty acids containing a single double bond are extremely common in biological material. The monounsaturated fatty acids of both bacteria and eukaryotic microorganisms usually have the *cis* configuration. However, in bacteria multiple-positional isomers are common; whereas in eukaryotic microorganisms, yeasts, fungi, and protozoa, palmitoleic and oleic acids (Fig. 10) are the most usual monoenoic acids. In algae we find various positional isomers of hexadecenoic acid. The blue-green algae (Table III) and the cryptomonads (Table V) contain only the $16:1\omega7$ isomer (palmitoleic acid). But in other algae the hexadecenoic acid fraction is usually a mixture of isomers. For example, $16:1\omega9$ and $16:1\omega5$ as well as $16:1\omega7$ have been identified in photoauxotrophic *Euglena* (see Bloch *et al.*, 1967), while $16:1\omega11$, $16:1\omega9$, $16:1\omega7$, and $16:1\omega5$ have been found in the dinoflagellate *Gyrodinium cohnii* (Harrington and Holz, 1968). Similar mixtures of 16:1-positional isomers have been found in red algae (Table IV), green algae (Table VIII), and some *Chrysophyta* (Table VI).

B. MULTIPLE MECHANISMS FOR THE BIOSYNTHESIS OF MONOUNSATURATED FATTY ACIDS

1. *Desaturase System I*

Experiments on animals had for some years been interpreted as evidence that unsaturated fatty acids were formed secondarily from their saturated analogs rather than by an independent synthetic pathway (for a review, see Green and Wakil, 1960). That this process might require molecular oxygen was indicated by the observation that the yeast *Saccharomyces cerevisiae* required oleic acid to grow under strictly anaerobic conditions (Andreasen and Stier, 1954). Both of these interpretations were shown to be correct when Bloch and his co-workers isolated an enzyme system from *S. cerevisiae* (Bloomfield and Bloch, 1960) and later from another yeast, *Candida utilis* (Yuan and Bloch, 1961; F. Meyer and Bloch, 1963), which directly desaturates palmitic to palmitoleic and stearic to oleic acids and requires the presence of molecular oxygen. Similar enzymes have been isolated from the bacterium *Mycobacterium phlei* (Fulco and Bloch, 1964), the phytoflagellate *Euglena gracilis* when grown under heterotrophic conditions (Nagai and Bloch, 1965), rat liver (Marsh and James, 1962; Holloway *et al.*, 1963), and hen liver (Jones *et al.*, 1969).

The basic properties of these isolated enzyme systems are similar in all cases (Table I) and such systems will be referred to as desaturase system I.

TABLE I

PROPERTIES OF ISOLATED ENZYMES OF FATTY ACID DESATURASE SYSTEM TYPE I

Source[a]	Intracellular Localization of Enzymes	Substrate, CoA	Product	Cofactor Requirements
Mycobacterium phlei (2)	Particulate $(100,000 \times g - 2$ hr)	18:0	$18:1\omega9$[b]	NADPH, O_2, FAD (or FMN) Fe^{2+}
Saccharomyces cerevisiae (1)	Microsomes	16:0	$16:1\omega7$ CoA	NADPH, O_2
Candida utilis	Microsomes			
Euglena gracilis (4) (hetero-trophic)	Particulate	18:0	$18:1\omega9$ (?)[c]	NADPH, O_2
Rat liver (3)	Microsomes			NADPH (or NADH), O_2
Hen liver (3)	Microsomes (chloroform-methanol ex-tracted)	16:0 18:0	$16:1\omega7$ (?)[c] $18:1\omega9$ (?)[c]	NADH (or NADPH) O_2, crude phospho-lipids

[a] References: (1) Bloomfield and Bloch (1960); (2) Fulco and Bloch (1964); (3) Jones *et al.* (1969); (4) Nagai and Bloch (1968).

[b] Intact cells produce $10–18:1\omega8$.

[c] (?), not determined whether CoA ester or free acid is the product.

Enzymes of this system are organized into particles, and such particles are usually associated with the microsomal fraction of cells. System I (Fig. 11) catalyzes the direct desaturation of long-chain acyl-CoA compounds to their 9,10-unsaturated analogs (Bloomfield and Bloch, 1960; Holloway *et al.*, 1963; Jones *et al.*, 1969). This system is specific for acyl-CoA compounds and will not utilize acyl-ACP compounds or free fatty acids as substrates (Bloomfield and Bloch, 1960; Marsh and James, 1962; Nagai and Bloch, 1968; Renkonen and Bloch, 1969). Molecular oxygen and NADPH (or NADH) are required cofactors (Bloomfield and Bloch, 1960; Holloway *et al.*, 1963; Jones *et al.*, 1969). It has been demonstrated that with the yeast system the requirement for molecular oxygen cannot be satisfied by the substitution of artificial electron acceptors (Bloomfield and Bloch, 1960).

Recently, Holloway and Wakil (1970) have fractionated the desaturase type I system of hen liver into several subparticles. They report that in this system oleic acid production requires three enzymes: a desaturase

$$\text{Stearoyl-CoA} \xrightarrow[\substack{\text{(particulate} \\ \text{enzyme complex)}}]{\text{O}_2;\ \text{NADPH}} \text{oleic acid (CoA)}$$

Fig. 11. Aerobic desaturation system I.

enzyme that is specific for stearyl-CoA, an NADH cytochrome b_5 reductase, and cytochrome b_5 (Holloway and Wakil, 1970).

2. *Desaturase System II*

The mechanism whereby plants synthesize monounsaturated fatty acids was a cause for speculation for a number of years. The biosynthesis of octadecenoic acid from ^{14}C-acetate could be demonstrated in cell-free extracts from plant tissues (Mudd and Stumpf, 1961; Stumpf and James, 1963), and it was found that this process, too, required molecular oxygen (Mudd and Stumpf, 1961; Stumpf and James, 1963). However, neither intact plant tissue (James, 1963) nor plant tissue extracts (Barron and Stumpf, 1962; Stumpf and James, 1963; Hawke and Stumpf, 1965) could convert exogenously supplied ^{14}C-palmitic or ^{14}C-stearic acids to ^{14}C-octadecenoic acid. Similar observations were made on algae, and therefore the existence of a plant pathway of monounsaturated fatty acid biosynthesis was postulated (for a review, see Erwin and Bloch, 1964). The anomalies of the plant pathway were finally resolved when Bloch and his co-workers showed that photoauxotrophically grown *Euglena* and spinach leaves both yield enzyme systems that desaturate stearate to oleate (Nagai and Bloch, 1965). While these "plant" systems share the same basic mechanism of the desaturase systems described for yeasts and animals (see Section III,B,1), they differ from them in several important respects. They will be referred to as desaturase system II.

Desaturase system II isolated from photoauxotrophic *Euglena* (Fig. 12) is specific for stearoyl-ACP and will not utilize stearoyl-CoA or free stearate; only the 18:1ω9 isomer is produced (Nagai and Bloch, 1965, 1968). Palmitoyl-ACP and other shorter-chain fatty acyl-ACP compounds are chain-elongated but not directly desaturated (Nagai and Bloch, 1965, 1967). Similar results have been obtained with spinach enzyme preparations (Nagai and Bloch, 1965, 1967, 1968). In both photoauxotrophic *Euglena* and spinach, desaturase system II is composed of fully soluble enzymes that are associated with the chloroplast (Nagai and Bloch, 1965). The photoauxotrophic *Euglena* system has been fractionated into three distinct enzymes: a desaturase that is specific for stearoyl-ACP, an NADPH oxidase, and ferredoxin. All three enzymes are required for the production of oleic acid (Nagai and Bloch, 1966, 1968).

Desaturase enzymes of the system II type are presumably present in other photosynthetic organisms reported to possess the plant pathway of monoenoic fatty acid biosynthesis. Apparently, in contrast to type I desaturase systems, it is characteristic of type II systems that the thioester bond of stearoyl-ACP produced during long-chain fatty acid synthesis is conserved (see Section II,B,1) and that there is free movement of stearoyl-ACP from the enzymes of the fatty acid synthetase system to those of the desaturase system. Such free exchange is made possible by the fact that both the fatty acid synthetase system and the desaturase type II system found in photoauxotrophic *Euglena* and in higher plants consist of soluble enzymes (rather than particulate completes, as in yeasts and animals; see Section II,B,2) and because both sets of enzymes are localized in the same cell compartment, the chloroplast. Organisms employing exclusively the type II desaturase system also appear to be deficient in a mechanism for converting either palmitoyl- or stearoyl-CoA to ACP derivatives. Hence in such organisms, exogenously supplied ^{14}C-palmitate or ^{14}C-stearate may be activated to their CoA derivatives and serve as substrates for lipid synthesis but cannot serve as direct precursors of unsaturated fatty acids. On the other hand, shorter-chain fatty acids can be converted to ACP derivatives and hence when exogenously supplied will serve as substrates for desaturation.

3. *Mechanism of Oxidative Desaturation*

The possible mechanisms of oxidative desaturation of fatty acids and the basis for some of the characteristic features of the enzyme systems have recently been reviewed (see Bloch, 1969) and will simply be summarized here.

A requirement for molecular oxygen and NADPH (or NADH) is characteristic of all type I and type II fatty acid desaturate systems isolated to date (see Table I), and they probably all share a common basic mechanism. This mechanism might involve oxygenase-type relations, in which case oxygenated intermediates such as hydroxystearic acids might be expected. However, all attempts to demonstrate that hydroxy acids are such intermediates have proved unsuccessful. For example, Lennarz and Bloch (1960) demonstrated that 9- or 10-hydroxystearic acids will substitute for oleic acid in supporting growth of yeasts under anaerobic conditions, which

$$\text{Stearoyl-ACP} \xrightarrow[\text{(soluble enzymes)}]{\substack{O_2;\ \text{NADPH};\\ \text{ferredoxin}}} \text{oleoyl-ACP}$$

Fig. 12. Aerobic desaturation system II.

might be expected if they were intermediates in oleic acid synthesis. However, subsequent investigations in Bloch's laboratory revealed that these dietary hydroxy acids are not converted to unsaturated fatty acids. Rather, their growth-supporting activity is due to their conversion to acetoxy and other derivatives, which in anaerobic yeasts substitute in turn for unsaturated fatty acids as precursors of membrane lipids (Light *et al.*, 1962). Furthermore, fatty acid desaturase system I preparations isolated from yeast and rat liver do not convert either 9- or 10-hydroxystearic acid or their CoA derivatives directly to unsaturated fatty acids (Light *et al.*, 1962; Marsh and James, 1962). Similarly, crude enzymes of the desaturase II type system isolated from photoauxotrophic *Euglena* fail to convert either 9- or 10-hydroxystearoyl-ACP compounds to oleic acid (Nagai and Bloch, 1968); instead they convert these compounds to their keto derivatives (Gurr and Bloch, 1966).

If fatty acid desaturation is an oxygenase-type reaction, unconjugated pteridines might be required, since they have been demonstrated to be cofactors in other biosynthetic oxygenase reactions (see Forrest and Van Baalen, 1970). That this might be the case is suggested by the observations of Kidder and Dewey (1963) that mixtures of oleic and linoleic acids partially replace the pteridine requirement of the zooflagellate *Crithidia fasciculata*. Again, however, it has not been possible to demonstrate that pteridines have any cofactor activity in fatty acid desaturase enzyme preparations (Fulco and Bloch, 1964). These cumulative failures, while somewhat discouraging, do not disprove the idea that oxygenase-type reactions constitute the basic mechanism of aerobic fatty acid desaturation, and the issue must be considered to be unresolved.

Another characteristic feature of all the aerobic fatty acid desaturase enzymes isolated to date is their high degree of geometrical and positional stereospecificity (Table I) since only *cis*-9,10-unsaturated fatty acids are produced. Indeed, Schroepfer and Bloch (1965) have shown that the desaturase enzymes of *Corynebacterium diptheriae* selectively remove a particular hydrogen atom from each pair of hydrogen atoms at carbon atoms 9 and 10 of stearic acid.

This specificity for producing only 9,10-monounsaturated fatty acids has been suggested by Scheuerbrandt and Bloch (1962) as a feature that distinguishes between aerobic desaturation and the anaerobic pathway of unsaturated fatty acid synthesis (see Section III,B,4). However, there are exceptions. Thus positional isomers of 9,10-monounsaturated fatty acids are found in organisms that synthesize unsaturated fatty acids via either of the two aerobic desaturase systems. But in some cases, such as the formation of vaccenic acid ($18:1\omega7$) in vertebrates (Harlan and Wakil,

1964) and in *Euglena* (Nagai and Bloch, 1967), these isomers probably arise via chain elongation of 9,10-unsaturated hexadecenoic acid rather than a direct desaturation of stearic acid. In other cases, direct desaturation of long-chain fatty acids to positional isomers of 9,10-monounsaturated fatty acids has been demonstrated with intact cells (although not with purified enzyme systems). For example, enzyme preparations from *Mycobacterium phlei* produce only palmitoleic acid, whereas intact cells desaturate palmitate to the 10,11 isomer of palmitoleic acid (Lennarz *et al.*, 1962b; Fulco and Bloch, 1964). Many species of *Bacilli* directly desaturate palmitic acid to *cis*-5-hexadecenoic acid (Fulco, 1967a, 1969), while other species desaturate palmitate to a mixture of palmitoleate and its 8,9 and 10,11 isomers (Fulco, 1967a, 1969). Similarly, James and his co-workers have reported that the eukaryotic green-algae *Chlorella vulgaris* desaturates a variety of long-chain fatty acids directly to either their 7,8- or 9,10-monounsaturated analogs (Howling *et al.*, 1968). At present, the reasons for this discrepancy between results of *in vivo* and *in vitro* studies are obscure.

4. Anaerobic Pathway of Monounsaturated Fatty Acid Biosynthesis

Molecular oxygen is an absolute requirement for the production of monounsaturated fatty acids via the direct desaturation type of mechanism, yet there are many microorganisms that can be cultivated on fat-free media in the complete absence of oxygen. Awareness of this discrepancy prompted Bloch and his co-workers to undertake a series of investigations of obligate anaerobic bacteria belonging to the genus *Clostridia* (Goldfine and Bloch, 1960; Scheuerbrandt *et al.*, 1961) and of the facultative anaerobic *Escherichia coli* (Bloch, 1962; Lennarz *et al.*, 1962b). These studies revealed the existence of a mechanism of monounsaturated fatty acid biosynthesis unrelated to the direct desaturation type. This anaerobic pathway produces monoenoic fatty acids via the β,γ dehydration of medium-chain-length β-hydroxy acids and the subsequent chain elongation of the resulting 3-enoates (Fig. 13). Both 9,10-monounsaturated fatty acids and their positional isomers can be produced depending on the chain length of the β-hydroxy acid that undergoes β,γ dehydration (Fig. 13).

The anaerobic pathway has been studied in detail in *Escherichia coli* and has been shown to be essentially a modification of the fatty acid synthetase system already discussed (Section II,B,1). During the formation of long-chain fatty acids in *E. coli*, growing acyl chains are diverted at the 10 carbon stage from the usual sequence leading to saturated fatty acids to a branch route that leads to monounsaturated fatty acids. This is accomplished by an enzyme, β-hydroxydecanoyl thioester dehydrase, that

1. Octanoyl-ACP + malonyl-ACP $\longrightarrow$ β-ketodecanoyl-ACP $\xrightarrow{\text{NADPH}}$

    β-hydroxydecanoyl-ACP $\xrightarrow{-H_2O}$ β,γ-decenoyl-ACP $\xrightarrow{+4\ \text{malonyl-ACP}}$

    vaccenic acid (ACP)

2. Decanoyl-ACP + malonyl-ACP $\longrightarrow$ β-ketododecanoyl-ACP $\xrightarrow{\text{NADPH}}$

    β-hydroxydodecanoyl-ACP $\xrightarrow{-H_2O}$ β,γ-dodecenoyl-ACP

    $\xrightarrow{+\ 3\,\text{malonyl-ACP}}$ oleic acid (ACP)

Fig. 13. Anaerobic pathway of unsaturated fatty acid biosynthesis.

converts the hydroxydecanoyl-ACP intermediate in the fatty acid synthetase reactions (Fig. 6) to *cis*-3-decenoyl-ACP rather than the usual
trans-2-decenoyl-ACP. The enoyl-ACP reductase of the fatty acid synthetase system cannot reduce the *cis*-3-thioester; hence the double bond is
preserved during subsequent condensation and reduction reactions, and
an unsaturated long-chain fatty acid is eventually produced. The β-hy-
droxydecanoyl thioester dehydrase of *E. coli* has been isolated and purified,
and its properties and mode of action have recently been reviewed in some
detail (see Bloch, 1969).

5. *Other Mechanisms for Synthesizing Monounsaturated Fatty Acids*

Nagai and Bloch (1965) have suggested that a third aerobic pathway
of monoenoic fatty acid biosynthesis may exist in euglenids, green algae,
and higher plants. In their proposed pathway, hexadecenoic and octadecenoic fatty acids would arise via β,γ desaturation of decanoic and dodec-
anoic acids followed by chain elongation of the β,γ acids (Fig. 14). This
pathway could produce several positional isomers of hexadecenoic acid,
and these have been detected in algae (Nagai and Bloch, 1965, 1967).
Attempts to obtain enzymic evidence for the pathway in either algae or
higher plants have been unsuccessful (see Nagai and Bloch, 1967; Stumpf
et al., 1967). Moreover, the reported ability of *Chlorella vulgaris* to desaturate palmitic acid directly to several positional isomers of palmitoleic
acid (see Section III,B,3) provides an alternative explanation of the origin
of the compounds identified by Nagai and Bloch in *Euglena*.

The aerobic β,γ desaturation pathway has also been suggested to operate
in vertebrates (Raju and Reiser, 1969; Johnson *et al.*, 1967). The evidence
for the existence of this pathway in higher animals is as follows. Sterculic
acid inhibits the formation of oleic acid from ^{14}C-labeled stearic acid in

$$10:0 \xrightarrow{\text{O}_2 \ (?)} 10:1\omega7 \xrightarrow{+3\text{C}_2} 16:1\omega7 \xrightarrow{+\text{C}_2} 18:1\omega7$$

$$12:0 \xrightarrow{\text{O}_2 \ (?)} 12:1\omega9 \xrightarrow{+2\text{C}_2} 16:1\omega9 \xrightarrow{+\text{C}_2} 18:1\omega9$$

Fig. 14. Proposed aerobic β,γ desaturation.

vertebrates (see Section III,C) but not its formation from ^{14}C-labeled acetate or from ^{14}C-labeled lauric acid (Raju and Reiser, 1969). Reiser and co-workers believe that sterculic acid directly inhibits an enzyme of the desaturase I system of vertebrates (see Section III,C) and that lauric acid (exogenously supplied or produced from acetate via the fatty acid synthetase system) undergoes β,γ aerobic desaturation and chain elongation of the resulting 12:1ω9 acid to oleic acid. Again the evidence is indirect, the site of action of sterculic acid inhibition is unknown (see Section III,C,1), and alternative explanations for the experimental results can be invoked.

C. INHIBITORS OF UNSATURATED FATTY ACID BIOSYNTHESIS

1. *Sterculic Acid*

Sterculic acid (Fig. 15), a cyclopropene fatty acid, appears to inhibit the direct desaturation of palmitic and stearic acids to their corresponding monoenes in higher animals. *In vivo* studies have demonstrated that feeding this acid to rats (Reiser and Raju, 1964) and chickens (Donaldson, 1967) results in a greatly reduced ability to convert ^{14}C-labeled palmitate and stearate to palmitoleate and oleate. *In vitro* experiments with whole rats show that while sterculic acid inhibits the incorporation of radioactivity from labeled palmitate and stearate into oleate, the *de novo* synthesis of radioactive oleate from radioactive acetate is not affected (Reiser and Raju, 1964; Raju and Reiser, 1969).

The green alga *Chlorella vulgaris* reacts to sterculic acid in a manner similar to rats and chickens; oleic acid synthesis from radioactive acetate is not affected, while oleic acid synthesis from radioactive stearate is almost completely inhibited in the presence of $3 \times 10^{-5} M$ sterculic acid (see James *et al.*, 1968).

In the same organisms sterculic acid is a relatively poor inhibitor of the desaturation of oleic acid to linoleic acid, effective inhibition requiring greatly increased amounts $(3 \times 10^{-3} M)$ of sterculic acid (James *et al.*, 1968).

Sterculate has a somewhat different effect on unsaturated fatty acid biosynthesis in higher plants. Normally, if leaf tissue is incubated with

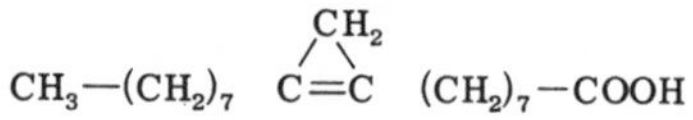

$$CH_3-(CH_2)_7 \quad \overset{\displaystyle \underset{/\diagdown}{CH_2}}{C=C} \quad (CH_2)_7-COOH$$

Sterculic acid

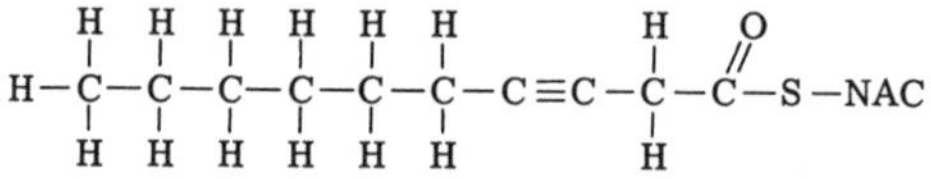

$$H-\overset{\displaystyle\overset{H}{|}}{\underset{\displaystyle\underset{H}{|}}{C}}-\overset{\displaystyle\overset{H}{|}}{\underset{\displaystyle\underset{H}{|}}{C}}-\overset{\displaystyle\overset{H}{|}}{\underset{\displaystyle\underset{H}{|}}{C}}-\overset{\displaystyle\overset{H}{|}}{\underset{\displaystyle\underset{H}{|}}{C}}-\overset{\displaystyle\overset{H}{|}}{\underset{\displaystyle\underset{H}{|}}{C}}-\overset{\displaystyle\overset{H}{|}}{\underset{\displaystyle\underset{H}{|}}{C}}-C\equiv C-\overset{\displaystyle\overset{H}{|}}{\underset{\displaystyle\underset{H}{|}}{C}}-\overset{\displaystyle\overset{O}{\|}}{C}-S-NAC$$

3-Decynoyl-NAC

Fig. 15. Inhibitors of unsaturated fatty acid biosynthesis.

radioactive acetate in the dark under anaerobic conditions, the radioactivity accumulates in the palmitate and stearate. Upon the admission of oxygen the label then appears in palmitoleic and oleic acids (James et al., 1968). The presence of sterculic acid has little or no effect on the aerobic transfer of radioactivity from stearate to oleate in such an experiment; the desaturation of stearoyl-ACP is apparently not sensitive to this acid (James et al., 1968). However, in its presence a significant inhibition of the transfer of radioactivity from palmitate to hexadecenoic acid is observed; the reason for this differential effect on palmitate and stearate desaturation is unknown (James et al., 1968).

The mechanism of sterculate action is obscure. Raju and Reiser (1967) reported that the addition of sterculic acid to rat liver enzyme preparations also inhibits the in vitro desaturation of ¹⁴C-labeled long-chain fatty acids, and they suggest that the inhibitor reacts with sulfhydryl groups in the desaturase complex (Raju and Reiser, 1967). However, recent observations of Pande and Mead (1970) have not supported this suggestion, and the in vitro inhibition observed with sterculic acid may be nonspecific. James and Gurr have suggested that sterculic acid acts not on the desaturase enzymes themselves but on the enzyme transferring the saturated long-chain fatty acid from CoA to ACP (James, 1968; Gurr, 1971). This explanation would appear to require that acyl-ACP compounds be the actual substrates for desaturation of stearate to oleate and of oleate to linoleate in microorganisms, plants, and animals. This is not supported by current evidence (see Section III,B,1).

2. Acetylenic Acid Derivatives

The in vivo production of unsaturated fatty acids in E. coli via the anaerobic pathway is inhibited by the acetylenic acid derivative 3-decynol-n-acetylcysteamine (Fig. 15). Growth of the bacterium is completely halted by addition of the compound to the culture medium; the inhibition is reversed by addition of oleic and vaccenic acids (see Bloch, 1969). The compound has been shown to be a specific inhibitor of the key enzyme in the anaerobic pathway, β-hydroxydecanoyl thioester dehydrase (see Bloch, 1969). The compound is specific for the anaerobic pathway and does not affect the aerobic desaturation reactions of yeasts and animal cells (see Bloch, 1969).

3. *Triparanol*

Triparanol, an inhibitor of cholesterol biosynthesis in vertebrates, appears to inhibit unsaturated fatty acid biosynthesis in the ciliated protozoan *Tetrahymena pyriformis* (Pollard et al., 1964; Shorb et al., 1965). Triparanol inhibition of cell division and growth in this ciliate is reversible by the addition of unsaturated fatty acids (Holz et al., 1962, 1963). Similar effects in experiments with triparanol have been reported with a variety of yeasts, algae, and protozoa (Rosenbaum et al., 1965; Aaronson and Bensky, 1965; Aarsonson et al., 1969).

D. COMPARATIVE STUDIES

1. *Methodology*

If a tissue or microbial population utilizes a direct desaturation mechanism for synthesizing monounsaturated fatty acids and is exogenously supplied with ^{14}C-labeled acetate, it will, in the absence of oxygen, incorporate the label exclusively into saturated fatty acid; in the presence of oxygen it will incorporate the label into both saturated and unsaturated fatty acids. However, in a tissue or microbial population utilizing the anaerobic mechanism for synthesizing monounsaturated fatty acids, the incorporation of exogenously supplied ^{14}C-labeled acetate into its unsaturated fatty acids will be indifferent to a presence or absence of molecular oxygen. Tissues and microbial populations employing the aerobic direct desaturation mechanism type I (but not those employing either the anaerobic mechanism or the aerobic direct desaturation mechanism type II) will incorporate radioactivity from exogenously supplied 1-^{14}C-labeled palmitic and stearic acids into their unsaturated fatty acids without prior breakdown of the exogenously supplied radioactive fatty acids and without reutilization of the label. If extensive degradation of the exogenously supplied radioactive palmitate and stearate occurs with reutilization of the label, this will be detected, since radioactivity will be found in cellular fatty acids of a shorter chain length than that provided by the experimenter.

2. *Survey*

This ability of suitably designed radioisotope experiments using intact cells to detect the type of mechanism of monounsaturated fatty acid biosynthesis present in a tissue or microbial population has facilitated comparative studies on the distribution of the different pathways among various groups of microorganisms, plants, and animals. The results of comparative investigations based upon the use of such radioisotope techniques on whole cells along with the results from a far more limited number of investigations based upon direct assay of isolated enzyme systems are summarized in Table II.

The anaerobic pathway is very common among bacteria, both among heterotrophs and photoauxotrophs; it is found not only in facultative and obligate anaerobes, but also in obligate aerobes such as pseudomonads (Table II). However, the mechanism appears to be confined to the bacteria and has never been detected in any eukaryotic organism (Table II).

The aerobic direct desaturation mechanism type I is the pathway of monounsaturated fatty acid biosynthesis most widely found in nature. It is present in some prokaryotes, including all blue-green algae; in some bacteria; and in most eukaryotes, including all yeasts and fungi, protozoa, higher animals, and several groups of eukaryotic algae, including the red algae, chrysomonads, and euglenids (Table II). It probably exists among the green algae but does not appear to be present in higher plants (Table II).

The aerobic direct desaturation mechanism type II has been found only among the euglenids, the green algae, and the higher plants (Table II).

There is at least one organism, *Euglena gracilis*, that possesses the genetic information for synthesizing the enzymes of both types of desaturase systems. Environmental conditions determine which type of enzyme will actually be synthesized. *Euglena* cultures grown as strict photoauxotrophs produce only the enzymes of desaturase system type II, while *Euglena* cultures grown as strict heterotrophs produce only the enzymes of desaturase system type I (Nagai and Bloch, 1965).

No organism has ever been demonstrated to possess the genetic information for synthesizing the enzymes of both the anaerobic mechanism and either of the two desaturase mechanisms. The possibility that facultative anaerobic bacteria might possess both the anaerobic pathway and a direct desaturation mechanism has been investigated in *Escherichia coli* and it was found that *E. coli* employs only the anaerobic mechanism regardless of whether it grows as an anaerobe or an aerobe (see Bloch *et al.*, 1961).

These findings are, of course, subject to the qualification that the number of organisms examined is relatively small, and more extensive studies

TABLE II

DISTRIBUTION OF THE SEVERAL PATHWAYS OF MONOUNSATURATED FATTY ACID
BIOSYNTHESIS

Group	Species[a]	Pathway of Unsaturated Fatty Acid Biosynthesis[b]	Methodology Employed[c]
Prokaryotic micro-organisms			
Bacteria			
Actinomycetes	*Mycobacterium phlei* (8, 24)	Des I[d]	W.C., Enz
	Corynebacterium diptheria (9)	Des I	W.C.
Spirochetes	*Treponema zuelzerae* (28)	AN	W.C.
Gram-positive eubacteria	*Sarcina* (2 species) (32)	AN	W.C.
	Micrococcus lysodeikticus (9)	Des I	W.C.
	Lactobacillus plantarum (2)	Des I	W.C.
	Bacillus megaterium (3 strains) (9, 10)	Des I[e]	W.C.
	B. subtilis, B. pumilus, B. macerans, B. stearothermophilus, B. brevis (3 strains), *B. coagulans, B. licheniformis* (2 strains) (12)	Des I	W.C.
	Clostridium butyricum (13) *C. kluyveri* (32)	AN	W.C.
Gram-negative eubacteria	*Escherichia coli* (2)	AN	W.C., Enz
	Pseudomonas (3 species) (9)	AN	W.C.
	Rhodospirillum rubrum (34)	AN	W.C.
	Rhodopseudomonas palustris, R. capsulatum, Chlorobium limicola (15)		
Cyanophyta (blue-green algae)	*Anabaena variabilis* (2) (30)	Des I	W.C.
	Anacystis nidulans (30)	Des I	W.C.
Eukaryotic micro-organisms			
Yeasts and fungi	*Saccharomyces cerevisiae* (3)	Des I	W.C., Enz
	Candida utilis (Torulopsis) (33) (26)	Des I	W.C. Enz
	Penicillium chrysogenum (1, 2)	Des I	W.C.
Algae			
Rhodophyta (red algae)	*Porphyridium cruentum* (7)	Des I	W.C.
Chrysophyta (chryso-monads)	*Ochromonas malhamensis* (7)	Des I	W.C.
	Poteriochromonas stipitata (7)	Des I	W.C.

TABLE II   (Continued)

DISTRIBUTION OF THE SEVERAL PATHWAYS OF MONOUNSATURATED FATTY ACID
BIOSYNTHESIS

Group	Species[a]	Pathway of Unsaturated Fatty Acid Biosynthesis[b]	Methodology Employed[c]
Euglenophyta (Euglenids)	*Euglena gracilis* Heterotrophic (16) (29)	Des I	W.C. Enz
	Photoauxotropic (29)	Des II	W.C., Enz
	Astasia longa (7)	Des I	W.C.
Chlorophyta (green algae)	*Chlorella pyrenoidosa* (2)	Des II	W.C.
	C. vulgaris (16)	Des I[f]	
	Scenedesmus D, *Ankistrodesmus braunii, Polytoma uvella, Chlamydomonas reinhardi* (7)	Des II	W.C.
Protozoa Sarcodina Soil amoebas	*Acanthamoeba* sp. (5)	Des I	W.C.
	Hartmannella rhysoides (7)	Des I	W.C.
Slime molds	*Dictyostelium discoideum* (4)	Des I	W.C.
Zoomastigophora [Kinetoplastidia (hemo- flagellates)]	*Crithidia fasciculata, C. onco- pelti, Blastocrithidia culicis, Leptomonas leptoglossi* (27)	Des I	W.C.
	Leishmania tarentolae (22, 27)	Des I	W.C.
	L. enrietti (23)	Des I	W.C.
	Trypanosoma lewisi (22)	Des I	W.C.
Ciliata (tetra- hymenids)	*Tetrahymena pyriformis, T. corlissii, T. setifera, T. paravorax, Glaucoma chattoni* (6)	Des I	W.C.
Metaphyta Pterophyta (ferns)	*Dryopteris filix-mas, Athyrium filix-foemina, Scolopendrium vulgarae* (14)	Des II	W.C.
Angiosperms	*Carpolbrutus chilense* (ice plant) (11)	Des II (?)[g]	W.C.
	Soybean seed cotyledons (19)	Des II (?)[h]	Enz
	Spinach (29)	Des II	Enz
	Barley seedlings (17)	Des II	Enz
	Ricinus communis (castor leaves) (20)	Des II	W.C.
Metazoa Invertebrates Nematode	*Turbatrix aceti* (vinegar eel), grown axenically (31)	Des I	W.C.
Vertebrates	Rat liver (18, 25)	Des I	Enz
	Hen liver (21)	Des I	Enz

might reveal a quite different pattern of distribution. An additional source of concern arises from the fact that the presence of the aerobic desaturation type II system in an organism has usually been established by the failure of the organism to desaturate exogenously supplied 1-^{14}C-labeled palmitic and stearic acids rather than by actual isolation and characterization of the desaturase enzymes themselves. On the basis of such failure to desaturate, *Euglena gracilis* and several green algae (Table II) were considered to contain only desaturase system II (Erwin *et al.*, 1964), yet *Euglena* when grown heterotrophically does contain the enzymes of desaturase system I (Nagai and Bloch, 1965) and hence should desaturate exogenously added palmitate and stearate. The failure of the intact *Euglena* cells to desaturate could not be due to a permeability barrier or a deficiency in fatty acid activation since exogenously supplied 1-^{14}C-labeled palmitate and stearate were metabolized and incorporated into the complex lipids of the cell (Erwin *et al.*, 1964). James and his associates have reported that intact *Euglena* cells will desaturate exogenously added palmitate and stearate if the cells are suspended in phosphate buffer (Harris *et al.*, 1965). Whole

---

[a] References: (1) Bennett and Quackenbush (1969); (2) Bloch *et al.* (1961); (3) Bloomfield and Bloch (1960); (4) Davidoff and Korn (1962b); (5) Davidoff and Korn (1963); (6) Erwin and Bloch (1963a); (7) Erwin *et al.* (1964); (8) Fulco and Bloch (1964); (9) Fulco *et al.* (1964); (10) Fulco (1967a, 1969); (11) Fulco (1965); (12) Fulco (1967a, 1969); (13) Goldfine and Bloch (1960); (14) Haigh *et al.* (1969); (15) Harris and James (1965); (16) Harris *et al.* (1965); (17) Hawke and Stumpf (1965); (18) Holloway *et al.* (1963); (19) Inkpen and Quackenbush (1969); (20) James (1963); (21) Jones *et al.* (1969); (22) Korn *et al.* (1965); (23) Korn and Greenblatt (1963); (24) Lennarz *et al.* (1962b); (25) Marsh and James (1962); (26) F. Meyer and Bloch (1963); (27) H. Meyer and Holz (1966); (28) H. Meyer and F. Meyer (1969); (29) Nagai and Bloch (1965, 1968); (30) Nichols *et al.* (1965); (31) Rothstein and Götz (1968); (32) Scheuerbrandt *et al.* (1961); (33) Yuan and Bloch (1961); (34) Wood *et al.* (1965).

[b] AN, anaerobic pathway; Des I, direct desaturation system I; Des II, direct desaturation system II.

[c] Enz, demonstrated with isolated enzyme system employing acyl-CoA and acyl-ACP substrates; W.C., demonstrated in whole cells (or tissue slices) by incorporation of radioactive acetate and intermediate and long-chain saturated fatty acids into monounsaturated fatty acids.

[d] Isolated enzyme system produces 16:1ω7 and 18:1ω9, but 16:1ω8 is the major product in whole cells.

[e] Cells produce only 16:1ω11 from 16:0.

[f] Direct desaturation of stearate obtained with whole cells incubated in phosphate buffer but not with cells grown on tryptone broth.

[g] Produces 14:1ω5 and 18:1ω9 from ^{14}C-14:0 but does not produce 18:1 from ^{14}C-18:0 and hence may be either Des I or Des II.

[h] Free fatty acids were employed as substrates in a crude enzyme system, and it is uncertain whether the system is of the Des I or II type.

cells of the green alga *Chlorella vulgaris* also will desaturate added palmitate and stearate only if the cells are suspended in phosphate buffer rather than their usual growth medium (Harris *et al.*, 1965). This suggests that *C. vulgaris*, like *Euglena*, appears to contain desaturase system I, an interpretation that is supported by the susceptibility of desaturase in *Chlorella* to inhibition by sterculic acid (see Section III,C,1).

This phenomenon may be more general, and some organisms, particularly other green algae reported to possess only desaturase system type II, may, like *Euglena*, also possess desaturase system type I enzymes whose activity is "repressed" by the particular experimental conditions employed.

3. Significance of Multiple Pathways

The existence of multiple pathways for the biosynthesis of monoenoic fatty acids raises several questions concerning their evolution and physiological significance. Bloch has suggested that the anaerobic pathway of unsaturated fatty acid synthesis represents the most primitive one, since it is widely believed that early forms of life probably arose under anaerobic conditions (Bloch, 1962). The widespread occurrence of this pathway among the prokaryotic bacteria is in agreement with this view. Bloch has further suggested that the aerobic direct desaturation mechanism was selected for during the evolution of prokaryotic microorganisms into eukaryotic microorganisms because of the ability of this type of mechanism to produce polyunsaturated fatty acids of the methylene-interrupted type that are so ubiquitous among the membrane lipids of eukaryotic cells (Bloch, 1962). From this viewpoint it is significant that the prokaryotic blue-green algae that have been suggested to be the progenitors of the eukaryotic line (see R. M. Klein and Cronquist, 1967) utilize aerobic desaturase system type I to produce their monounsaturated fatty acids (Table II), and many, but not all, of these organisms contain large amounts of polyunsaturated fatty acids (see Section V,B,1). Furthermore, while bacteria do not normally contain polyenoic acids, a bacterium containing desaturase system type I has been shown to be capable of producing such compounds (see Section IV,B,1).

The origin of the aerobic desaturation mechanism among bacteria presents some difficulties since it seems unlikely that it could have arisen by modification of the anaerobic pathway. If the aerobic desaturase mechanism does, in fact, utilize oxygenase reactions, then it may have arisen by the conversion of an oxygenase-dependent catabolic pathway to serve a biosynthetic function. A number of bacteria are known to possess pathways for catabolizing long-chain alkanes. The reaction sequence is usually initiated by the introduction of molecular oxygen into the alkanes, and in some cases the initial attack is in the middle of the molecule. For example, in *Nocardia salmonicolor* hydrocarbon degradation is initiated by an oxygen-dependent insertion of a double bond into the substrate molecule. Hexadecane is converted to a mixture of *cis*-7-hexadecene and *cis*-8-hexadecene, while octadecane is converted to *cis*-9-octadecene (Abbott and Casida, 1968). The unsaturated alkanes cannot be used directly for unsaturated fatty acids, nor do these bacteria produce their unsaturated fatty acids via the direct desaturation pathway (Abbott and Casida, 1968). In *Micrococcus cerificans* hexadecane is first converted to palmitic acid, then desaturated to palmitoleic acid, and then further degraded for carbon and energy (see Abbott and Casida, 1968).

The aerobic desaturase type II systems have been found only in a restricted group of organisms, namely in higher plants; in green algae, which are commonly considered to be the stem group from which higher plants originated; and in the euglenids, which many authorities believe are either an offshoot of the green algae themselves or an offshoot of evolutionary progenitors of the green algae (see R. M. Klein and Cronquist, 1967). The type II mechanism almost certainly represents an adaptation of the desaturase type I system. The aerobic desaturase type II system consists of soluble enzymes and some representatives of the groups shown to have fatty acid synthetase systems of the soluble type, which are also localized in the chloroplast (see Section II,B,2). This suggests that during the course of evolution both systems have been placed in the chloroplast and their enzymes modified so as to facilitate their utilization of photosynthetically produced reductants, energy sources, and CO_2 fixation products. Evidence that the operation of both the fatty acid synthetase system and the desaturase system of spinach may be coupled to photosynthesis has recently been reported (Stumpf and Boardman, 1970).

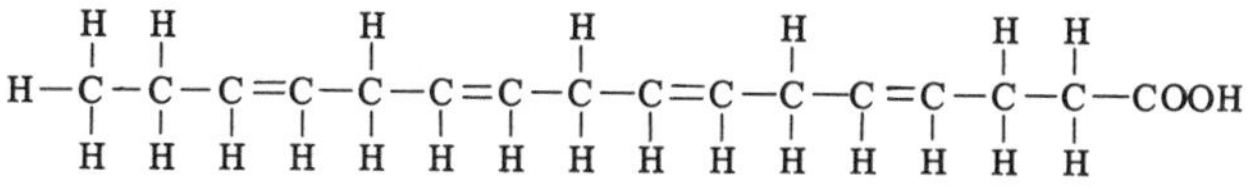

All *cis*-4, 7, 10, 13-hexadecatetraenoic acid (16:4ω3)

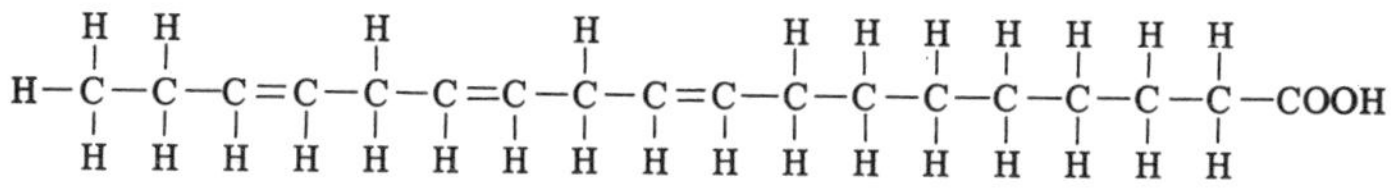

α-Linolenic acid
(all *cis*-9, 12, 15-octadecatrienoic acid; 18:3ω3)

Fig. 16. Some ω3 polyunsaturated fatty acids.

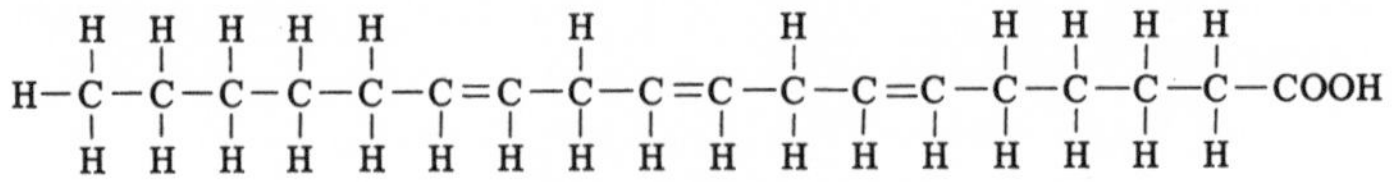

$$\begin{array}{c} \text{H}\quad\text{H}\quad\text{H}\quad\text{H}\quad\text{H}\qquad\quad\text{H}\qquad\qquad\text{H}\qquad\quad\text{H}\quad\text{H}\quad\text{H}\quad\text{H} \\ |\quad|\quad|\quad|\quad|\qquad\quad|\qquad\qquad|\qquad\quad|\quad|\quad|\quad| \\ \text{H}-\text{C}-\text{C}-\text{C}-\text{C}-\text{C}-\text{C}=\text{C}-\text{C}-\text{C}=\text{C}-\text{C}-\text{C}=\text{C}-\text{C}-\text{C}-\text{C}-\text{COOH} \\ |\quad|\quad|\quad|\quad|\quad|\quad|\quad|\quad|\quad|\quad|\quad|\quad|\quad|\quad|\quad|\quad| \\ \text{H}\quad\text{H}\quad\text{H}\quad\text{H}\quad\text{H}\quad\text{H}\quad\text{H}\quad\text{H}\quad\text{H}\quad\text{H}\quad\text{H}\quad\text{H}\quad\text{H}\quad\text{H}\quad\text{H}\quad\text{H}\quad\text{H} \end{array}$$

γ-Linolenic acid
(all *cis*-6, 9, 12-octadecatrienoic acid; 18:3ω6)

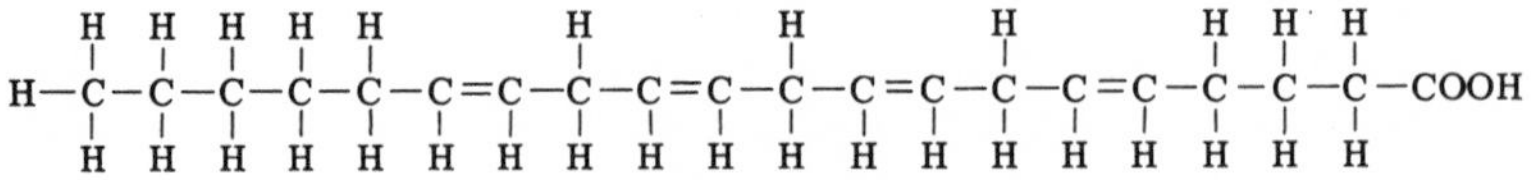

$$\begin{array}{c} \text{H}\quad\text{H}\quad\text{H}\quad\text{H}\quad\text{H}\qquad\quad\text{H}\qquad\qquad\text{H}\qquad\qquad\text{H}\qquad\quad\text{H}\quad\text{H}\;\;\text{H} \\ |\quad|\quad|\quad|\quad|\qquad\quad|\qquad\qquad|\qquad\qquad|\qquad\quad|\quad|\quad| \\ \text{H}-\text{C}-\text{C}-\text{C}-\text{C}-\text{C}-\text{C}=\text{C}-\text{C}-\text{C}=\text{C}-\text{C}-\text{C}=\text{C}-\text{C}-\text{C}=\text{C}-\text{C}-\text{C}-\text{COOH} \\ |\quad|\quad|\quad|\quad|\quad|\quad|\quad|\quad|\quad|\quad|\quad|\quad|\quad|\quad|\quad|\quad|\quad| \\ \text{H}\quad\text{H}\quad\text{H}\quad\text{H}\quad\text{H}\quad\text{H}\quad\text{H}\quad\text{H}\quad\text{H}\quad\text{H}\quad\text{H}\quad\text{H}\quad\text{H}\quad\text{H}\quad\text{H}\quad\text{H}\quad\text{H} \end{array}$$

Arachidonic acid
(all *cis*-5, 8, 11, 14-eicosatetraenoic acid; 20:4ω6)

Fig. 17. Some ω6 polyunsaturated fatty acids.

IV. Biosynthesis of Polyunsaturated Fatty Acids

A. TYPES OF COMPOUNDS

Fatty acids containing multiple double bonds are characteristically found in the polar lipids of eukaryotic microorganisms in abundance and wide variety, their structures differing in the length of the carbon chain, the degree of unsaturation, and the location of the double bonds. Generally, these fatty acids are all *cis* in configuration and have multiple double-bond systems of the methylene-interrupted type. They can be assigned to one of two principal categories, the ω3 and the ω6 families. The ω designation categorizes polyenoic acids according to their terminal portion, the terminal portion being the part of the fatty acid from the terminal methyl group to the closest double bond (see Schlenk and Sand, 1967; Marcel *et al.*, 1968). Thus α-linolenic acid and related ω3 compounds all have methylene-interrupted double-bond systems that terminate near the methyl end of the molecule (Fig. 16). Arachidonic acid and α-linolenic acid, which belong to the ω6 family (Fig. 17), have, on the other hand, a stretch of six carbon atoms between the termination of their methylene-interrupted double-bond system and the methyl end of the molecule. This distinction appears to be of considerable biological significance.

B. BIOSYNTHESIS OF POLYUNSATURATED FATTY ACIDS

1. *Serial Desaturation*

The typical polyunsaturated fatty acids found in eukaryotic cells are synthesized via the serial desaturation and chain elongation of oleic acid.

Direct evidence that this mechanism for producing polyenoic fatty acids operates *in vivo* in the tissues of higher plants (Simmons and Quackenbush, 1954; Harris and James, 1965) and animals (for a review, see Mead, 1960, 1968) has been obtained by the employment of radioisotope techniques. Similar studies have demonstrated that polyunsaturated fatty acids arise via serial desaturation of monoenoic fatty acids in the following eukaryotic microorganisms: a yeast, *Candida utilis* (Yuan and Bloch, 1961); a fungus, *Penicillium chrysogenum* (Bennett and Quackenbush, 1969); protozoa, including a soil amoeba, *Acanthamoeba* sp. (Korn, 1964a), several ciliates of the genus *Tetrahymena* (Erwin and Bloch, 1963a), slime molds (see Korn *et al.*, 1965), and a number of parasitic zooflagellates (see Korn *et al.*, 1965; H. Meyer and Holz, 1966); and algae, including the euglenids, *Euglena gracilis* and *Astasia longa* (Hulanicka *et al.*, 1964; Erwin *et al.*, 1964; Nichols and Appleby, 1969), a red algae, *Porphyridium cruentum* (Nichols and Appleby, 1969), several chrysomonads (Erwin *et al.*, 1964; Nichols and Appleby, 1969), several green algae, including *Ankistrodesmus braunii*, *Scenedesmus* D (Erwin *et al.*, 1964), and *Chlorella vulgaris* (Harris *et al.*, 1965; Harris and James, 1965), and a related colorless alga, *Polytoma uvella* (Erwin *et al.*, 1964). The polyunsaturated fatty acids synthesized by these organisms are of the methylene-interrupted type.

In contrast, among prokaryotic microorganisms only the blue-green algae generally contain polyunsaturated fatty acids. The blue-green algae also synthesize their polyunsaturated fatty acids via the serial desaturation of oleic acid (Nichols and Wood, 1968). Most bacteria employ the anaerobic pathway for producing their monoenoic fatty acids. Because of the manner in which double bonds are introduced in the anaerobic mechanism, this mechanism could not be employed to produce polyenoic acids of the methylene-interrupted type (see Section III,E). However, even bacteria that synthesize their monoenoic fatty acids by a direct desaturation mechanism do not usually synthesize polyunsaturated fatty acids. The only known exception is *Bacillus licheniformis*, which under certain environmental conditions can be induced to synthesize small amounts of polyunsaturated fatty acids, primarily 5,10-hexadecadienoic acid (Fulco, 1970).

2. Cell-Free Systems

Little is known about the enzymology of polyenoic fatty acid biosynthesis, although the mechanism appears to be basically similar to that employed for monoenoic fatty acid formation. Enzyme preparations that convert monoenoic fatty acids to dienoic fatty acids have been isolated from four organisms: the yeast *Candida utilis* (F. Meyer and Bloch, 1963), the green alga *Chlorella vulgaris* (Harris and James, 1965), safflower seeds

(McMahon and Stumpf, 1964), and rat liver (Holloway *et al.*, 1963; Nugteren, 1962). The rat-liver-isolated enzyme system also desaturates each of the intermediates (Fig. 21) between linoleic acid and arachidonic acid (Nugteren, 1962; Stoffel and Ach, 1964).

The enzyme systems for producing dienoic fatty acids generally resemble the desaturase type I systems of monounsaturated fatty acid synthesis. They employ the CoA thioester of oleic acid as the substrate and require molecular oxygen and either NADH or NADPH (F. Meyer and Bloch, 1963; Holloway *et al.*, 1963; McMahon and Stumpf, 1964; Nugteren, 1962; Harris and James, 1965). Furthermore, they are particulate enzyme systems. In yeast, the enzyme system for polyenoic fatty acid synthesis is localized in a 100,000 *g* particulate fraction (F. Meyer and Bloch, 1963); the vertebrate system has been described as microsomal (Holloway *et al.*, 1963). The enzymes from *Chlorella* (Harris and James, 1965) and safflower seeds (McMahon and Stumpf, 1964) appear to be localized in the chloroplasts.

However, the biosynthesis of polyenoic fatty acids may require additional cofactors. Biosynthesis of linoleic acid by cell-free particulate preparations from *Candida utilis* requires the presence of an unidentified supernatant factor, whereas, monoenoic fatty acid biosynthesis does not (F. Meyer and Bloch, 1963). On the basis of experiments with intact cells, I have suggested that sterols may be precursors of cofactors required for polyunsaturated fatty acids in the ciliated protozoan *Tetrahymena setifera* (Erwin *et al.*, 1966).

Some recent studies on algae suggest the possibility that an alternative mechanism for producing polyunsaturated fatty acids may exist, namely one in which the substrates are intact lipids. The evidence is derived from experiments on *Chlorella* and *Euglena*. Upon incubation of *Chlorella vulgaris* with ^{14}C-acetate, all the lipids of the alga become labeled; three lipids, phosphatidyl glycerol, phosphatidyl choline (PC), and monogalactosyl diglyceride (MGDG), incorporate the label very rapidly. Following this rapid uptake, the total radioactivity of these three lipids remains constant, but the distribution of the ^{14}C among different fatty acid residues of these lipids changes with time. Initially, ^{14}C is found only in the palmitic and stearic acid residues; then the ^{14}C label appears in the monoenoic fatty acid residues at the same time as it declines in the saturated fatty acid residues; finally, it accumulates in the trienoic fatty acid residues. Assay of the radioactivity of individual molecular species of one of these lipids, MGDG, reveals that the specific activities of individual fatty acid residues are different depending on the molecular species from which they are isolated. Thus after a 2-hour period the only dienoic acids containing ^{14}C were those found in molecular species of MGDG containing an average of five double bonds per molecule. These labeling patterns are difficult to understand if the polyunsaturated fatty acids found in MGDG and the other two rapidly labeled lipids are first produced by serial desaturation reactions and then incorporated into the lipids. However, the patterns would be predictable if further desaturation of the fatty acids of these lipids takes place after *de novo* synthesis of the lipid molecules from saturated or monounsaturated fatty acids (see Nichols *et al.*, 1967; Nichols, 1968; Nichols and Moorhouse, 1969; Safford and Nichols, 1970).

Other experiments on *Chlorella* support this interpretation. When exposed to ¹⁴C-oleic acid, *Chlorella* rapidly incorporated 70% of the molecule into a single lipid, phosphatidyl choline; after 30 minutes of incubation, ¹⁴C-labeled linoleic acid is found but exclusively in the form of ¹⁴C-linoleoyl-PC (see Gurr, 1971). When a micellar dispersion of ¹⁴C-labeled oleoyl-PC prepared by incubating *Chlorella* with ¹⁴C-oleic acid under nitrogen in the dark is incubated in the light with an *in vitro*-enzyme (chloroplast) system obtained from *Chlorella*, ¹⁴C-linoleoyl-PC is produced. A comparison of the relative rates of *in vitro* desaturation of ¹⁴C-oleoyl-PC and ¹⁴C-oleoyl-CoA suggests that the former is the better substrate (see Gurr, 1971).

Also of significance is the fact that Bloch and co-workers (Renkonen and Bloch, 1969) have demonstrated that enzyme preparations of *Euglena* employ oleoyl-ACP for synthesis of monogalactosyl diglyceride but use oleoyl-CoA for synthesis of phospholipids. Together these experimental observations suggest that the polyunsaturated fatty acids of the lipids of *Euglena* and *Chlorella* (and related algae) may be produced by either of

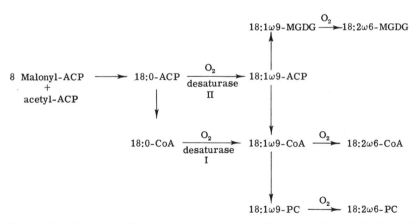

Fig. 18. Possible alternative mechanisms of oleic acid desaturation in *Euglena* and *Chlorella*.

two alternative mechanisms (see Fig. 18). In the first mechanism oleoyl-CoA is produced either directly via the desaturase system type I or indirectly by transfer from oleoyl-ACP produced via the desaturase system type II. Oleoyl-CoA is then desaturated to linoleoyl-CoA, which in turn is a substrate for the production of various polyunsaturated fatty acids via further desaturation and chain elongation. Alternatively, oleoyl-CoA and oleoyl-ACP may first be employed to synthesize oleoyl-PC and oleoyl-MGDG, respectively, as suggested by the findings of Renkonen and Bloch (1969). Subsequent desaturations would then take place on the intact lipids (Fig. 18). The evidence for the latter type of mechanism is indirect and may have an alternative explanation. However, such a mechanism is not without precedent—the formation of cyclopropane fatty acids from oleoyl-containing phospholipids (see Section II,D) would be an analogous reaction.

3. Pathways of Polyunsaturated Fatty Acid Biosynthesis

While all eukaryotic cells synthesize their polyunsaturated fatty acids from oleic acid by the same enzymic mechanism, they often employ different metabolic pathways, for further desaturation of oleic acid can proceed in two different directions. The $\omega 3$ type of desaturation usually produces linoleic acid and α-linolenic acids from oleate by the introduction of additional double bonds between the 9,10 position and the methyl end of the molecule (Fig. 19). Some algae (see Bloch et al., 1967) also contain an $\omega 3$ type of desaturation that produces a 16 carbon atom series of polyenes (Fig. 19). In some organisms α-linolenate accumulates; in others this acid undergoes a carboxyl type of desaturation coupled with chain elongation (Fig. 20). However, all the final products belong to the $\omega 3$ family of polyunsaturated fatty acids (see Section IV,A).

The $\omega 6$ family of polyunsaturated fatty acids are also synthesized from linoleic acid via chain elongation and further desaturation. However, in this type all subsequent double bonds are introduced exclusively toward the carboxyl end of the molecule (Fig. 21). Most organisms that produce $\omega 6$ polyenoic fatty acids exclusively and hence are specialized for carboxyl-directed desaturation retain the ability to produce linoleic acid from oleic acid (Fig. 21). Where this is not so, as in vertebrates, linoleic acid becomes a dietary requirement or essential fatty acid (for reviews, see Aees-Jorgensen, 1961; Alfin-Slater and Aftergood, 1968; Mead, 1968). In the absence of dietary essential fatty acid, vertebrates carry out a carboxyl-directed desaturation sequence that starts with oleic acid (Fulco and Mead, 1959, 1960; Holloway et al., 1963) and produces an $\omega 9$ polyenoic acid, $20:3\omega 9$ (Fig. 22).

Synthesis of α-linolenic acid:

$$18:1\omega9 \xrightarrow{O_2} 18:2\omega6 \xrightarrow{O_2} 18:3\omega3$$

Synthesis of hexadecatetraenoic acid:

$$16:1\omega9 \xrightarrow{O_2} 16:2\omega6 \xrightarrow{O_2} 16:3\omega3 \xrightarrow{O_2} 16:4\omega3$$

Fig. 19. $\omega3$ pathway of polyunsaturated fatty acid biosynthesis.

18:4ω3

↗ O_2

18:3ω3

↓ $+C_2$

$$20:3\omega3 \xrightarrow{O_2} 20:4\omega3 \xrightarrow{O_2} 20:5\omega3 \xrightarrow{+C_2} 22:5\omega3 \xrightarrow{O_2} 22:6\omega3$$

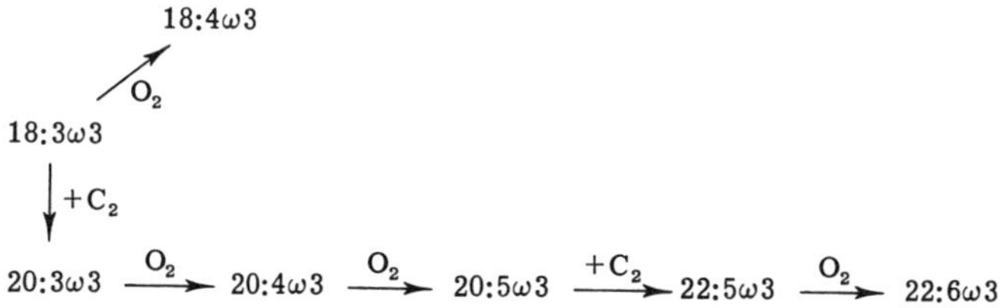

Fig. 20. Further desaturation of α-linolenic acid.

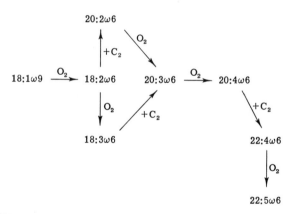

Fig. 21. $\omega6$ pathway of polyunsaturated fatty acid biosynthesis.

$$18:1\omega9 \xrightarrow{O_2} 18:2\omega9 \xrightarrow{+C_2} 20:2\omega9 \xrightarrow{O_2} 20:3\omega9$$

Fig. 22. Formation of $\omega9$ polyunsaturated fatty acids in essential acid-deficient vertebrates.

The $\omega6$ type of polyenoic fatty acid biosynthesis can produce arachidonic acid via either of two routes: desaturation of linoleic to γ-linolenic acid followed by chain elongation (Fig. 21) or, alternatively, chain elongation of linoleic to 11,14-eicosadienoic acid followed by carboxyl-directed desaturation (Fig. 21). The first route is the major one in vertebrates (Marcel et al., 1968), the chrysomonad Ochromonas danica, and the red alga Porphyridium cruentum (Nichols and Appleby, 1969). The second route is the predominante one in the euglenids Euglena gracilis and Astasia longa (Hulanicka et al., 1964; Nichols and Appleby, 1969), the soil amoeba Acanthamoeba sp. and Hartmannella rhysoides, and the slime mold Physarum polycephalum (Korn et al., 1965).

C. Environmental Factors Influencing Unsaturated Fatty Acid Biosynthesis

The fatty acid composition of a microbial or tissue culture population is extremely flexible and readily often dramatically modified in response to environmental parameters such as media composition and pH, length of incubation, aeration, incubation temperature, and degree of illumination.

1. Media Composition

Requirements for dietary fatty acids and other lipids are not uncommon among microorganisms. The older literature has been reviewed (Hutner and Holz, 1962), and current studies with bacteria, yeasts, and fungi that require fatty acids are discussed in detail in Chapter 6.

However, even non-lipid-requiring microorganisms and tissue cells may respond dramatically to the presence of lipids in crude or synthetic media. De novo synthesis of fatty acids from ^{14}C-acetate is almost completely suppressed in tissue culture cells (Greyer et al., 1962; Bailey, 1966) and in ciliated protozoa (Erwin, 1971) by the presence of either free fatty acids or complex lipids in the culture media. Furthermore, microorganisms may take up from the culture media and incorporate into their neutral and polar lipids fatty acids that are structurally quite different from those normally synthesized by the organisms. For example, 20:2$\omega6$, 20:3$\omega6$, and 20:4$\omega6$ are incorporated into the polar lipids of the ciliate Tetrahymena pyriformis, and the soil amoeba Acanthamoeba sp. incorporates $\omega3$ polyenoic acids into its polar lipids when the culture media supply these compounds (Lees and Korn, 1966). These fatty acids are not found in Tetrahymena and Acanthamoeba when the organisms are grown on lipid-free media (Table XII). Similarly, Escherichia coli will incorporate polyunsaturated fatty acids into its lipids but neither synthesizes nor requires such compounds (Silbert et al., 1968). The implication of these and similar observations are clear. The fatty acid composition of a microorganism may be considered a reliable index of its biosynthetic capacities only if that organism is cultured axenically in lipid-free media. Likewise, the results of radioisotope trace experiments on fatty acid synthesis in a microorganism must be interpreted with extreme caution if the organism is grown on lipid-containing media, and such experiments are virtually meaningless if the cultures are heavily contaminated with other microorganisms.

Even when microorganisms are cultured on lipid-free media, variations in media composition may produce large-scale changes in the fatty acid spectra of an organism. The effects of the presence of leucine, isoleucine, or valine or their short-chain fatty acid derivatives on the fatty acid composition of microbes have already been mentioned (Section II,D). Effects of other alterations in the media composition have also been observed, and a few examples will suffice. The nitrogen content of the culture media influences the proportion of saturated to unsaturated fatty acids in green algae (see Shaw, 1966) and bacteria (Marr and Ingraham, 1962). The ratio of saturated to unsaturated fatty acids varies inversely with sugar concentration in *Escherichia coli* (Marr and Ingraham, 1962) and in the fungus *Blakeslea trispora* (Dedyukhina and Bekhtereva, 1969). The choice of substrate and the purity of sucrose employed in culture media profoundly influence fatty acid composition in euglenids (see J. A. Erwin, 1968). The addition of urea or potassium monobasic phosphate to cultures of *Blakeslea trispora* increased the organisms' content of $18:2\omega6$ and $18:3\omega6$ acids (Dedyukhina and Beklterreva, 1969).

2. *Aeration*

Increased oxygen tension has been reported to elevate unsaturated fatty acid content in plants (James, 1969), the ciliate *Tetrahymena* (Koch and Scherbaum, 1967), the dinoflagellate *Gyrodinium cohnii* (Harrington and Holz, 1968), and cryptomonad flagellates (D. H. Beach, 1961). These results are not remarkable in the light of the requirement for molecular oxygen for unsaturated fatty acid biosynthesis in the organisms. However, the increased content of cyclopropane acids at the expense of unsaturated fatty acids observed in the bacteria *Escherichia coli* and *Pseudomonas fluorescens* at low oxygen tension (Crowfoot and Hunt, 1970) must have a different explanation.

3. *Light and Carbon Dioxide Tension*

In many photosynthetic algae (see Section V,F,1) $\omega3$ polyunsaturated fatty acid synthesis is depressed and $\omega6$ polyene synthesis enhanced by the absence of light. Increased light intensity (D. H. Beach, 1961; Constantopoulos and Bloch, 1967b) also enhances the synthesis of $\omega3$ polyenoic acids.

Increased carbon dioxide tension increases $18:3\omega3$ and particularly $16:4\omega3$ acids in *Euglena* (Hulanicka *et al.*, 1964) and in cryptomonads (D. H. Beach, 1961). A deficiency of manganese decreases $\omega3$ polyenoic fatty acid biosynthesis in *Euglena* (Constantopoulos, 1970), presumably via its effect on chlorophyll synthesis. All these effects are related to morphological and physiological changes in the photosynthetic organelles and their unique lipid composition in these algae and will be discussed further in Section V and more fully in Chapter 5.

4. *Temperature*

The degree of unsaturation of an organism's fatty acids usually varies inversely with the growth temperature (for a general review of temperature effects on microorganisms, see Farrell and Rose, 1967). Psychrophilic microorganisms, those capable of growth at $0°–5°C$ and with growth optima below $20°C$, generally have more highly unsaturated fatty acids than their mesophilic counterparts, those that grow in the $20°–40°C$ range (Kates and Hagen, 1964; Sumner *et al.*, 1969); thermophilic microorganisms, those that grow above $45°C$, generally have more highly saturated fatty acids than mesophiles (Sumner *et al.*, 1969). "True" thermophiles, those capable of growing above $60°C$, are all prokaryotic organisms (Brock, 1967). The thermophilic blue-green algae lack polyunsaturated fatty acids (Table III), while the thermophilic bacteria usually lack even monounsaturated fatty acids (see Daron, 1970). Furthermore, when the same organism is grown at two different temperatures it usually synthesizes more unsaturated fatty acids at the lower temperature. This phenomenon has been observed in bacteria (Marr and Ingraham, 1962; Kates and Hagen, 1964; Bishop and Still, 1963; Daron, 1970), blue-green algae (Holton *et al.*, 1964), eukaryotic algae (D. H. Beach, 1961; Patterson, 1970), yeasts (Kates and Baxter, 1962; F. Meyer and Bloch, 1963; Brown and Rose, 1969), fungi (Sumner and Morgan, 1969; Sumner *et al.*, 1969), and protozoa (Erwin and Bloch, 1963a; Koch and Scherbaum, 1967; Greenblatt and Wetzel, 1966), as well as in metazoa, invertebrates (see Lewis, 1962; Farkas and Herodek, 1964), and vertebrates (see Knipprath and Mead, 1968; Caldwell and Vernberg, 1970). These results have usually been given an ecological interpretation. The changing degree of unsaturation of an organism's fatty acids represents a physiological adaptive response to changing environmental temperature that facilitates survival of the organism at different temperatures (e.g., see Knipprath and Mead, 1968; Caldwell and Vernberg, 1970).

However, what at first glance appears to be a very simple and universal phenomenon turns out to be quite complex, the apparent simplicity arising from a failure to consider the complexities of the available experimental data and the deficiencies in many of the experiments themselves. First, decreasing the incubation temperature increases the oxygen availability since oxygen solubility in water increases significantly with temperature decrease. Oxygen tension is a significant factor in the relative amounts of saturated and unsaturated fatty acids produced (see Section IV,C,2); indeed, James has shown that in nonphotosynthetic tissue from higher plants the increased unsaturated fatty acids obtained at lower temperatures is a direct result of increased oxygen tension (James, 1969). Yet most experiments on the effects of temperature on fatty acid composition have ignored simultaneous changes in oxygen availability.

Second, if the alteration in the degree of unsaturation of an organism's fatty acids were strictly an adaptive response, one might expect to find a linear correlation between growth temperature and the degree of unsaturation. In many studies on the effects of temperature on fatty acid unsaturation only two different growth temperatures were employed, and the data from these experiments do not permit us to decide whether or not the effects are linear with temperature change. In some cases where fatty acid composition has been studied over a range of temperatures, the alterations observed in the ratio of saturated to unsaturated fatty acids have not been gradual but abrupt. For example, in many bacteria the decreased saturated to unsaturated fatty acid ratio observed with decreasing temperature is actually a reflection of the monounsaturated to cyclopropane fatty acid ratio. In the bacterium *Serratia marcesens* the monounsaturated fatty acid to cyclopropane acid ratio was the same at 37° and 30°C, changing in favor of the monounsaturated components only at 20°C (Bishop and Still, 1963). In the blue-green alga *Anacystis nidulans*, which contains only saturated and monounsaturated fatty acids, there is no change in the ratio of saturated to unsaturated acids between 26° and 35°C, monounsaturated fatty acid biosynthesis decreasing at 41°C (Holton *et al.*, 1964).

Furthermore, to speak of a high or low saturated to unsaturated fatty acid ratio is meaningless when the organisms contain a variety of unsaturated fatty acids not all of which respond to temperature shifts in the same manner. For example, the bacterium *Bacillus licheniformis* contains two desaturase enzymes; one functions only at low temperatures and produces 16:1ω11, while the other is temperature-independent and produces 16:1ω6 acids (Fulco, 1970). In eukaryotes the response to temperature changes is even more complex. In cryptomonad flagellates the synthesis of 18:4ω3 and 20:5ω3 is enhanced by a reduction of the temperature from 25° to 15°C, but 18:3ω3 synthesis is inhibited (D. H. Beach, 1961). In the green alga *Chlorella sorokiniana* the highest degree of unsaturation is observed at 22°C, decreasing at both lower and higher temperatures. Dienoic acid predominates at 14°C, trienoic at 22°C, and saturated at 48°C (Patterson, 1970). Again, while the overall saturation of the fatty acids of goldfish decreased at 10°C as compared to 35°C, many polyenoic acids were more predominant at 35°C; thus the content (in both triglycerides and phospholipids) of 16:1, 18:1, 20:3, and 20:5 acids decreased at 35°C, but the content of 18:2, 20:4, and 22:6 acids increased at 35°C (Knipprath and Mead, 1968).

Finally, in some organisms, changes in the ratio of saturated to unsaturated fatty acids produced with elevated temperature are not adaptive but pathological. For example, in the ciliate *Tetrahymena pyriformis* the ratio of saturated to unsaturated fatty acid increases between 25° and 35°C with the principal effect being exerted on monounsaturated fatty acids (Erwin and Bloch, 1963a). At higher temperatures (37°–40°C) a disruption of polyunsaturated fatty acid synthesis occurs. No change in the fatty acid composition of the membrane phospholipids occurs over the entire temperature range; rather there is a steady decrease in the amount of phospholipid synthesized. As a result, at the upper temperature, 40°C, the ciliates become morphologically abnormal and eventually die. These effects at 40°C are prevented by supplying polyunsaturated fatty acids in the culture medium (Erwin, 1970). Similar changes in ability to desaturate with concomitant pathological effects have been observed with the zooflagellate *Leishmania enrietti* (Greenblatt and Wetzel, 1966).

The above data suggest the following interpretation of the effects of growth temperature on the fatty acid composition of organisms. Different unsaturated fatty acids (particularly in eukaryotic organisms) are probably synthesized by different enzymes that have different temperature optima. Similarly, the systems that produce the desaturase enzymes themselves may have different temperature optima; F. Meyer and Bloch (1963), for example, have demonstrated that the ability of enzyme systems isolated from the yeast *Candida utilis* to produce different unsaturated fatty acids depends on the temperature at which the yeast was grown. Hence the effects of growth temperature on the fatty acid composition of different organisms will be complex and variable. These effects may, in some cases, be of adaptive value to the organism but are not necessarily so, and indeed may in some cases be detrimental. From the limited data available it would seem that the operation (or synthesis) of desaturase enzymes producing polyunsaturated fatty acids does not operate above the mesophilic temperature range. This may be one reason eukaryotic microorganisms (whose membrane lipids probably require polyunsaturated fatty acids) are never found in truly thermophilic environments and all true thermophiles are prokaryotic cells (see Brock, 1967).

5. Culture Age

As with increasing temperature, a general pattern is observed: With increasing culture age the ratio of saturated to unsaturated fatty acids increases. In bacteria this actually represents an increased cyclopropane to unsaturated fatty acid ratio (Marr and Ingraham, 1962; Law et al., 1963; Crowfoot and Hunt, 1970). A general decrease of unsaturated fatty acids with culture age has been found in protozoa (Erwin and Bloch, 1963b; Koch and Scherbaum, 1967; H. Meyer and Holz, 1966), dinoflagellates (Harrington and Holz, 1968), cryptomonads (D. H. Beach, 1961) and other algae (see Shaw, 1966), yeasts (Suomalainen and Keranen, 1968), and fungi (Sumner et al., 1969; Shaw, 1966).

Again, however, the apparent universality of the phenomenon does not necessarily indicate a simple underlying mechanism. First, there are exceptions to the pattern. In cryptomonads, 18:3ω3 concentration decreases with culture age, but 18:4ω3 and 20:5ω3 concentrations increase (D. H. Beach, 1961); in the chrysomonad *Ochromonas danica* there is a general increase in unsaturated fatty acids with culture age (Gellerman and Schlenk, 1965), and this is also true of the fungus *Cephalosporium* (Huber and Redstone, 1967). Second, there would again be a tendency for the oxygen to become limiting in older culture. Finally, in bacteria the changes take place in the fatty acids of the phospholipids of the cell membrane, but in the protozoan *Tetrahymena pyriformis* (Erwin and Bloch, 1963a; Erwin, 1970; Jonah and Erwin, 1971) the changes simply represent an accumulation of storage triglyceride, which contains large amounts of saturated and monounsaturated acids. The fatty acid composition of polar lipids obtained from membranes isolated from *T. pyriformis* is identical in young and old cultures (Jonah and Erwin, 1971). Similarly, old cultures of *Euglena* produce a storage wax rich in saturated acids (Rosenberg, 1963b), and similar accumulation of storage lipids rich in saturated and monounsaturated fatty acids and poor in polyunsaturates appears to account for the phenomenon in many green algae and fungi (see Shaw, 1966).

V. Phyletic Distribution of Polyunsaturated Fatty Acids

A. Notes on Methodology

1. Origin of the Biological Material

The fatty acid composition of a microbial population is readily modified in response to environmental parameters (see Section IV,C). Information from an analysis of the fatty acid composition of a microorganism may be considered a reliable reflection of its biosynthetic capacities only if the cells employed for analysis were grown axenically on lipid-free, well-defined synthetic media and preferably under several different sets of carefully controlled environmental conditions. I have, where possible, indicated the methodology employed in obtaining the biological material used for fatty acid analysis. Emphasis has been placed on results obtained with axenic cultures grown on lipid-free media.

2. *Identification of Fatty Acid Components of Microorganisms*

Gas-liquid chromatography permits the rapid separation and identification of a mixture of long-chain fatty acids (as their methyl esters) and requires very small amounts of material. However, as with most chromatographic procedures, unequivocal identification of compounds by GLC alone is rarely possible, even when several different stationary phases are employed. Several accessory techniques (also essentially "micro") have been developed that when used in conjunction with GLC permit the identification of most naturally occurring long-chain fatty acids. Mixtures of methyl esters of fatty acids can be separated on the basis of their degree of unsaturation by silica gel TLC employing mercuric acetate adducts or impregnating the gel with silver nitrate (argenation TLC). The number of carbon atoms in an unsaturated fatty acid may be confirmed by subjecting the compound to catalytic hydrogenation and identifying the products by GLC. Positional isomers of polyunsaturated fatty acids whose number of carbon atoms and double bonds have been determined can usually be identified by GLC (provided that reference compounds are available) since such isomers usually have distinctly different retention times on the polar stationary phases in general use. However, positional isomers of monoenoic fatty acids usually have identical or almost identical retention times on all stationary phases now in use. The position of the double bonds in a monoenoic fatty acid can be established by degradation of the compound and GLC analysis of the products employing either permanganate oxidation or ozonalysis. These methods can also be used to confirm the identification of positional isomers of polyunsaturated fatty acids. For a discussion of these techniques, see Ackman (1969), Holman and Rahm (1966), and Privett (1966).

In this section assessment of the degree of confidence that can be placed on the identification of fatty acids reported in the literature has been made on the basis of which techniques were employed and has been indicated for literature citations wherever possible. Reports of fatty acid analysis predating the development of modern techniques have generally been ignored. For coverage of some of the earlier literature, see reviews by Shorland (1962), Kates (1964), and Shaw (1966).

B. PROKARYOTIC ALGAE AND BACTERIA

1. *Bacteria*

The almost ubiquitous presence of polyunsaturated fatty acids of the methylene-interrupted type in the polar lipids of eukaryotic cells stands in sharp contrast to their absence in bacteria, for bacteria lipids do not contain polyunsaturated fatty acids except as fortuitous contaminants.

2. *Cyanophyta*

Morphologically, the blue-green algae resemble the bacteria; physiologically, they resemble the eukaryotic algae in that they possess the oxygen-evolving type of photosynthesis. Biochemically, the fatty acid spectrum of the blue-green algae is intermediate between that characteristic of bacteria and that characteristic of eukaryotic microorganisms.

The fatty acid composition of 26 species of blue-green algae is given in Table III. In addition, Stanier and his co-workers have examined the fatty acid content of about 40 strains of unicellular Cyanophyta usually assigned to the order Chroococcales and about 15 strains of various filamentous Cyanophyta. (Stanier and co-workers have abandoned the traditional botanical nomenclature for the various groups of Cyanophyta and have distributed their strains of unicellular Cyanophyta among three typological groups, I, II, and III, which were constructed on the basis of both morphological and biochemical criteria.) All strains were cultured axenically, and the identity of the positional isomers of their polyunsaturated fatty acids was confirmed by degradation studies. The results of these studies, however, are at this time available only in summarized form (Kenyon and Stanier, 1970; Stanier *et al.*, 1971).

Some blue-green algae resemble the bacteria in that they do not contain any polyenoic fatty acids; others contain only the dienoic acid, linoleic acid (Table III, Kenyon and Stanier, 1970; Stanier *et al.*, 1971). Many species, however, resemble higher algae in that they contain large amounts of 18 carbon atom trienoic fatty acids (Table III, Kenyon and Stanier, 1970; Stanier *et al.*, 1971). A few species of blue-green algae contain small amounts of a 16:2 acid, but more highly unsaturated fatty acids of the 16 carbon atom series often found in eukaryotic algae are absent (Table III). Similarly, polyenoic fatty acids containing more than three double

TABLE III
Fatty Acids of Algae: Cyanophyta

Organism[a]	Saturated (total)[c]	16:1	18:1	16:2	18:2	ω3	18:3[d]	ω6
Order Chroococcales								
Agmenellum quadruplicatum (6)[e]	40	15	16	4	14		5	
Anacystis cyanea (7)[f]	28	20	7	0	10		5	
A. marinus (6)[e]	56	36	4	0	0	0		0
A. montanus (7)[f]	21	3	25	0	18		19	
A. nidulans (1)[g]	49	39	10	0	0	0		0
(5)[e]	40	33	4	0	0	0		0
(6)[g]	53	34	13	0	0	0		0
Chroococcus turgidus (7)[f]	28	12	7	4	16		26	
Coccochloris elabens (6)[g]	53	12	13	0	17		3	
Synechococcus cedrorum (2)[e]	4	39	10	0	0	0		0
Order Chamaesiphonales								
Myxosarcina chroococoides (4)[e]	42	9	7	1	9	33		0
Order Oscillatoriales								
Anabaena cylindrica (4)[e]	50	6	6	6	24	11		0
A. flos-aquae (4)[e]	41	6	5	4	37	11		0
A. variabilis (5)[e]	39	15	14	0	14	17[m]		0
Chlorogloea fritschii (4)[h]	48	5	14	0	17	16		0
Grown photoautotrophically (2)[i]	44	17	14		13	12		
Grown heterotrophically (2)[i]	41	19	26		13	0		
Hapalosiphon laminosus (2)[i,j]	57	24	18	0	0	0		0
Lyngbya aestuarrii (7)[f]	26	12	8	4	17		18	
L. lagerhaimii (6)[e]	45	15	31	0	7	0		0
Microcoleus chthonoplastes (6)[g]	51	13	14	0	5		18	
Nostoc sp. (7)[f]	41	10	20	1	8		7	
N. muscorum A (2)[e]	36	15	7		10		21	
N. muscorum G (6)[e]	35	20	16	0	14		11	
Oscillatoria sp. (2)[h]	32	24	26		10		7	
O. williamsii (6)[e]	42	24	11	14	4	0		0
Plectonema terebrans (6)[e]	40	13	20	5	11		6	
Spirulina platensis (4)[k]	46	10	5	0	12	0		21
(7)[k]	47	6	10	0	16		13	
Trichodesmium erythaeum (6)[l]	45	4	3	0	4		19	
Order Stigonematales								
Mastigocladus laminosus (4)[e]	39	43	17	0	2	0		0

bonds or more than 18 carbon atoms also found in many groups of eukaryotic algae do not appear to exist among the blue-greens.

The 18 carbon atom triene of *Anabaena cylindrica*, *A. variabilis*, *A. flos-aquae*, *Chlorogloea fritschii*, and *Myxosarcina chroococcoides* and a number of other "species" of Cyanophyta (Stanier *et al.*, 1971) has been shown by degradation studies to be the ω3 acid, α-linolenic acid (Levin *et al.*, 1964 Nichols and Wood, 1968). However, a number of blue-green algae, including the filamentous form *Spirulina platensis* (Table III, Nichols and Wood, 1968) and several strains of unicellular forms belonging to Stanier's typological group II, have been shown to contain substantial amounts of the ω6 isomer, γ-linolenic acid (Stanier *et al.*, 1971). The identity of the isomer was confirmed by the results of degradation studies (Nichols and Wood, 1968; Stanier *et al.*, 1971).

Holton and co-workers (Holton *et al.*, 1968) have suggested that the presence of polyunsaturated fatty acids in a blue-green alga is correlated roughly with its degree of morphological and physiological complexity.

---

[a] References: (1) Holton *et al.* (1964); (2) Holton *et al.* (1968); (3) Levin *et al.* (1964); (4) Nichols and Wood (1968); (5) Oró *et al.* (1967); (6) Parker *et al.* (1967); (7) Schneider *et al.* (1970).

[b] To the nearest whole number; 0 denotes present in amounts <1% or absent.

[c] *Trichodesmium erythaeum* contained primarily 10:0 (27%), 14:0 (21%), and 16:0 (17%) plus small amounts of 12:0, 15:0, and 18:0; *A. marinus* contained primarily 14:0 (21%) and 16:0 (32%) plus small amounts of 15:0 and 18:0. In all other blue-green algae the saturated fatty acids consisted primarily of 16:0 plus only small amounts of 12:0, 14:0, 15:0, and 18:0. Branched-chain fatty acids have been reported only for *A. nidulans* [Ref. (5)], and these include *i*-13:0 (2%), *i*-14:0 (2%), and *i*-16:0 (2%).

[d] The position of the double bonds in 18:3 was not specified in these reports.

[e] Grown axenically on lipid-free chemically defined media (usually mineral) as photoautotrophs under constant illumination at 24°–30°C and gassed with 0.5–1% CO_2 (in air).

[f] Grown as described in footnote *e*, but unialgal-bacterial contamination was described as varying from very slight to none.

[g] Grown as described in footnote *e*, but grown at 35°–40°C.

[h] Grown as described in footnote *e*, but in air.

[i] Grown axenically at 35°C on a lipid-free chemically defined medium (mineral medium plus sucrose) and gassed with 0.5% CO_2 (in air). Grown as a heterotroph in the dark in one experiment and photosynthetically under constant illumination in another experiment.

[j] Thermophilic.

[k] Grown as described in footnote *f*, but the medium contained small amounts of soil extract.

[l] Not cultured; collected from a natural bloom.

[m] While not specified in this report, the position of the double bonds has been established by other authors; see Levin *et al.* (1964).

TABLE IV

FATTY ACIDS OF ALGAE: RHODOPHYTA

Fatty Acids (wt. %)[a,b][f]	Class Bangiophyceae, Order Porphyridiales[c]				Class Florideophyceae[d]				
	Porphyridium sp. (1)[e]	Porphyridium cruentum (2)[e]	(5)[e]	(4)[e]	Order Ceraminales Ceramium rubrum (3)[e]	Laurencia obtusa (5)[e]	Rhodomela sufusca (3)[e]	Order Gigantinales Gracilaria conferviodes (5)[e]	Plocacium coccineum (3)[e]
Saturated (total)[f]	39	33	31	26	37	65	34	31	40
16:1	0	7	2	2[i]	7[i]	4	5	3	7
18:1	2[g]	4	4	3	14[i]	12[c]	15[i]	16[c]	6[i]
18:2ω6	8	18	15	16	1	4	2	2	3
18:3ω3	0	3	0	0	2	1	1	1	0
18:4ω3	0	0	2	0	1	0	1	1	1

Fatty acid								
20:3	0	0	2	1[k]	0	1[k]	0	7[k]
20:4ω6	24	33	36	5	5	14	46	12
20:5ω3	34[h]	11	17	17	9	24	1	22
22:6	0	0	0	1	0	1	0	1

[a] See Table II.

[b] The identity of the fatty acids was based primarily upon GLC data in Refs. (1) and (4), argenation TLC, and GLC retention time and cochromatography plus GLC of reduction products in Ref. (2), while in Refs. (3) and (5) the identity of the positional isomers was confirmed by degradation studies.

[c] The species of *Porphyridium* are unicellular forms and were cultured axenically as photoautotrophs in a lipid-free chemically defined medium.

[d] The species of Florideophyceae are multicellular forms that were not cultured but collected from the sea.

[e] References: (1) Ackman *et al.* (1968); (2) Erwin *et al.* (1964); (3) Klenk *et al.* (1963); (4) Nichols and Appleby (1969); (5) Wagner and Pohl (1965); (6) Pohl *et al.* (1968).

[f] Primarily 10:0 with smaller amounts of 14:0 and 18:0 plus small amounts of 12:0, 15:0, 19:0, 20:0, and 21:0. None of these workers reported branched-chain fatty acids in red algae.

[g] 18:1ω9 from degradation studies.

[h] Polyunsaturated eicosenoic acids reported but not further identified.

[i] Small amounts of *trans*-16:1ω13 also detected (localized in the phosphatidyl glycerol fraction where it forms 32% of the total fatty acids in this lipid).

[j] 16:1ω7 (from degradation studies).

[k] 20:3ω6 (from degradation studies).

[l] Mixture of 18:1ω9 and 18:1ω7 (from degradation studies).

Unicellular forms of the order Chroococcales that lack such complexity may or may not contain polyunsaturated fatty acids (Table III, Stanier *et al.*, 1971), but all of the 15 morphologically more complex strains of the *Calothrix, Anabaena, Chlorogloea, Oscillatoria,* and *Spirulina* types examined by Kenyon and Stanier (1970) plus 18 other species of varying degrees of complexity listed in Table III contain polyunsaturated fatty acids. The sole exception, *Hapalosiphon laminosus,* is normally an inhabitant of hot springs (Holton *et al.*, 1968), and its fatty acid composition may reflect adaptation to a thermophilic mode of life.

C. Eukaryotic Algae

1. *Rhodophyta*

The red algae are considered by many taxonomists to be the most primitive of the eukaryotic algae; they are generally multicellular, macroscopic, nonmotile marine forms. They can be divided into two classes, the Bangiophyceae and the Florideophyceae. Our knowledge of the fatty acids of members of the class Bangiophyceae is generally limited to *Porphyra columbina* plus a few species of the genus *Porphyridium*; the latter are unicellular forms that can be cultured axenically in the laboratory. Several species belonging to various orders of the class Florideophyceae have been analyzed for their fatty acid composition by Klenk and co-workers (Klenk *et al.*, 1963) and Pohl and co-workers (Wagner and Pohl, 1965; Pohl *et al.*, 1968); all these species were collected wild from the sea. Red algae for which good quantitative fatty acid data are available are listed in Table IV.

The two major polyunsaturated fatty acids of the red algae are the ω6 acid, arachidonic acid, and the ω3 acid, eicosapentaenoic acid, both 20 carbon polyenoic acids (Table IV). Shorter- and longer-chain polyenoic acids are not significant components of the lipids of these algae. The *Porphyridium* species contain substantially more linoleic acid than the species of the class Florideophyceae. This difference might represent a real taxonomic distinction or may simply be a reflection of the fact that only the *Porphyridium* species were cultured axenically.

2. *Pyrrophyta*

Three classes of unicellular flagellated algae are grouped together to form the division Pyrrophyta (R. M. Klein and Cronquist, 1967). There are two classes of dinoflagellates, the Desmophyceae and the Dinophyceae, that are largely marine; the class Cryptophyceae (or cryptomonads), is found in both marine and freshwater environments.

Our knowledge of the fatty acid composition of the dinoflagellates and

TABLE V

FATTY ACIDS OF ALGAE: PYRROPHYTA

Organism[a]	Saturated (total)	16:1	18:1ω9	20:1	18:2ω6	16:poly	18:3ω3	18:4ω3	20:5ω3	22:6ω3	Minor Acids
Class Desmophyceae											
Order Prorcentrales											
Exuviella sp. (7)[c]	39[j]	2[l]	5	1	9	3[o]	2	4[r]	4	10[u]	14[x]
Prorocentrum micans (3)[d]	24[k]	23[l]	6	0	0	26[o]	6[a]	0	12	0	1[x]
Class Dinophyceae											
Order Gymnodiniales											
Amphidinium carterii (7)[c]	17[j]	1[l]	2	2	1	2[o]	3	15[r]	20	22[u]	8[x]
(1)[c]	47[j]	0	8	0	0	0	0	10	7	25	3[x]
Gymnodinium nelsoni (7)[c]	42[j]	5[l]	16	0	1	0	1	2[r]	5	18[u]	6[x]
Order Peridiniales											
Glendodinium sp. (7)[c]	31[j]	4[l]	5	0	5	0	6	16[r]	2	18[u,v]	2[x]
Gonyaulax catanella (7)[c]	50[j]	3[l]	7	0	3	2[o]	6	7[r]	1	9[u]	4[x]
G. polyedra (8)[e]	38[j]	1[l]	3	0	2	0	3	14	14	23	2[x]
G. tamarensis (7)[c]	42[j]	4[l]	5	0	4	0	5	7[r]	3	14[u,v]	3[x]
Gyrodinium cohnii (6)[f]	48[k]	1[l]	14	0	0	0	0	0	0	30[u]	0
Peridinium triquetum (7)[c]	35[k]	5[l]	2	0	9	0	3	8[r]	2	16[u,v]	6[x]
P. trochoideum (7)[c]	45[k]	8[l]	9	0	3	2[o]	2	5[r]	1	11[u,v]	5[x]
(2)[c]	53[k]	4[l]	7	0	0	4[o]	3	7	13	0	2[x]
Class Cryptophyta											
Unidentified cryptomonad (1)[c]	17[j]	1	3[x]	1	11	0	23[a]	16	14	1[v]	4[y]
Chilmononas paramecium (2)[a]	37[k]	0	9	0	12	0	27	0	6	3	6[z]
(5)[h]	28[j]	2	10	0	8	0	40	0	10[s]	0	2[z]

TABLE V (Continued)

FATTY ACIDS OF ALGAE: PYRROPHYTA

Organism[a]	Satu-rated (total)	16:1	18:1ω9	20:1	18:2ω6	16:poly	18:3ω3	18:4ω3	20:5ω3	22:6ω3	Minor Acids
Chromoonas sp. (2)[c]	19[j]	2[m]	5	0	3	1[p]	23[k]	23	14	6[w]	14[x]
Cryptomonas sp. WH (2)[c]	10[k]	2[m]	5	0	0	2[p]	7	44	16	10	1[z]
C. appendiculata (3)[d]	18[j]	12	3	10	4	3[p]	12	13	10[t]	0[v,w]	6[aa]
C. malculata (3)[d]	20[j]	7	4	17	0	4[p]	6	16	17[t]	0[v,w]	4[aa]
C. ovata (2)[i]	11[i]	4[m]	2	0	0	2[p]	17[q]	34	12	7[w]	3[z]
C. ovata var. *palustris* (4)[i]	23[j]	9	16	0	26	1[p]	5	0	0	0[v]	20[bb]
Hemiselmis brunescens (2)[d]	14[i]	3	2	18	0	3[p]	8	31	14	0[v,w]	5[aa]
H. rufescens (3)[d]	21[i]	10	2	14	1	4[p]	7	17	8[t]	0[t,v]	6[aa]
H. virescens (2)[c]	31[j]	5[m]	7	0	3	0	22[q]	16	7	12[t]	3[z]
Rhodomonas lens (2)[c]	32[k]	5[m]	10	0	2	0	16	13	13	5[t]	2[z]

[a] References: (1) Ackman *et al.* (1968); (2) D. H. Beach *et al.* (1970); (3) Chuecas and Riley (1969); (4) Collins and Kalnins (1969); (5) Erwin *et al.* (1964); (6) Harrington and Holz (1968); (7) Harrington *et al.* (1970); (8) Patton *et al.* (1966).

[b] See Table III.

[c] Marine organisms grown axenically in lipid-free chemically defined media at 15°–25°C; photosynthetic forms grown under constant illumination as photoautotrophs or photoauxotrophs.

[d] Same as in footnote *c* but *not* axenic-unialgal cultures.

[e] Not cultured, but collected from the sea.

[f] Same as in footnote *c*, but this organism is a nonphotosynthetic form grown heterotrophically in the dark.

[g] Freshwater, nonphotosynthetic form; grown axenically at 25°C in the dark on a lipid-free chemically defined medium (minerals plus vitamins) with lactate as the carbon source.

h Same as in footnote g, but employing acetate as the carbon source.

i Freshwater, photosynthetic form; grown axenically at 15°–20°C as a photoheterotroph on a lipid-free chemically defined medium under constant illumination.

j Primarily 16:0 with smaller amounts of 14:0 and 18:0 plus trace amounts of 12:0, 17:0, 20:0, and 24:0.

k Primarily 14:0 and 16:0 with smaller amounts of 18:0 plus trace amounts of 12:0, 17:0, 20:0, and 24:0; G. cohnii contained larger amounts of 12:0 (8%).

l The sum of several positional isomers; in G. cohnii the ω5, ω7, ω9, ω11, and ω13 isomers of 16:1 have been identified (by degradation studies); see Harrington and Holz (1968).

m Identified (by degradation studies) as 16:1ω7; several of the cryptomonads in addition contain small amounts of trans 16:1ω13. These include Cryptomonas sp. WH (3%), C. ovata (1%), H. virescens (2%), and R. lens (1%); unidentified cryptomonad present.

n Also contains 18:1ω7 (7%) plus trace amounts of 18:1ω5.

o Sum of 16:2 and 16:3ω4.

p Sum of 16:2ω7, 16:2ω4, 16:3ω5, and 16:4ω1.

q These organisms also contain 18:3ω6: P. micans (2%), Chroomonas sp. (1%), C. ovata (5%), H. rufescens (2%), and unidentified cryptomonad (trace).

r In addition, these organisms contain presumed cis, trans isomers of 18:4ω3 (2–7%).

s Polyunsaturated eicosenoic acids not further identified.

t These organisms also contain small amounts of 20:4ω3: C. appendiculata (6%), C. malculata (1%), and H. rufescens (3%).

u In addition, these organisms contain presumed cis, trans isomers of 22:6ω3 (1–5%).

v These organisms also contain small amounts of 22:5ω3 (1–3%).

w These organisms also contain small amounts of 22:4ω6 (1–2%).

x Minor components consist of 13:0 and 19:0; branched chain, 14:0 and 15:0; 14:1, 14:2, 20:4, 22:4, and 24:4; 16:3ω3 and 16:4ω3 and 18:4ω6.

y Include 15:0 and 17:0, 14:1 and 17:1, and 17:2.

z Not specified.

aa Include 13:0 and 15:0; branched chain, 15:0 and 17:0; 19:1 and 22:1; 20:2, 20:3, and 22:3.

bb Includes shorter-chain fatty acids, 6:0, 8:0, 9:0, and 10:0; 13:0, 15:0 and 19:0; 14:1, 15:1, and 17:1; 15:2, 17:2, 19:2, and 20:2.

cryptomonads stems primarily from the studies of Holz and his co-workers, who have analyzed the fatty acids of 18 species, which were cultured axenically in lipid-free synthetic media under carefully controlled environmental conditions (Harrington and Holz, 1968; Holz, 1969; D. H. Beach *et al.*, 1970; Harrington *et al.*, 1970). They have carefully identified the fatty acids, separating them according to degree of saturation using argenation TLC, identifying the separated compounds by GLC, ascertaining chain length by GLC of the reduction products of the fatty acids, and confirming the identity of positional isomers by degradation studies. Their results, along with other reports from the literature, are summarized in Table V.

The fatty acid spectra of photosynthetic marine cryptomonads and dinoflagellates are very similar when both groups are grown axenically on lipid-free synthetic media (Table V). Aside from variable amounts of linoleic acid, only ω3 polyunsaturated fatty acids are synthesized. These are primarily 18:4ω3, 20:5ω3, and 22:6ω3 (Table V); α-linoleic acid is usually a minor acid in the dinoflagellates but is present in substantially higher amounts in the cryptomonads (Table V). Polyunsaturated 16 carbon atom fatty acids were absent or present in only trace amounts. The heterotrophic marine dinoflagellate *Gyrodinium cohnii* contained only a single polyenoic fatty acid, 22:6ω3 (Table V). Deviations from this general pattern have been reported (Table V), but only with organisms not grown in axenic culture in chemically defined media.

Of the two freshwater photosynthetic cryptomonads examined, one, *Cryptomonas ovata*, resembles the marine forms, while the other, *Cryptomonas ovata* var. *palustris*, lacks the 18:4ω3, 20:5ω3, and 22:6ω3 acids characteristic of the marine forms (Table V). The one heterotrophic, freshwater cryptomonad examined, *Chilomonas paramecium*, also lacks 18:4ω3 and has a generally lower content of 20:5ω3 and 22:6ω3; it does contain large amounts of α-linolenic acid (Table V). To date, no freshwater dinoflagellates have been analyzed for fatty acid composition.

3 Chrysophyta

The algae of division Chrysophyta consist of four classes (Cronquist, 1971), the Chloromonadophyceae, the Xanthophyceae, the Chrysophyceae, and the Bacillariophyceae. There are no data on the fatty acid composition of members of the first of these classes, and only two species of the second class the Xanthophyceae have been examined, *Olisthodiscus* sp. and *Monodus subterraneus*. Eicosapentaenoic acid is the major polyunsaturated fatty acid of these two algae when they are grown axenically in chemically defined media (Table VI). Small amounts of linoleic and arachidonic acids

TABLE VI

FATTY ACIDS OF ALGAE: CHRYSOPHYTA

Organism[a]	Saturated (total)	16:1	18:1	18:2ω6	16:2 + 16:3	18:3 ω6	18:3 ω3	18:4ω3	20:3	20:4ω6	20:5ω3	22:6ω3
Class Xanthophyceae												
Order Chloramoebales												
Olisthodiscus sp. (1)[c,d]	30[i]	4	1	1	2	0	7	0	1[o]	1	22	0
Order Mischococcales												
Monodus subterraneus (8)[c]	27[i]	24	9	4	0	0	0	0	1	5	29	0
(7)[c]	25[i]	27	5	4	0	0	0	0	0	5	31	0
Class Chrysophyceae												
Order Ochromonadales												
Coccolithus huxlyi (2)[d,e]	25[i]	28	10	2	11[m]	0	1	0	0	0	17	0[o]
Coccolithus huxlyi and Cricosphaera carteri (5)[c,d]						+	2–10	2–18				6–7
Order Chromulinales												
Monochrysis lutheri (1)[d]	29[j]	18	3	1	4[m]	0	0	0	1[o]	1	22	0
(2)[d,e]	22[j]	20	6	2	22[m]	0	0	1	2[p]	0	19	0[o]
Pseudopedinella sp. (2)[d,e]	15[j]	24	3	2	24[m]	0	0	0	1[p]	0	27	0[o]
Class Bacillariophyceae: Centrales												
Order Eupodiscales												
Cyclotella cryptica (6)[d,f]	22[c]	35	1	1	13	0	2[n]	0	0	0	19	0
(1)[c,d]	37[i]	42	1	0	7[m]	0	0	0	0	0	8	2
Skeletonema costatum (1)[c,d]	29[k]	17	0	1	12[m]	0	0	3	0	0	13	2
Thalassiosira fluviatilis (1)[c,d]	32[j]	45	0	0	10[m]	0	0	0	0	0	8	2

TABLE VI (Continued)

Organism[a]	Saturated (total)	16:1	18:1	18:2ω6	16:2 + 16:3	18:3 ω6	18:3 ω3	18:4ω3	20:3	20:4ω6	20:5ω3	22:6ω3
Order Rhizosolenisles												
Lauderia borealis (2)[d,e]	22[i]	21	2	1	15[m]	0	0	0	0	0[r]	30	0[e]
Cricosphaera carteri (2)[d,e]	18[j]	21	3	3	20[m]	0	0	2	0	1[r]	20	0
C. elongata (2)[d,e]	19[i]	21	2	2	17[m]	0	1	1	0	0	28	0
Ochromonas danica (8)[g]	28[j]	1	14	20	0	14	4	4	5	6	0	0[e]
(4)[g]	29[j]	+	8[l]	16	0	10	2	+	5[p]	11	+	0[e]
O. malhamensis (3)[g]	14[i]		5[l]	23	0	3	15	0	?[q]	8	?[q]	0
Poteriochromonas stipitata (3)[g]	24[i]	3	7[l]	18	0	1	3	0	?[q]	4	?[q]	0
Syracosphaera carterae (1)[c,d]	26[i]	2	8	10	5[m]	0	8		0	0	4	0[e]
Order Isochrysidales												
Dicrateria inornata (2)[d,e]	18[i]	8	17	5	4[m]	0	13	20	1[p]	0	8	0
Isochrysis galbana (2)[d,e]	27[i]	5	15	11	5[m]	1	14	17	1[p]	0	3	0
Prymnesium parvum (2)[d,e]	22[i]	10	25	18	2[m]	0	11	2	1[p]	0	4	0[e]
Order Biddulphiales												
Chaeroceros septentrionale (2)[d,e]	22[i]	20	4	1	13[m]	0	1	0	2[p]	1[r]	21	0[e]
Ditylium brightwelli (2)[d,e]	25[i]	30	6	2	15[m]	0	1	0	1[p]	0[r]	11	0[e]
Class Bacillariophyceae: Pennales												
Order Fragilariales												
Asterionella japonica (2)[d,e]	19[i]	21	3	1	13[m]	0	0	0	2[p]	7[r]	20	0[e]
Order Bacillariales												
Cylindrotheca fusiformis (6)[d,f]	32[i]	25	3	3	5	2[n]	2[n]		0	9	17	0

Fatty Acids (wt. %)[b]

Nitzschia angularis (6)[d,f]	30[i]	31	3	2	6	1[n]	0	0	0	21	0
N. thermalis (6)[d,f]	34[i]	46	1	1	5	0	0	4	7	0	
Order Naviculales											
Navicula pelliculosa (6)[f,h]	14[i]	31	6	4	21	3[n]	0	0	5	15	0
Order Phaeodactyles											
Phaeodactylum											
$\quad$tricornutum (6)[d,f]	20[i]	31	2	1	11	1[n]		0	0	26	0
$\quad$(2)[d,e]	20[i]	27	5	1	18[m]	0	0	0[r]	18	0[s]	
$\quad$(1)[c,d]	37[i]	47	1	1	6[m]	0	0	0	7	1	

[a] References: (1) Ackman et al. (1964, 1968); (2) Chuecas and Riley (1969); (3) Erwin et al. (1964); (4) Haines et al. (1962) and Gellerman and Schlenk (1965); (5) Holz (1969); (6) Kates and Volcani (1966); (7) Koelensmid et al. (1962); (8) Nichols and Appleby (1969).

[b] See Table III; +, present.

[c] These photosynthetic algae were cultured axenically at 18°–24°C in lipid-free chemically defined media under constant illumination.

[d] Marine species.

[e] These photosynthetic algae were grown in unialgal cultures in Erd-Schreiber medium at 20°C under constant illumination.

[f] These photosynthetic diatoms were cultured axenically at 18°–20°C under constant illumination in media containing 1 gm/liter of bactotryptone peptone.

[g] These freshwater chrysomonads have functional chloroplasts but require organic carbon sources for growth in the light and hence were grown photoheterotrophically under axenic conditions at 20°–24°C.

[h] Freshwater species.

[i] Primarily 16:0 plus small amounts of 14:0, 18:0, and 20:0.

[j] Primarily 14:0 and 16:0 plus small amounts of 18:0 and 20:0.

[k] Contains 14:0 (33%) plus 16:0 (16%) plus traces of 18:0 and 20:0.

[l] Shown to be 18:1ω9 (by degradation studies).

[m] Identified from GLC data as a mixture of 16:2ω4 and 16:2ω6 plus 16:3ω4 and 16:3ω6.

[n] Identified from GLC data alone; positional isomer not specified.

[o] Reported to be 20:3ω3.

[p] Reported to be 20:3ω6.

[q] Total unidentified polyunsaturated eicosenoic acids other than 20:4ω6 reported for O. malhamensis (28%) and P. stipitata (27%).

[r] Reported to contain in addition small amounts of 20:4ω3.

[s] Contains small amounts of one or more of the following acids: 22:2, 22:3, 22:4, and 22:5.

are also present, but neither γ-linoleic acid nor the ω3 acids, 18:4ω3 and 22:6ω3, are found (Table VI).

More is known about the fatty acid composition of the third class, the Chrysophyceae. The fatty acids of several freshwater chrysomonads that grow in axenic culture on lipid-free synthetic media have been analyzed, namely *Ochromonas danica, O. malhamensis,* and *Poteriochromonas stipitata.* These chrysomonads are unicellular, flagellated forms possessing unusual nutritional versatility. They contain chloroplasts and grow in the light as photoheterotrophs. They grow well in the dark as strict heterotrophs on dissolved nutrients, and they are also facultative phagotrophs (see M. B. Allen, 1969). This physiological versatility is reflected biochemically in their fatty acid composition. All three species contain substantial amounts of linoleic acid plus more highly unsaturated polyenoic acids of both the ω3 family, including 18:3ω3, 18:4ω3, and 20:5ω3, and the ω6 family, including 18:3ω6, 20:3ω6, and 20:4ω6 (Table VI). *Ochromonas danica* also synthesizes some docosenoic polyenoic acids of the ω6 family, 22:4ω6 and 22:5ω6 (Gellerman and Schlenk, 1965). Polyenoic 16 carbon atom fatty acids have not been detected.

The fatty acids of a number of marine chrysomonads grown in axenic culture in lipid-free synthetic media have been analyzed by Holz and co-workers (Holz, 1969) and Ackman and his associates (1964, 1968). Chuecas and Riley (1969) have studied the fatty acids of several other marine chrysomonads grown in unialgal culture. The results of these studies are summarized in Table VI. In addition, Holz (1969) has reported that seven other species of marine chrysomonads, *Chrysochromulina kappa, Dicrateria inornata, Isochrysis galbana, Monochrysis lutheri, Pavlova gyrans, Prymnesium parvum,* and *Stichochrysis immobilis,* grown axenically in lipid-free chemically defined media, contain 18:3ω3 (2–11%), 18:4ω3 (2–9%), 20:5ω3 (1–9%), and 22:6ω3 (1–10%) but no ω6 polyunsaturated fatty acids, except for trace amounts of γ-linolenic acid.

Marine chrysomonads generally have similar fatty acid spectra that differ from those of freshwater chrysomonads in several respects. Unlike freshwater forms, marine chrysomonads synthesize little of the ω6 acids; they lack arachidonic acid and contain only trace amounts of γ-linolenic ac d (Table VI). Furthermore, they contain higher concentrations of 20:5ω3 and usually, but not always, have significantly higher amounts of 18:4ω3 than freshwater forms (Table VI). Many marine forms also have significant amounts of 22:6ω3, at least when grown in axenic culture on synthetic media. This fatty acid pattern is reminiscent of that of the cryptomonads and dinoflagellates. Unlike freshwater forms, some marine chrysomonads also have significant amounts of 16:2 and 16:3 fatty acids

(Table VI); however, these are not of the ω3 type found in euglenids and green algae but rather mixtures of 16:2ω4, 16:2ω6, 16:3ω4, and 16:3ω6.

The fourth class of the division Chrysophyta, the Bacillariophyceae (or diatoms), consists primarily of unicellular forms that possess a silicified wall and are widely found in both marine and freshwater habitats. Ackman and co-workers have grown several marine diatoms axenically on lipid-free synthetic media (Ackman *et al.*, 1964, 1968) and examined their fatty acids. Kates and Volcani have grown one freshwater and five marine diatoms axenically in media containing small amounts of yeast extract and examined their fatty acids (Kates and Volcani, 1966), and Chuecas and Riley (1969) have analyzed the fatty acids of several marine diatoms grown in unialgal culture. The results of these studies are summarized in Table VI. The diatoms contain only small amounts of linoleic acid, little or no 18:3 fatty acids, and no 18:4 fatty acids (Table VI). Some, but not all, species contain an eicosatetraenoic acid, which is probably arachidonic acid (Table VI). The major polyenoic fatty acid of the marine diatoms is 20:5ω3, but substantial amounts (5–13%) of polyenoic 16 carbon atom compounds of the type found in marine chrysomonads are also present in diatoms (Table VI). The one freshwater diatom examined has a fatty acid spectrum similar to that of the marine forms (Table VI).

4. *Euglenophyta*

The euglenids are of considerable interest since they constitute a well-defined group sharing characteristics of both the plant and animal kingdoms. Most members of the group are freshwater unicellular flagellates that lack a cell wall. Their nutrition and metabolism spans the entire plant-animal spectrum. The previously mentioned (Section II,B,2) ability of *Euglena gracilis* to grow as a strict photoauxotroph, a photoheterotroph, or a strict heterotroph (lacking differentiated chloroplasts) represents an excellent example of this plant-animal ambiguity. *Euglena* can be permanently "cured" of its plantlike propensities by treatment with streptomycin or other chemical and physical agents (see Schiff and Epstein, 1968); these *Euglena* mutants grow well as heterotrophs. Naturally occurring non-chloroplast-containing counterparts of *Euglena* also exist (i.e., *Astasia longa*); they, too, grow well as strict heterotrophs. Finally, another group of euglenids exists represented by *Peranema trichophorum*, which are not only strict heterotrophs but also phagotrophs.

As in the case of the chrysomonads the nutritional and metabolic versatility of these flagellates is reflected in their fatty acid composition. *Euglena gracilis* is capable of synthesizing virtually all the 16, 18, 20, and 22 carbon atom polyenoic fatty acids of both of the ω3 and ω6 type found in nature (Korn, 1964b) (Table VII). Which fatty acids are actually

TABLE VII

FATTY ACIDS OF ALGAE: EUGLENOPHYTA

	Relative Amount (wt. %)[a]						
	Euglena gracilis						Astasia longa (2)[b,d]
	Strain Z				Variety bacillaris[c]		
Fatty Acids	Photo-auxo-troph (2)[d,e]	Photo-hetero-troph (3)[d,f]	Photo-hetero-troph (4)[d,g]	Hetero-troph (2)[d,h]	Wild type (1)[d]	Bleached mutant (1)[d,i]	
Saturated total	32[l]	36[n]	34[p]	34[l]	34[l]	12[q]	37[l]
16:1[j]	6	2	6	16	4	3	4
18:1[k]	10	1	2	7	8	5	7
16:2ω6	3	3	9	0			0
18:2ω6	4	3	12	7	3	2	1
20:2ω6	0	2	1	3			5
16:3ω3		7	5	0	0	0	0
18:3ω3	32	9	15	2	16	1	0
20:3ω6	0	2	8	1	?	?	4
20:3ω3	—[m]	2		—[m]	?	?	—[m]
16:4ω3	16	3	0	0	0	0	0
18:4ω3	0	0	2	0	0	0	0
20:4ω6	3	3	0	10	↑	↑	18
20:4ω3	0	2	0	3			0
20:5ω3	2	4	0	10	42	65	4
22:5ω6	↑	4	2–5	13			0
22:5ω3	2	1					0
22:6ω3	↓	2[o]			↓	↓	0

[a] See Table III.

[b] Cells were grown heterotrophically on an acetate containing lipid-free synthetic medium and harvested in the log phase.

[c] Cells were grown and harvested as in footnote h but under constant illumination.

[d] References: (1) J. Erwin and Bloch (1963b); (2) Hulanicka et al. (1964); (3) Korn (1964b); (4) Rosenberg (1963a).

[e] Cells were grown axenically on mineral-vitamin medium at 24°–25°C with 5% CO_2 (in air) under constant illumination and harvested in the log phase.

[f] Cells were grown in the light at 24°–25°C on a lipid-free chemically defined medium containing organic carbon sources and harvested after 7–10 days.

[g] Cells were grown in an organic medium in the light and harvested in the stationary phase.

[h] Cells were grown on the same medium as in footnote f but at 24°–25°C, incubated in the dark, and harvested after 4–5 days (log-phase cells).

[i] Derived from wild type by treatment with streptomycin.

synthesized depends on whether the organism is growing as a photosynthetic "plant" or as a heterotrophic "animal" (Erwin and Bloch, 1962, 1963b; Hulanicka et al., 1964; Rosenberg, 1963a; Rosenberg and Pecker, 1964; Rosenberg et al., 1965). α-linolenic acid and polyunsaturated 16 carbon atom acids of the ω3 family constitute the bulk of the polyenoic fatty acids of the strict photoauxotroph. These acids are found in only trace amounts in the dark-grown heterotrophs in which polyenoic 20 and 22 carbon atom fatty acids of both the ω3 and ω6 types predominate (Table VII). Photoheterotrophically grown *Euglena* contains intermediate amounts of the various types of acids. Streptomycin-bleached *Euglena* and the natural heterotroph *A. longa* resemble dark-grown normal *Euglena* in their fatty acid composition (Table VII). The fatty acids of the phagotrophic *Peranema trichophorum* have not yet been studied.

5. *Chlorophyta*

The Chlorophyta consist primarily of a single large class, the Chlorophyceae (the green algae). A second small class, the Charophyceae, appear to be a highly modified offshoot of the Chlorophyceae and are sometimes included in the Chlorophyta. Table VIII lists the fatty acids of a single member of the Charophyceae, *Nitella*, which was collected from the wild. The polyunsaturated fatty acids of *Nitella* consist primarily of linoleic and α-linolenic acids; arachidonic acid is also present in small amounts.

Considerably more data are available on the fatty acids of the green algae, particularly for members of the orders Volvocales and Chlorococcales (see Table VIII). Many species can be readily cultured in the laboratory on lipid-free media. For a number of the species, the identity of positional isomers of polyenoic fatty acids has been confirmed by degradation studies.

The Volvocales, considered to be the most primitive of the green algae (Cronquist, 1971), are flagellated forms inhabiting both marine and freshwater environments. All but two of the volvocids examined are marine forms, and all but one are photosynthetic. The volvocids contain a wide variety of polyunsaturated fatty acids. All contain linoleic acid. A number

---

[i] Mixture of 16:1ω7 and 16:1ω9.

[k] Primarily 18:1ω9; small amount of 18:1ω7.

[l] Primarily 16:0 plus 14:0 plus small amounts of 12:0, 18:0, and i-15:0.

[m] Small amounts included in value for 20:4ω6.

[n] Primarily 13:0, 14:0, and 16:0 with small amounts of 15:0, 17:0, and 18:0.

[o] Also reported: 17:1 (2%), 15:4ω2 (1%), 19:4ω5 (6%), 21:4ω5 (1%), and 21:5ω5 (2%).

[p] Primarily 14:0, 16:0, and 19:0 plus small amounts of 12:0, 15:0, and 18:0.

[q] Primarily 16:0 with small amounts of 12:0, 14:0, and 18:0.

TABLE VIII

FATTY ACIDS OF ALGAE: CHLOROPHYTA

Fatty Acids (wt. %)[b]

Organism[a]	Saturated (total)[c]	16:1	18:1	16:2	18:2ω6	16:3	16:4	18:3 ω6	18:3 ω3	18:4ω3	20:4ω6	20:5ω3	22:6ω3
Class Chlorophyceae													
Order Volvocales													
Chlamydomonas sp. strain 285 (2)[c]	20	7	7	1	28	0	1	1	0	17	0	0	0
Chlamydomonas sp. strain 430 (2)[c]	22	2	17	7	10	7	0	3	29	1	0	0	0
Chlamydomonas reinhardi (3)[d]	26	3	24	?	5	+[p]	+[p]	6	31	0		2[p]	
Dunaliella primolecta (2)[c]	19	10	6	8	6	7	6	2	10	7	10	0	0[s]
D. salina (12)[e]	42	15	11	0	8	0	0	0	19	0	0	0	0
D. tertiolecta (1)	18	0	3	2	5	4	17[a]	6	37	0	0	0	0[s]
(2)[c]	20	10	8	3	6	5	7	1	8	8	10	0	0[s]
Chlorella pyrenoidosa (6)[f]	17	3	14	0	11	4	13	1	31	5	0	0	0
(5)[g]	20	3[m]	46	3	10	7[a]	+		12	0	0	0	0
(10)[h]	23	2	17	6	19	0	0	←— 26[t] —→			0	0	0
C. sorokiniana (8)[d]	34	3	4	6	21	13	0	←— 18[t] —→		0	0	0	0
C. variegata (2)[i]	19	5	6	14	23	6	0	0	28		0	0	0
C. vulgaris (7)[j]													
Photoautotrophic	30	8[m,n]	2	7	34	2[p]	+[p]		20	0	0	0	0
Heterotrophic	21	6[m]	11	6	54	0	0		7	0	0	0	0

Species									←16[t]→		←17ʳ→	
Coelastrum microsporum (10)[h]	14	1	30ᵘ	5	13	0	0	0	7	0	0	0
Scenedesmus D₃ (3)[l]	26	3`	8	?	+	+ᵖ	0	0	0	0	0	0
S. obliquus (5)[g]	37	2	8	+	7	+ᵠ	15	0	2	0	0	0
(13)[g]	12	2	5ᵛ	0	–	0	59ʷ	0	0	0	0	0
S. quadricauda (10)[h]	13	4	18ᵘ	0	7	0	7–20	5–7	2–6	0	0	0
Dunaliella salina,[k] D. Viridis,[k] and D. tertiolecta (4)						3–5			12–38	2–6	3–6	0
Heteromastix rotunda (2)[c]	22	16	2	3	3	3	2	0	4	9	1	30	0ᵍ
Halosphaera viridis (1)[e]	23	1	5ᵛ	0	7	0	3ᵃ	1	7	6	0	0	4
Platymonas (Tetraselmis) sp. (1)[k]	21	3ᵐ	14	0	3	1	15ᵃ	1	19	0	1	8	1
Platymonas convoluta and Prasinocladus maximus (4)[k]							9		13		0	3–6	0
Platymonas tetrathele (13)[k]	31	4	13ᵛ	2	12	4ᵃ	14ᵃ	0	16	8	2	4
Polytoma uvella (3)[z]	24	1	8	0	5	0	0	9	3	0	17ʳ	
Order Chlorococcales												
Ankistrodesmus braunii (11)[f]	33		54		+				13			
S. quadricauda (6)[f]	14	2	6	1	14	2	19	1	34	0	0	0
Tetraedrom sp. (10)[h]	20	3	34	3	11	0	0	←18[t]→				
Order Siphonocladales												
Valonia utricularis (13)[e]	46	3	18ᵛ	1	8	5ᵃ	0	0	10	6	2	0
Order Codiales												
Codium elongatum (13)[e]	37	10	18ᵛ	3	10	2ᵃ	0	0	16	5	5	4	0
Codium fragile (5)	30	2	11	1	6	12	0	0	27	2	5	5ᶻ	0

TABLE VIII (Continued)

Organism[a]	Saturated (total)	16:1	18:1	16:2	18:2ω6	16:3	16:4	18:3 ω6	18:3 ω3	18:4ω3	20:4ω6	20:5ω3	22:6ω3
Order Caulerpales													
Halimeda tuna (13)[e]	32	6	26[v]	2	13	3[a]	0	0	5	2	2	4	2
Enteromorpha sp. (6)[e]	21	2	9	3	18	20	14[q]	←—— 18[t] ——→		17	—	3[x]	0
Enteromorpha compressa (5)[e]	23	1	8[u]	1	5	2	15	0	26	9	—	8[x]	0
Ulva fascinata (13)[e]	24	4	21[v]	1	8	0	3[a]	0	14	21	1	0	0[w]
Order Zygnematales													
Spriogyra sp. (9)[f]	30	6	12	1	7	4	8	←—— 17[t] ——→		4	—	4[x]	0
Class Charophyceae													
Order Charales													
Nitella sp. (6)[f]	23	2	2	?	31	7	?	0	17	0	6	0	0

Fatty Acids (wt. %)[b]

[a] References: (1) Ackman *et al.* (1968, 1970); (2) Chuecas and Riley (1969); (3) Erwin and Bloch (1963b); (4) Holz (1969); (5) Klenk *et al.* (1963); (6) Koelensmid *et al.* (1962); (7) Nichols (1965); (8) Patterson (1970); (9) Schlenk and Gellerman (1965); (10) Schneider *et al.* (1970); (11) Williams and McMillan (1961); (12) Williams (1965); (13) Wagner and Pohl (1965) and Pohl *et al.* (1968).

[b] See Table III; +, present.

[c] Marine photosynthetic alga, grown in unialgal culture on a nonsynthetic (Erd-Schreiber) medium at about 20°C under constant illumination.

[d] Freshwater photosynthetic alga grown axenically in a lipid-free chemically defined medium in the presence of organic carbon sources (under constant illumination).

[e] Marine photosynthetic alga not cultured but collected from the sea.

[f] Freshwater photosynthetic alga grown axenically at 22°–26°C on a lipid-free chemically defined medium either as a photoautotroph or as a photoauxotroph under constant illumination.

[g] Freshwater photosynthetic alga; conditions of culture axenic but not clearly specified.

[h] Same as in footnote f but unialgal, not axenic; bacterial contamination reported as being slight to none.

[i] Same as in footnote d, but the medium contained small amounts of bactotryptone (medium essentially lipid-free).

[j] This freshwater photosynthetic alga was grown under two different sets of conditions: as a photoauxotroph as in footnote f or as a heterotroph cultured in a lipid-free chemically defined medium (containing organic carbon sources) in the dark.

[k] Marine photosynthetic alga grown axenically at 15°–20°C on a lipid-free chemically defined medium, either as a photoautotroph or as a photoauxotroph (under constant illumination).

[l] Freshwater photosynthetic alga; not cultured but collected wild.

[m] Shown to be $16:1\omega7$ plus $16:1\omega9$ (by degradation studies).

[n] Also contains small amounts of $trans$-$16:1\omega13$.

[o] Saturated fatty acids usually primarily $16:0$ with smaller amounts of $14:0$ and $18:0$ plus trace amounts of $12:0$, $15:0$, $17:0$, and $20:0$.

[p] Shown to be $16:3\omega3$ and $16:4\omega4$ (by degradation studies) by other authors (see Bloch *et al.*, 1966; Constantopolous and Bloch, 1967b).

[q] Shown to be $16:3\omega3$ and $16:4\omega4$ (by degradation studies).

[r] Polyunsaturated eicosenoic acids not further identified.

[s] Also contains $20:3\omega6$ (1–2%), $20:4\omega5$ (0–4%), $22:5\omega6$ (0–2%), and $22:5\omega3$ (4–6%).

[t] Identified by GLC only; positional isomer not specified.

[u] Also contains $trans$-$18:1\omega9$ (7–14%).

[v] Shown to be $18:1\omega9$ (by degradation studies).

[w] Shown to be $16:4\omega3$ (by degradation studies), but value includes small amounts of $18:2\omega6$.

[x] Values include small amounts of $20:4\omega6$.

[y] Also contains $22:5$.

[z] Freshwater nonphotosynthetic alga grown axenically at 24°–26°C in the dark on a lipid-free simple chemically defined medium containing acetate as the carbon source.

of ω3 polyenoic acids of the 16, 18, and 20 carbon atom series have been found, including 16:3ω3, 16:4ω3, 18:3ω3, 18:4ω3, and 20:5ω3 (Table VIII); however, 22:6ω3 and other 22 carbon fatty acids are rare (Table III). The volvocids also usually contain ω6 polyenoic acids, primarily γ-linolenic acid (Table VIII). One exception to this pattern is the freshwater heterotrophic volvocid *Polytoma uvella*, which has very little α-linoleic acid and no other ω3 acids. It does have γ-linolenic acid and a greatly increased content of polyenoic 20 carbon atom fatty acids, but positional isomers have not been characterized (Table VIII).

The green algae of the order Chlorococcales are nonmotile forms that generally occupy a freshwater habitat. Fatty acid analysis has been performed only on freshwater forms. The fatty acid spectra of the members of the Chlorococcales are considerably simpler than those of the volvocids. The algae of the order Chlorococcales usually synthesize only linoleic acid and ω3 polyenoic acids (Table VIII). The ω3 acids consist primarily of α-linolenic acid plus 16:3ω3 and 16:4ω3. 18:4ω3 is relatively rare, and polyenoic fatty acids of more than 18 carbon atoms are conspicuous by their absence (Table VIII).

The fatty acid composition of members of other orders of green algae has not been extensively investigated. In Table VIII data are given for one species of the order Siphonocladales, two species of the order Codiales, one species of the order Caulerpales, and three species of the order Uvales. All seven algae are macroscopic marine forms and were collected from the wild. Their fatty acid composition more closely resembles that of members of the Volvocales than that of the Chlorococcales. However, unlike the Volvocales they do not contain γ-linolenic acid, and in those forms that contain any ω6 polyenoic acid it is arachidonic acid (Table VIII).

6. *Phaeophyta*

The brown algae represent a morphologically complex group, ranging from small filamentous forms to those having a complex thallus up to a 150 ft long; they generally occupy a marine habitat. The algae that have been analyzed for fatty acid content were all collected wild from the sea. Klenk and co-workers reported that three species of the common sea weed *Fucus* (*F. serratus*, *F. platycarpus*, and *F. vesiculosus*) contain the following fatty acids as major components of their lipids: 14:0 (10–12%), 16:0 (25–26%), 16:1 (2%), 18:1ω9 (16–19%), 18:2ω6 (5–9%), 18:3ω3 (6–8%), 18:4ω3 (6–7%), 20:4ω6 (10–11%), and 20:5ω3 (8%). Polyenoic 16 and 22 carbon atom fatty acids are present only in trace amounts. The identity of the positional isomers was confirmed by degradation studies (Klenk *et al.*, 1963). These results are in qualitative agreement with those of other workers (see Paquot *et al.*, 1970).

A similar fatty acid composition has been found in four other species of brown algae, *Dictoyopteris polypodoides*, *Styptocaulon scoparium*, *Taomia atomaria*, and *Undaria pinnatifida*, by Pohl and co-workers (Pohl *et al.*, 1968). They reported finding the following fatty acids: 14:0 (4–8%), 16:0 (13–28%), 16:1 (4–9%), 18:1ω9 (8–15%), 18:2ω6 (6–12%), 18:4ω3 (7–9%), 20:4ω6 (7–9%), and 20:5ω3 (3–10%). Again the identity of the positional isomers were confirmed by degradation studies, and again neither polyenoic 16 carbon atom compounds or polyenoic 22 carbon atom fatty acids were detected. Earlier, Wagner and Pohl had reported the occurrence of 16:1, 16:3, 18:1, 18:2, 18:3, 18:4, 20:3, 20:4, and 20:5 in *Calpomenia sinuosa*, *Cutleria multifida*, *Lessonia fusca*, *Padina pavonia*, and *Sargassum linifolium*. Positional isomers were not designated, and no quantitative data were given (Wagner and Pohl, 1965).

D. YEASTS AND FUNGI

1. *Fungi*

The fungi are neither unicellular nor multicellular but rather are *coencytic*—they consist of a mycelium, a mass of multinucleate cytoplasm enclosed in a usually branched system of tubes (Stanier *et al.*, 1963). The mycelia of the lower fungi or Phycomycetes generally contain no cross walls. The mycelia of the higher fungi do contain numerous cross walls, which, however, are incomplete and permit cytoplasmic, but not nuclear, exchange between compartments. The higher and lower fungi are also distinguished from one another by having different types of asexual spore formation; in the former it is exogenous, and in the latter, endogenous (Stanier *et al.*, 1963). The lower fungi have been grouped by some authors into three classes; two of the classes, the Chytridomycetes and the Oomycetes, are generally aquatic and have life cycles containing a flagellated form (Stanier *et al.*, 1963; Cronquist, 1971); the third class, the Zygomycetes, are terrestrial and lack a flagellated stage (Stanier *et al.*, 1963; Cronquist, 1971). The higher fungi in turn are divided into two classes, the Ascomycetes and the Basidiomycetes (Stanier *et al.*, 1963; Cronquist, 1971), on the basis of the type of spore formation characterizing the sexual phase of their life cycle.

The fatty acids of 30 different species of lower and higher fungi have been examined by Shaw (1965). Shaw has reviewed his own findings along with the observations of others (see Shaw, 1966), and I have relied heavily on this review for literature references prior to 1965.

a. Fatty Acids of the Lower Fungi. In his 1966 review Shaw suggested that the polyunsaturated fatty acid patterns of fungi were simple and

TABLE IX

Fatty Acids of the Fungi: Phycomycete Group

Organism[a,b]	Incubation temp. (°C)	Saturated (total)	16:1	18:1	18:2ω6	18:3ω6	20:2	20:3	20:4ω6	20:5ω3	22:6ω3
Class Chytridomycetes											
Order Chytridiales											
Phlyctochytrium punctatum (1)	25	40[l]	2	29[v]	13	4[w,x]	0	0	3[y]	1	0
Dermocystidium sp. (1)[d,e,f]	20	35[l]	19	4[v]	2	0[w]	24	2	2	5	2
Class Oomycetes											
Order Peronosporales											
Phytophthora infestans (7)	22–23	38[m]	3	31	5	0	0	2	10	0	8[z]
Pythium sp. (2)		26[m]	1	36[v]	14	1	5	3	2	10	
P. acanthicum (2)		36[m]	2	31[v]	12	1	2	3	3	9	
P. debaryanum (5)		37[l]	8	21	16	5	0	1	4	0	7[aa]
(7)	22–23	30[l]	6[u]	14	20	2	0	2	12	0	19[z]
Order Saprolegniales											
Schizochytrium aggregatum (1)[d,e]	20	43[l]	1	18[v]	3	0	2	2	4	4	11[bb]
Thraustochytrium roseum (1)[d,e]	20	37[l]	1	35[v]	3	1	0	0	2	6	34[bb]
T. aureum (1)[d,e]	20	35[l]	1	6[v]	2	0	0	0	5	6	34[bb]
Saprolegnia litoralis (5)	25	35[n]	4	30	14	3	0	0	0	0	0
Class Zygomycetes											
Order Enteromophthorales											
Basidiobolus haptosporus (7)	22–23	27[o]	14	23	15	22	0	0	0	0	0

Organism											
B. meristosporus (7)	22–23	20[o]	9	20	24	23	0	0	0	0	0
B. ranarum (7)	22–23	21[o]	10	30	18	23	0	0	0	0	0
Condiobolus denaesporus (8)	22–23	64[p]	6	15	2	2	0	1	9	0	0
C. osmoides (7)	22–23	32[q]	12	42	6	2	0	0	14	0	0
C. thromboides (7)	22–23	21[o]	15	39	6	2	0	1	6	0	0
Delacroixia cornata (7)	25	53[r]	2	27	7	6	0	5	5	0	0
Entomophthora sp. (7)[g]	22–23	22[o]	18	38	4	2	0	0	12	0	0
Entomophthora sp. (7)[h]	22–23	64[h]	0	5	1	2	0	3	14	0	7[z]
Entomophthora sp. (7)[i]	22–23	68[h]	1	4	3	5	0	3	13	0	3[z]
E. apiculata (7)	22–23	52[m]	2	25	8	4	0	8	2	0	0
E. conglomerata (7)	22–23	54[h]	16[u]	34	5	3	0	1	13	0	0
E. coronata (5)	25	28[m]	9	35	7	6	0	0	0	0	0
(4)[d]	25	55[m]	1	21[v]	11	2	0	0	5	0	0
(7)	22–23	56[m]	2	25	6	4	0	5	4	0	0
E. exitialis (7)	22–23	40[m]	0	17	6	4	0	6	27	0	0
E. ignobilis (7)	22–23	29[n]	6	39	5	2	0	0	19	0	0
E. megasperma (7)	22–23	75[s]	1	7	2	2	0	1	9	0	3[z]
E. muscae (7)	22–23	24[l]	17[u]	38	5	2	0	0	14	0	0
E. obscura (7)	22–23	66[t]	0	3	0	0	0	1	4	0	24[z]
E. thaxteriana (7)	22–23	17[l]	17[u]	36	6	2	0	2	19	0	0
E. tipula (7)	22–23	27[l]	18[u]	30	6	3	0	1	12	0	0
E. virulenta (7)	22–23	19[u]	19[u]	38	4	3	0	0	14	0	0
Order Mucorales											
Choanephora cucuribitarum (5)	25	34[l]	10	25	16	15	0	0	0	0	0
Helicostylum pyriforme (5)	25	38[l]	2	44	8	9	0	0	0	0	0
Mortierella renispora (2)	28	25[l]	1	21	8	6	2	6	6	27	6
Mucor sp. (6)[j]	48	19[l]	3	49	14	8	0	0	0	0	0
		22[l]	3	55	15	4	0	0	0	0	0
M. globosum (4)	25	43[l]	8	26	8	16	0	0	0	0	0

TABLE IX (Continued)

FATTY ACIDS OF THE FUNGI: PHYCOMYCETE GROUP

Organism[a,b]	Incubation temp. (°C)	Saturated (total)	Fatty Acids (wt. %)[c]								
			16:1	18:1	18:2ω6	18:3ω6	20:2	20:3	20:4ω6	20:5ω3	22:6ω3
M. hiemalis (6)[d]	25	27[l]	3	33	19	19	0	0	0	0	0
M. javanicus (6)	25	32[l]	5	27	13	14	0	0	0	0	0
M. miehei (6)[i]	25	29[l]	3	48	16	4	0	0	0	0	0
	48	40[l]	3	48	10	2	0	0	0	0	0
M. mucedo (6)[d]	25	33[l]	3	26	25	20	0	0	0	0	0
M. pusillus (6)[d,i]	25	31[l]	3	40	20	6	0	6	0	0	0
	48	32[l]	3	42	19	4	0	0	0	0	0
(4)	48	27[l]	1	59	11	1	0	0	0	0	0
M. racemosus (6)	25	23[l]	3	37	17	19	0	0	0	0	0
M. ramannianus (6)	25	24[l]	3	28	14	31	0	0	0		0
M. strictus I (6)[d,k]	10	32[l]	2	33	12	20	0	0	0	0	0
	20	25[l]	3	45	14	13	0	0	0	0	0
M. strictus II	10	38[l]	2	35	12	15	0	0	0	0	0
	20	29[l]	3	41	15	12	0	0	0	0	0
Phycomyces blakesleeanus											
(5)	25	39[l]	2	32	20	5	0	0	0	0	0
(3)		31[l]	2	30	35	2	0	0	0	0	0
Rhizopus sp. (6)[d,i]	25	27[l]	2	30	26	12	0	0	0	0	0
	48	42[l]	3	35	22	5	0	0	0	0	0
R. arrhizus (5)	25	28[l]	4	40	16	10	0	0	0	0	0
R. nigricans (7)	22–23	19[l]	3	30	19	28	0	0	0	0	0
R. stolonifer (5)	25	34[l]	4	3	8	16	0	0	0	0	0

[a] References: (1) Ellenbogen *et al.* (1969); (2) Haskins *et al.* (1964); (3) Jack (1966); (4) Mumma *et al.* (1970); and Mumma and Bruszewski (1970); (5) Shaw (1965); (6) Sumner and Morgan (1969) and Sumner *et al.* (1969); (7) Tyrrell (1967); (8) Tyrrell (1968).

[b] All organisms were grown axenically, except where otherwise noted; all were grown in media containing salts, glucose peptone, or yeast extract.

[c] See Table III.

[d] Grown on lipid-free chemically defined media.

[e] Marine organisms.

[f] Lower Phycomycete of uncertain taxonomic position.

[g] From aphids (all species of *Entomophthora* are parasitic on insects).

[h] From spruce worm.

[i] From sugar cane froghopper.

[j] Thermophilic strains; fatty acid values for the same fungus grown at different temperatures are for cultures of equivalent growth.

[k] Psychrophilic; fatty acid values for the same fungus grown at different temperatures are for cultures of equivalent growth.

[l] Primarily 16:0, smaller amounts of 14:0 and 18:0.

[m] Primarily 14:0, smaller amounts of 16:0 and 18:0; no branched-chain compounds.

[n] Equal amounts of 14:0, 16:0, and 18:0; no branched-chain compounds.

[o] Primarily 16:0, smaller amounts of 14:0 and 18:0.

[p] 14:0 (15%), 16:0 (10%), 18:0 (3%); i-14:0 (20%), a-15:0 (13%), i-16:0 (2%).

[q] Primarily equal amounts of 14:0 and 16:0, smaller amounts of 12:0 and 18:0; no branched-chain compounds.

[r] Primarily 14:0, small amounts of 12:0, 13:0, 15:0, 16:0, and 18:0; no branched-chain compounds.

[s] Primarily 12:0 and 14:0 with smaller amounts of 10:0, 16:0, and 18:0.

[t] Primarily 18:0, smaller amounts of 14:0 and 16:0.

[u] Contains small amounts of 16:2.

[v] 18:1ω9 (from degradation studies).

[w] Also contains 18:3ω3 (1–2%).

[x] Also contains 18:4ω3 (1%).

[y] Also contains 20:4ω3 (3%).

[z] Polyenoic acids greater than 20 carbon atoms; not further identified.

[aa] Probably 22:3, not 22:6ω3 in this case.

[bb] Also contains small amounts of 22:2ω6, 22:4ω6, 22:5ω6, and 22:5ω3 (from GLC retention times, no degradation data).

TABLE X

FATTY ACIDS OF THE HIGHER FUNGI

Organism[a,b]	Fatty Acid (wt. %)[c]						
	16:0[d]	16:1	18:0	18:1	18:2ω6	18:3ω3	18:OH
Class Ascomycetes[e]							
Order Taphrinales							
Taphrina deformans (17)	21	8	5	52	7	2	0
Order Eurotiales							
Aspergillus flavus (17)	13	0	18	37	32	0	0
A. fresenius (18)	16	0	3	20	53	0[m]	0
A. niger (17)	14	2	8	28	36	9	0
A. niger (16)	16	1	7	21	38	16	0
Penicillium atrovenetum (22)	15	1	5	31	43	1	0
P. chrysogenum (17)	13	0	12	19	43	6	0
P. chrysogenum (2)[f]	14	0	4	6	54	21	0
P. chrysogenum (16)	12	0	6	11	65	6	0
P. duponti (16)[g]	25	0	11	42	22	0	0
P. javanicum (4)	20	1	17	25	32	1	0
P. lanosum (17)	27	2	6	42	23	0	0
P. notatum (17)	20	3	6	14	54	2	0
P. roqueforti (27)[f]	12	1	4	11	50	17	0
Order Sphaeriales							
Butryosphaeria ribis (17)	32	0	21	15	20	17	0
Chaetomium globosum (17)	19	2	8	17	46	7	0
(16)	31	10	10	9	36	1	0
C. thermophile (16)[g]	58	3	4	8	27	0	0
Mycosphaerella musicola (17)	16	2	6[k]	17	49	5[m]	0
Neurospora crassa	18	3	8[k]	11	42	8	0
Stemphylium dendriticum (7)	19	2	4	17	52	6	0
Order Diaporthales							
Glomerella cingulata (11)	44	2	6	26	20	1	0
Order Hypocreales							
Claviceps sp. (15)[h]	34	2	7	39	14	0	+[n,o]
C. gigantea (15)[h]	17	3	3	56	20	0	+[n,o]
C. paspali (15)[h]	19	7	2	57	15	0	+[n,o]
C. purpurea (15)[h]	20	7	4	23	14	0	33[o]
(20)[h]	33	4	5	21	12	0	22[o]
C. sulcata (15)[h]	11	1	5	13[k]	5	0	64[p]
Fusarium sp. (8)	21	1	11	23	30	13	0
F. aquaeductuum (5)	11	0	4	25	58	1	0
F. aquaeductuum var. medium (5)	12	0	6	29	47	2	0
F. moniliforme (17)	14	0	11	30	42	1	0

TABLE X   (Continued)

FATTY ACIDS OF THE HIGHER FUNGI

Organism[a, b]	Fatty Acid (wt. %)[c]						
	16:0[d]	16:1	18:0	18:1	18:2ω6	18:3ω3	18:OH
F. solani (8)	19	0	14	24	37	5	0
F. sambucinum (5)	12	0	7	33	39	4	0
Nectria ochroleuca (17)	19	5	12	10	43	9	0
Order Helotiales							
Botrytis cinerea (17)	19	1	4	11	16	42	0
(8)	17	1	5	23	44	10	0
Order Pezizales							
Pyronema domesticum (17)	14	0	21	29	35	0	0
Class Basidiomycetes[e]							
Order Agaricales							
Agaricus campestris (9)[h]	12	0	6	3	64	0	0
Amantia muscaria (19)	10	2	8	40	53	0	0
Auricularia Auricula-jadae (25)[i,j]	19	0	4	32	42	1	0
Clitocybe illudens (3)[h]	19	2	2	43	35	0	0
Collybia sp. (17)[h]	13	2	2	6	54	17	0
Coprinus comatus (11)	23	2	10	20	42	3	0
Corticium solani (17)	13	2	7	22[k]	28	15	0
Daedaleopsis confragosa (25)[i]	22	4	23	7	18	5	0
D. tricolor (25)[i]	21	9	11	16	34	0	0
Exobasium vexans (17)	16	3	6	23	34	13	0
Fomes sp. (17)[h]	12	2	3	5	70	4	0
F. annosus (6)[f]	21	0	37	10	29	0	0
F. fomentarius (10)[i,j]	35	3	10	17	4	1	0
Polyporus sulphureus (25)[i,j]	2	2	3	35	30	0	0
Rhizoctonia lamellifera (17)	21	3	10	12	30	14	0
Trametes orientalis (10)[i,j]	19	0	0	42	15[l]	0	0
Tricholoma nudum (12)	25	1	10	33	30	1	0
Order Tremellales							
Stilbum zacalloxanthum (17)	22	4	10	12	30	14	0
Order Ustillaginalles							
Ustilago scitaminea (17)	20	3	7	30[k]	32	2	
Class Deuteromycetes[e]							
(Fungi Imperfecti)							
Beauveria bassiana (21)	20	0	5	33	40	2	0
B. tenella (14)	24	0	9	26	24	2	0
Cephalosporium acremonium, corda (26)	7	6	0	75	12	0	0
Cephalosporium subverticallutum (17)	22	3	7	28	35	3	0

TABLE X   (Continued)

FATTY ACIDS OF THE HIGHER FUNGI

Organism[a,b]	Fatty Acid (wt. %)[c]						
	16:0[d]	16:1	18:0	18:1	18:2ω6	18:3ω3	18:OH
Cylindrocarpon radicicola (17)	22	2	8	25[k]	26	9	0
Epicoccum nigrum (5)	15	3	7	18	34	4	0
Hirsutella gigantea (17)	19	1	5	18	57	0	0
Humicola brevis (16)	29	2	4	20	41	4	0
H. grisea (16)	15	0	2	31	34	19	0
H. grisea var. *thermoidea* (16)[g]	29	0	2	4	29	0	0
H. insolens (16)[g]	30	0	1	37	32	0	0
H. lanuginosa (16)[g]	21	0	5	65	6	0	0
H. nigrescens (16)	21	0	4	29	34	12	0
Isaria farinosa (21)	17	1	6	41	34	2	0
Malbranchea pulchella (16)	11	0	11	27	51	0	0
M. pulchella var. *sulfurea* (16)[g]	26	0	8	35	31	0	0
Metarrhizium anisopliae (21)	21	1	9	23	47	0	0
Microsporum gipseum (24)	17	0	8	10	64	0	0
Pithomyces chartarum (7)	32	0	3	37	10	5	0
Pullularia pullulans (13)	31	3	8	46	11	0	0
Sporotrichium exile (16)	17	2	9	8	58	0	0
S. thermophile (16)[g]	28	0	7	28	35	0	0
Stibella sp. (16)	20	1	2	13	58	5	0
S. thermophila (16)	43	2	14	25	14	0	0
Trichophyton mentagrophytes (1)	24	0	11	17	45	0	0
T. rubrum (23)	24	0	17	13	52	0	0

[a] References: (1) Audette *et al.* (1961); (2) Bennett and Quackenbush (1969); (3) Bentley *et al.* (1964); (4) Coots (1962); (5) Foppen and Gribanovski-Sassu (1968) and Gribanovski-Sassu and Foppen (1968); (6) Gunasekaran *et al.* (1970); (7) Hartman *et al.* (1960, 1962); (8) Haskins *et al.* (1964); (9) Hughes (1962); (10) Ishida and Mitsuhashi (1970); (11) Jack (1966); (12) Leegwater *et al.* (1961); (13) Merdinger *et al.* (1968); (14) Molitoris (1963); (15) Morris (1967); (16) Mumma *et al.* (1970); (17) Shaw (1965, 1966); (18) Stone and Hemming (1968); (19) Talbort and Vining (1963); (20) Thiele (1964); (21) Tyrrell (1969); (22) Van Etten and Gottlieb (1965); (23) Wirth and Anand (1964); (24) Wirth *et al.* (1964); (25) Yokokawa (1969, 1970); (26) Huber and Redstone (1967); (27) Kubeczka (1968).

[b] Except where otherwise noted, all organisms were cultured axenically on crude (but usually lipid-poor) media.

[c] See Table III.

[d] All the fungi usually also contain small amounts of 12:0 and 14:0.

divided sharply along taxonomic lines, the lower fungi synthesizing linoleic and γ-linolenic acids and the higher fungi synthesizing linoleic and α-linolenic acids (Shaw, 1966). However, at that time the fatty acids of only a few species of lower fungi had been analyzed. Since then considerably more work has been done on this group. The fatty acid composition of approximately 50 different species belonging to all three classes of lower fungi is summarized in Table IX. Almost all these fungi were grown axenically either in lipid-poor crude media or in lipid-free chemically defined media.

Shaw's hypothesis is still applicable to one order of Zygomycete fungi, the order Mucorales. With a single exception, all species of Mucorales synthesize only two polyunsaturated fatty acids, linoleic and γ-linolenic acids (Table IX). However, the remaining species of the Zygomycetes and members of the other two classes of lower fungi show more complex fatty acid spectra. Many species synthesize not only linoleic and γ-linolenic acids but also significant amounts of other ω6 polyenoic acids, particularly arachidonic acid and a 20 carbon triene that is probably 20:3ω6 (Table IX). It is striking that none of the lower fungi synthesize α-linolenic acid or 18:4ω3 (Table IX). Indeed, not one of the species belonging to the class Zygomycetes synthesizes any ω3 polyenoic fatty acid (Table IX). However, a chytrid and some members of the Oomycetes do synthesize ω3 polyunsaturated fatty acids, particularly 20:5ω3 (Table IX).

b. Fatty Acids of the Higher Fungi. The fatty acid composition of approximately 80 species of higher fungi is summarized in Table X. From one point of view the literature on the fatty acids of higher fungi is difficult to

---

[e] The higher fungi are classified according to the type of spore formation characteristic of the sexual stage of their life cycle; where this has not yet been described, the organism is consigned to the Deuteromycetes (Fungi imperfecti). The sexual stages of many of the organisms listed under the Ascomycetes have not been described, but the organisms are placed there because they are obviously related to forms whose sexual stage has been demonstrated to characterize them as Ascomycetes.

[f] These organisms were cultured axenically on lipid-free synthetic media.

[g] Thermophilic fungi grown at 45°C or higher; all others are mesophilic fungi cultured at 22°–30°C.

[h] Not cultured but collected from the wild or from infected plants.

[i] Origin of the material not determined.

[j] Fatty acids only of the ether-extractable oils.

[k] Also contains small amounts of 20:1.

[l] Contains significant amounts of 20:2 (10%).

[m] Also contains trace amounts of 20:2, 20:3, 22:2, and 22:3.

[n] Present, but quantitative data not provided.

[o] Identified as ricinoleic acid.

[p] Identified as 9,10-dihydroxystearic acid.

summarize and frustrating to interpret. The fungi listed in Table X were rarely grown on lipid-free synthetic media, often were cultured on crude media that might be expected to be rich in lipid, and sometimes were simply collected from the wild. Furthermore, the relative amounts of different fatty acids in fungi clearly differ greatly with culture conditions and age, and these conditions were sometimes difficult to ascertain. And while in most cases the fatty acid composition given represents that of the mycelia of the fungi, this is not always true in the case of mushrooms. Finally, while most data are from an analysis of the fatty acids of the total lipids of the fungi, some represent an analysis of only the fatty acids of the hexane or ether extractable oils.

Yet in spite of such variations in approach, the results of the poly-unsaturated fatty composition of the higher fungi reported by various authors are surprisingly similar, at least qualitatively, and support the original contention of Shaw in regard to this group (1965, 1966), for, as suggested by Shaw, the higher fungi synthesize only two types of poly-unsaturated fatty acids, linoleic and α-linolenic acids (Table X). In a few cases small amounts of dienoic and trienoic 20 and 22 carbon atom compounds are found (Table X), but these are probably simply chain elongation products of linoleic and α-linolenic acids.

As previously mentioned (Section II,A), Ascomycete species belonging to the genus *Claviceps* synthesize large amounts of the hydroxy analog of linoleic acid, ricinoleic acid. These same species form no trienoic acids, and possibly the ricinoleic acid serves as a substitute in the lipids of these fungi (Table X).

2. *Yeasts*

Lodder (1970) has defined a yeast as a "microorganism in which the unicellular form is conspicuous and which belongs to the fungi." Lodder and his associates have further restricted the category so as to exclude those fungi that have a yeastlike phase in their life cycle. Hence the yeasts consist of a small group of unicellular organisms that are clearly related to the higher fungi but that may be considered degenerate since they do not form mycelia.

The yeasts listed in Table XI are typical, and most were grown on lipid-free synthetic media. Their fatty spectra are rather simple. Polyunsaturated fatty acids are limited to linoleic and α-linolenic acids, and thus the pattern closely resembles that of the higher fungi. However, the content of polyenoic fatty acids, particularly that of α-linolenic acid, is highly variable and very subject to oxygen availability and temperature (Table XI). Indeed, since oxygen soon becomes limiting with age in most yeast cultures (unless forcibly aerated), much of the quantitative variation reported by different authors is probably due to differences in culture technique, growth yields, and times of harvesting. Some yeasts, particularly of the genus *Saccharomyces*, contain only monounsaturated fatty acids (Table XI) and thus constitute the one group of eukaryotic organisms whose membranes lack polyunsaturated fatty acids.

E. PROTOZOA

Since the phytoflagellates have been included with the algae, the protozoa discussed here are the animallike members of the phylum. The Sarcodina (amoebalike and related protozoa) and the Zoomastigophora (the zooflagellates) belong to a single subphylum, the Sarcomastigophora (see Honigberg *et al.*, 1964). The ciliated protozoa constitute the bulk of a second subphylum, the Ciliophora (Honigberg *et al.*, 1964). Our present knowledge of the fatty acids of protozoa is limited to data obtained from the analysis of a small number of organisms all of which belong to one or the other of these two subphyla.

1. Sarcomastigophora

a. Sarcodina. Two closely related soil amoeba plus one true (noncellular) slime mold, all three of which can be grown axenically, have been analyzed for fatty acid content. The slime mold *Physarum polycephalum* was grown on a lipid-poor crude medium, while the two soil amoebas, *Acanthamoeba* sp. and *Hartmannella rhysoides*, were grown in lipid-free synthetic media. The two soil amoebas synthesize very large amounts of oleic acid (45–64%). Their polyenoic fatty acids consist exclusively of linoleic acid plus ω6 polyenoic acids containing 20 carbon atoms, 20:2ω6, 20:3ω6, and 20:4ω6 (Table XII). No ω3-type polyunsaturated fatty acids are synthesized. *Physarum polycephalum* has a very similar fatty acid composition but contains more linoleic acid and less oleic acid than the soil amoebas (Table XII).

A cellular slime mold *Dictyostelium discoideum* when grown on autoclaved *Escherichia coli* does not contain any fatty acids with methylene-interrupted double-bond systems but does synthesize several unique fatty acids. These are all *cis*-5,9-hexadecadienoic acid, all *cis*-5,9-octadecadienoic acid, and all *cis*-5,11-octadecadienoic acid (Davidoff and Korn, 1962a); they have been found in all developmental stages (myxamoebas, pseudoplasmodia, and mature fruiting bodies) of these organisms (see Korn *et al.*, 1965).

b. Zoomastigophora. The zooflagellates are potentially of great interest since many zoologists believe that the metazoa evolved from a colonial zooflagellate (see Hanson, 1958). Unfortunately, fatty acid analysis has

TABLE XI

FATTY ACIDS OF YEASTS

Organism[a]	Fatty Acid (wt. %)[b]							
	<16:0[c]	16:0	16:1	18:0	18:1	18:2ω6	18:3ω3	>20:0
Ascomycetous yeasts[d]								
Lipomyces lipofer (4)	0	14	2	7	70	7	0	0
Hansenula anomala (2)	0	12	10	0	28	24	19	0
H. anomala (8)[e]	5[k]	35	2	36	3	0	3	6[o,p]
H. anomala (Candida pelliculosa) var. anomala (2)	0	14	9	1	31	19	25	1
H. anomala var. schneigii (2)	0	12	3	0	28	25	14	1
Metschnikowia (Candida) pulcherrima (8)[e]	17[k]	20	0	18	0	1	0	20[o]
Pichia fermentans (8)[e]	35[k]	23	0	20	0	1	0	6[o]
P. membranaefaciens (2)	0	15	16	2	25	25	18	0
Saccharomyces bailii (S. acidifaciens) (8)[e]	21[k]	20	0	21	0	19	0	5[o]
S. bayanus (S. oviformis) (8)[e]	17[k]	39	0	← 26[n] →		0	1	5[o]
S. cerevisiae (6)								
Aerobic growth[f]	1	13	37[m]	5	44[m]	0	0	
Anaerobic growth[f]	24	60	0	15	2	0	0	
S. cerevisiae (7)	2	14	48	6	29	0	0	
(2)	0	8	35	0	30	0	1	
S. cerevisiae var. ellipsoidus (8)[e]	60[k]	15	0	17	0	0	1	0
Saccharomycodes ludwigi (8)	6	12	9	← 12[n] →		12	9	44[o]
Schizosaccharomyces pombe (10)	0	8	0	0	89	0	0	0
S. pombe var. liquefaciens (1)	8	16	5	7	37	2	0	0
(8)	19[k]	25	9	4	28	11	0	3[o,q]
Yeastlike genera of Ustilaginales								
Leucosporidium (Candida) scottii (5)[g,h]	+	15	2[m]	3	17[m]	34	28	0

Asporogenous yeasts not belonging to Sporobolomycetaceae

Candida sp. (5)[a,h]	+	14	18[m]	1	14[m]	40	11	0
C. krusei (2)	0	13	13	0	25	24	23	0
C. lipolytica (5)[o]								
10°C	+	13	13[m]	1	33[m]	37	0	0
25°C	+	22	11[m]	9	38[m]	18	0	0
C. mycoderma (2)	0	13	13	0	25	24	23	0
C. petrophillium (9)[o]	0[k]	7	9	1	34	45	0	0
C. rugosa (2)	14	13	11	2	33	30	9	0
C. (Torulopsis) utilis (3)[i]								
15°C	0[k]	16	8	+	26	33	17	0
30°C	0[k]	12	2	1	33	48	3	0
Kloeckera apiculata (8)[j]	24	16	0	48	0	13	0	0
Rhodotorula glutinis (5)[p]	+[l]	9	1	14	58	2	0	0

[a] References: (1) Baraud et al. (1970); (2) Bracco and Muller (1969); (3) Brown and Rose (1969); (4) Jack (1966); (5) Kates and Baxter (1962); (6) Light et al. (1962); (7) Longley et al. (1968); (8) Maurice and Baraud (1967); (9) Mizuno et al. (1966); (10) White and Hawthorne (1970).

[b] See Table III.

[c] Sum of 10:0, 12:0, and 14:0.

[d] Usually classified as members of the order Endomycetales.

[e] Values calculated from authors' separate analytical data for the fatty acids of the hexane- and methanol-extractable lipids of these yeasts using the ratio of the reported amounts of the two types of lipid extracts from each yeast.

[f] Grown under strict anaerobic conditions; alternatively, aerobically—in the presence of air but not aerated or stirred.

[g] These yeasts were grown axenically on media containing either peptone or yeast extract plus sugar. All other yeasts were grown on a lipid-free chemically defined media.

[h] Psychrophilic yeasts, grown at 10°C. All other yeasts are mesophilic and were grown at 25°–30°C (except where otherwise indicated)

[i] Values for hexane-extractable lipids only.

[j] Grown under high O_2 tension.

[k] Also contains small amounts of 13:0, 15:0, 17:0, and 19:0 (1–9% total).

[l] Contains straight-chain fatty acids lower than 16 carbon atoms and also branched-chain 15:0 (7%) and 17:0 (3%).

[m] Shown by degradation studies to be 16:1ω7 and 18:1ω9.

[n] Sum of 18:0 and 18:1.

[o] Sum of 20:0, 21:0, 22:0, 24:0, and 26:0.

[p] Also contains 22:2 (6%).

[q] Also contains 22:1 (9%) and 22:2 (4%).

been confined to a single group (order Kineplastida) of vertebrate blood parasites and related invertebrate parasites; the latter can be cultured axenically in lipid-free synthetic media. Our knowledge of the fatty acids of these organisms is due to Korn and his associates and to Meyer and Holz. With one exception the results from the two laboratories are in agreement. Six species of *Crithidia* plus *Blastocrithidia culicis* and *Leptomonas leptoglossi* when cultured in lipid-free chemically defined media synthesize primarily two polyunsaturated fatty acids, linoleic and γ-linolenic acids (Table XII). Small amounts of 20 and 22 carbon atom ω6 acids, including 20:3ω6 and 22:5ω6 (but not arachidonic acid), are also synthesized (Table XII). These organisms do not contain ω3 polyunsaturated fatty acids. Furthermore, Korn has reported that *Trypanasoma lewisi* obtained from rat blood or grown on blood agar has little capacity for synthesizing fatty acids *de novo* from ^{14}C-acetate but could synthesize linoleic and γ-linolenic acids from stearate via oleate (Korn *et al.*, 1965). Again, ω3 polyenoic acids are not synthesized.

Korn and co-workers have reported that another parasitic flagellate, *Leishmania tarentolae*, of the order Kineplastida, does synthesize substantial amounts of an ω3 polyunsaturated fatty acid, α-linolenic acid, and does not synthesize γ-linolenic acid (Table XII). However, Meyer and Holz report that *L. tarentolae* resembles the other flagellates of the order Kineplastida in that it does synthesize γ-linolenic acid (11%) and other ω6 polyenoic acids and does not synthesize α-linolenic acid or other ω3 polyenoic acids (Table XII). This conflict is puzzling since both groups employed the same organism grown on the same lipid-free synthetic medium and both groups definitively identified the 18 carbon polyenoic acid detected in the organism (Korn *et al.*, 1965; H. Meyer and Holz, 1966). Korn has furthermore suggested on the basis of radioisotope experiments that *Leishmania enrietti*, another hemoflagellate, when grown on a crude lipid-containing medium can also synthesize the ω3 acid, α-linolenic acid (Korn and Greenblatt, 1963).

2. *Ciliophora*

The ciliated protozoa are a large, well-defined taxonomic group. Unfortunately, only two groups of ciliates, the tetrahymenids and several species of *Paramecium*, can be cultured axenically. Fatty acid analysis has been carried out on five species of tetrahymenid ciliates, namely *Tetrahymena pyriformis*, *T. corlissi*, *T. paravorax*, *T. setifera*, and the closely related genus *Glaucoma chattoni*. All these ciliates grow on synthetic media; one of them, *G. chattoni*, requires a lipid for optimal growth (Holz *et al.*, 1961).

TABLE XII

FATTY ACIDS OF THE PHYLUM PROTOZOA

Organism[a,b]	Saturated (total)	19:0Δ[e]	16:1ω7	18:1ω9	18:2ω6	20:2ω6	18:3ω6	18:3ω3	20:3ω6	20:4ω6	22:5ω6	Other
					Fatty Acids (wt. %)[c,d]							
Subphylum Sarcomastigophora[f]												
Superclass Mastigophora												
Class II: Zoomastigophora[g]												
Order Kinetoplastida												
Crithidia sp. (5)[h]	25[n]	0[s]	0	21	20	0	21	0	1	0	13	1[z]
C. fasciculata (6)	21[n]	7	1	26	10	0	17	0	1[u]	←	11[w]	1
C. oncopelti (6)	24[n]	7	0	27	10	0	18	0	1[v]	←	10[w]	0
C. luciliae (6)	34[n]	7	0	17	13	0	17	0	2[v]	←	11[v]	1
C. acanthocephali (6)	25[n]	13	0	23	21	0	13	0	0	0	5[w]	1
Crithidia sp. (from *Arilus cristitus*) (6)	22[n]	13	0	23	20	0	16	0	1[v]	←	5[w]	0
Blastocrithidia culicis (6)	23[o]	0	0	23	18	0	29	0	0	0	8[w]	0
Leptomonas leptoglossi (6)	21[n]	4	0	10	70	0	0	0	0	0	0	0
Leishmania tarentolae (6)	15[p]	11	0	25	6	0	11	0	9[v]	←	0	1
(5)	18[n]	0	1	30	23	0	0	18	1	0	2[w]	5[u]
Superclass Sarcodina												
Class Rhizopodea												
Order Amoebida												
Acanthamoeba sp. (4)	25[q]	0	4	45	5	6	0	0	6	8	0	2[y]

TABLE XII   (Continued)

FATTY ACIDS OF THE PHYLUM PROTOZOA

Organism[a,b]	Saturated (total)	19:0Δ[e]	16:1ω7	18:1ω9	18:2ω6	20:2ω6	18:3ω6	18:3ω3	20:3ω6	20:4ω6	22:5ω6	Other
					Fatty Acids (wt. %)[c,d]							
Hartmannella rhysoides (2)	15[q]	0	4	64	2	5	0	0	3	4	0	2[y]
Order Eumycetozoida												
Physarum polycephalum (5)	13[q]	0	3	35	34	3	0	0	1	3	0	1[v]
Subphylum Ciliophora												
Class Ciliatea												
Order Hymenostomatida[i]												
Tetrahymena pyriformis (1)[j]	17[r]	0	12	9	18	?[t]	38	0		?[t]	0	0
T. corlissi ThX (1)[k]	14[r]	0	3	12	26	?[t]	31	0		?[t]	0	0
T. setifera HZ-1 (1)[k]	20[r]	0	3	4	13	?[t]	47	0		?[t]	0	0
T. paravorax RP (1)[k]	45[r]	0	3	1	16	0[t]	33	0		?[t]	0	0
Glaucoma chattoni A (3)												
Oleate grown[l]	18[r]	0	17	35	15	0	4	0		0	0	0
Lecithin grown[m]	41[r]	0	27	15	4	0	10	0		0	0	0

[a] References: (1) Erwin and Bloch (1963a); (2) Erwin et al. (1964); (3) Erwin (1971); (4) Korn (1963); (5) Korn et al. (1965); (6) H. Meyer and Holz (1966).
[b] *Physarum polycephalum* was grown axenically on a fatty-acid-poor crude medium. All the other protozoa were grown axenically

in stationary culture at 24°–28°C on chemically defined organic media and the cells were harvested in the log phase of growth (*T. pyriformis* was grown at 35°–37°C). Except where noted, the culture media were lipid-free.

c In identifying fatty acids all authors employed GLC retention times (compared with those of authentic standards) after prior separation of the fatty acids (as methyl esters) by argenation TLC according to the degree of unsaturation and also after hydrogenation (to confirm the number of carbon atoms); identification of positional isomers was confirmed by degradation studies.

d See Table III.

e The cyclopropane acid, 9,10-methyleneoctadecanoic acid.

f The phylum Protozoa is divided into four subphylums, two of which are listed here: the Ciliophora, containing the ciliated protozoa, and the Sarcomastigophora, containing the flagellated protozoa and the amoebalike organisms that form pseudopodia (Honigberg *et al.*, 1964).

g The Mastigophora or flagellated protozoa are divided into two classes; class I, the Phytomastigophora (phytoflagellates), have been grouped with the algae (Tables V–VIII).

h Data are only for the fatty acid composition of the phospholipid fraction, which accounted for approximately 80% of the total lipids of this organism.

i Subclass Holotrichia.

j Mating type II, Syngen 1.

k Media contained small amounts of cholesterol (200 μg/100 ml), required for growth by these organisms.

l *Glaucoma chattoni* requires a lipid for good growth; oleic acid (5 mg/100 ml).

m Synthetic *SN* 1,2-dipalmitoyl glyceryl phosphoryl choline (5 mg/100 ml).

n Primarily 18:0 plus small amounts of 14:0 and 16:0 plus small or trace amounts of 12:0, 15:0, 17:0, and 19:0 plus *i*-15:0, *i*-17:0, and *i*-19:0.

o Essentially equal amounts of 16:0 and 18:0 plus smaller amounts of 14:0, 17:0, and *i*-15:0 plus 7% *i*-17:0.

p Primarily 16:0 (11%) and *i*-17:0 (20%) with small amounts of 14:0, 17:0, and 18:0.

q *Acanthamoeba*, 14:0 (9%), 16:0 (9%), 18:0 (6%); *H. rhysoides*, 14:0 (4%), 16:0 (3%), 18:0 (8%); *P. polycephalum*, 16:0 (8%), sum of 14:0, 18:0, and branched chain (7%).

r Primarily 16:0 and 14:0, smaller amounts of 18:0, plus small amounts of 12:0, 13:0, 15:0, 17:0, and 20:0; *i*-15:0, *i*-17:0, and *i*-19:0. *Tetrahymena paravorax* contains substantial amounts of *i*-15:0 (7%), *i*-17:0 (10%), and *i*-19:0 (4%). *Glaucoma chattoni* contains very large amounts of 16:0 (27%) when grown on palmitate containing lecithin.

s Contains an unidentified fatty acid that from its retention time could be 19:0∆ (9%).

t Contains three unsaturated eicosenoic acids; not further identified (but no arachidonic acid); *T. pyriformis* (<1%), *T. corlissi* (4%), and *T. setifera* (<1%).

u 18:4ω3 (1%), 20:3ω3 (1%), 20:4ω3 (2%), and 22:5ω6 (1%).

v Polyunsaturated eicosenoic acids; not further identified.

w Polyunsaturated docosenoic acids; not further identified.

x 22:4ω6.

y 21:1.

These ciliates synthesize only two polyunsaturated fatty acids, linoleic acid and the ω6 acid, γ-linolenic acid (Table XII). These acids are primarily components of the phospholipids of the ciliates (Erwin and Bloch, 1963a). The total fatty acid composition of *G. chattoni* is influenced by the type of lipid supplement supplied in the culture medium (Table XII), but here, too, the phospholipids of the ciliate are rich in linoleic and γ-linolenic acids (Erwin, 1971).

The fatty acids of the one other group of ciliates that can be cultured axenically, members of the genus *Paramecium*, have not been analyzed. Katz and Keeny (1967) have reported on the fatty acid composition of the phospholipids of a nonaxenic mixed sample of holotrichous ciliates drawn from the rumen of a cow. The only polyenoic acids found were 18:2 and 18:3; the positional isomers were not identified.

F. PHYLOGENETIC AND PHYSIOLOGICAL IMPLICATIONS

1. *Polyunsaturated Fatty Acids and the Phylogenetic Relationships of Eukaryotic Microorganisms*

Basically there are two types of theories that attempt to account for the origin of eukaryotic microorganisms from prokaryotic ancestors and to define the phylogenetic relationships of eukaryotic algae, protozoa, and fungi. I shall refer to them as the *classic* and the *radical* types. The classic type of theory is epitomized in concepts set forth by Klein and Cronquist (1967) and Klein (1970); the radical theory has been most thoroughly developed by Margulis (formerly Sagan, see Sagan, 1967; Margulis, 1968).

Klein and Cronquist propose that the first primitive eukaryotic cells were derived from the blue-green algae, which in turn were derived from an ancestor of modern photosynthetic bacteria. The basic step in the evolution of the eukaryotic cell from the prokaryotic would be the invagination of the exterior limiting membrane of the prokaryotes to produce internal membrane systems. A tendency toward invagination of the limiting membrane to give internal cytoplasmic membranes is already seen in bacteria. In the blue-green algae this tendency is even more pronounced, and in these organisms intracytoplasmic membranous vesicles exist that are derived from the limiting membrane of the cell and that contain the enzymes for respiration and photosynthesis. According to the classic theory, the enclosure of these scattered cytoplasmic vesicles along with a small portion of the cellular DNA by limiting membranes would produce discrete entities-primitive chloroplasts and mitochondria. A similar enclosure by a limiting membrane of the bulk of the cellular DNA would produce a primitive nucleus, the net result being a primitive eukaryotic cell. Klein and Cronquist propose that such a hypothetical primitive eukaryote in turn produced, on the one hand, the red algae, and, on the other hand, a volvocid type of green algae (R. M. Klein and Cronquist, 1967; R. M. Klein, 1970). They suggest that subsequent eukaryotic evolution consisted essentially of the diversification of the volvocid line into more advanced algal groups and that heterotrophic and animallike protists were derived secondarily from algal forms via the loss of the photosynthetic apparatus (Klein and Cronquist, 1967).

The alternative type of theory as developed by Margulis postulates a radically different mode of origin of the eukaryotic cell. A brief summary of the theory of Margulis follows. Both aerobic bacteria and motile bacteria akin to modern spirochetes were ingested by a large phagocytotic anaerobic prokaryote and survived to become intracytoplasmic symbionts. The aerobic endosymbionts evolved into mitochondria, and the spirochetelike endosymbionts gave rise to the complex flagella of the modern eukaryotes, conferring upon these organelles their characteristic "2 + 9" substructure; their DNA has been utilized to form the basal bodies of cilia and flagella, the centromeres of chromosomes, and the centrioles of the modern eukaryote (see Sagan, 1967). The enveloping of the genome of the host cell by a membrane would produce a nucleus, and this would complete the formation of a hypothetical primitive eukaryotic cell. Further evolution of the eukaryotic line consisted essentially of the differentiation of the complex flagellar system and the development of mitosis via utilization of the microtubule elements found in flagella to form the mitotic spindle (see Sagan 1967). Indeed, the development of mitosis and the role played by the 2 + 9 homolog originating from the ancestral motile spirochetelike endosymbiont traces out the major pathway of the evolution of eukaryotic microorganisms. Margulis has constructed a tentative phylogeny of the major eukaryotes by arranging the various groups according to the function of their 2 + 9 homolog, the presence or absence of flagella (or cilia), and the type of mitosis found in each group. In contrast to R. M. Klein and Cronquist, Margulis designates the heterotrophic line as the ancestral one and suggests that the most "primitive" present-day eukaryotes are amoeboflagellated protozoa such as *Tetramitus* and that modern algal groups arose secondarily and independently of one another from different ancestral animallike protists via the ingestion of simple prokaryotic algae, the establishment of symbiotic relationships, and the subsequent evolution of the symbionts into chloroplasts (Sagan, 1967; Margulis, 1968).

The basic phylogenetic relationships postulated by Klein and Cronquist are summarized in Fig. 23. Those proposed by Margulis are summarized in Fig. 24. In both cases the schemes are abbreviated, extending for the most part only to major taxonomic groups; groups for which fatty acid data are lacking are deleted. Since Klein and Cronquist did not consider in detail the phylogeny of the protozoa and their relationships to higher animals, these relationships, as schematized in Fig. 23, are drawn from other authorities but are generally characteristic of classic schemes of phylogenetic relationships (see Hutner and Provasoli, 1951; Dougherty, 1955; Dougherty and Allen, 1960; Hanson, 1958). For the benefit of the nonbiologist, it must be emphasized that Klein and Cronquist and Margulis are not claiming that any presently existing species is directly ancestral to any large group of species that originated in the distant geological past. When a group is said to be phylogenetically related to or derived from a second group, this is simply a shorthand way of stating that the first group was derived from an ancestor of the second group and that the members of the second group retain many of the essential features of the ancestral form.

I have superimposed upon the phylogenetic schemes of Klein and Cronquist and Margulis the polyunsaturated fatty acids characteristic of major taxonomic groups. The result is a striking agreement between the patterns of polyunsaturated fatty acid distribution and the classic type of phyletic scheme. The variety of polyunsaturated fatty acids found in nature and the specificity of the enzyme systems required to synthesize them make these compounds very useful biochemical indicators of phylogenetic relationships.

Since polyunsaturated fatty acids of the methylene-interrupted type are ubiquitous in eukaryotic cells but very rare or absent in prokaryotic bacteria, it appears reasonable to assume that the acquirement of the ability to synthesize these compounds was essential for the evolution of the eukaryotic cell. If this assumption is correct, one should expect that the evolutionary history of polyunsaturated fatty acids might be similar to that observed with other important biological innovations. The initial appearance in organisms of a few simple types of polyunsaturated fatty acids would be followed by the appearance of an increasing complexity of structural types. This in turn would be followed by a secondary simplification as different organisms became specialized for producing a given type of polyunsaturated fatty acid.

The fatty acid spectra of the blue-green algae—the presumed progenitors of the initial eukaryotes and the volvocid green algae, and the stem line of all modern eukaryotes in the scheme of Klein and Conquist—conform to this expected mode of development. While some Cyanophyta lack polyunsaturated fatty acids, others possess substantial amounts of linoleic acid and either $\omega 3$ or $\omega 6$ polyenoic acids. However, the range of poly-unsaturated fatty acids found is very limited; only 18 carbon trienes are formed (Fig. 23). The utilization of these compounds may have been linked to evolutionary advances taking place in the photosynthetic membranes of the blue-green algae. In the volvocid green algae an increase in the complexity of structural types is found. (I have included in the order Volvocales species for which some authors would erect separate orders; while this is important from a taxonomic viewpoint, it is immaterial from a phyletic viewpoint—particularly since all the species share a common fatty acid pattern.) They synthesize the entire series of $\omega 3$ and $\omega 6$ acids found in nature with great variation in the number of carbon atoms and degree of unsaturation (Fig. 23).

From the volvocids two major evolutionary paths can be traced; one of these leads through the Chlorococcales to other groups of green algae (not shown in Fig. 23; see Table IX for their fatty acid composition) and ultimately to the metaphytes, terminating in the angiosperms or flowering plants (Fig. 23). This path involves a simplification of the fatty acid pattern of the volvocids; the $\omega 6$ acids are gradually lost and become vestigial in the higher metaphytes, the angiosperms. Synthesis of $\omega 3$ acids is accentuated but is also gradually simplified so that in the angiosperms' α-linolenic acid becomes the predominant if not the exclusive polyenoic fatty acid formed.

The second major evolutionary line leads from the volvocid algae to the algae of the Chrysophyta; the latter have lost the ability to synthesize the $\omega 3$ 16 carbon polyenoic series but have retained the capacity to form the other $\omega 3$ and $\omega 6$ polyenes. From the Chrysophyta this line leads to the two major groups of strict heterotrophic organisms. One group of heterotrophs, consisting of the fungi and yeasts, is either saprophytic or parasitic; these heterotrophs have a rigid cell wall and thus are totally dependent on the uptake of dissolved nutrients. Here, too, there is a progressive loss of the ability to synthesize various polyenoic fatty acids so that more "advanced" members of the group synthesize only linoleic and α-linolenic acids (Fig. 23). It should be noted that the suggestion that the higher fungi arose via the Zygomycetes (R. M. Klein and Cronquist, 1967) is difficult to reconcile with the fact that the Zygomycetes synthesize only $\omega 6$-type polyenoic fatty acids (Fig. 23; Table IX).

The second group of strict heterotrophic eukaryotic microorganisms derived from the Chrysophyta, the protozoa, have no cell wall, are generally phagotrophic, and are the presumed ancestors of the metazoa. This animal line may have arisen from the nutritionally unspecialized freshwater chrysomonads that combine photosynthesis, heterotrophy, and phag-

Higher plants: angiosperms; ω3 (18:3)   Higher animals: vertebrates; Lin; ω6

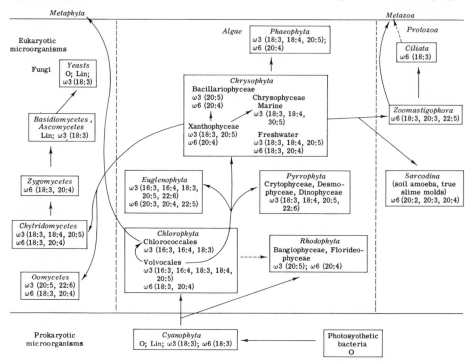

Fig. 23. Classic scheme of the origin and phylogenetic relationships of eukaryotic microorganisms and the distribution of polyunsaturated fatty acids. Adapted from R. M. Klein and Cronquist (1967) and Dougherty (1955). Key for fatty acid data: O, some or all members of the group do not synthesize any polyunsaturated fatty acids of the methylene-interrupted type. Lin, some or all members of the group synthesize only linoleic acid. ω3, some or all members of the group synthesize significant amounts of ω3 polyunsaturated fatty acids; the carbon number and degree of unsaturation of the most common compounds synthesized by this group is given in brackets. ω6, some or all members of the group synthesize significant amounts of ω6 polyunsaturated fatty acids; the carbon number and degree of unsaturation of the most common compounds synthesized by this group is given in brackets. Lin, linoleic acid cannot be synthesized and is a required nutrient. →, most probable phylogenetic relationships. ⟶, alternative phylogenetic relationship.

otrophy in the same organism. The animal line synthesizes only ω6 polyunsaturated fatty acid from linoleic acid by introducing additional double bonds only toward the carboxyl end of the molecule. This restriction on the direction of desaturation culminates in an inability to convert oleic to linoleic acid in the vertebrates (Mead, 1968), leaving these animals dependent on a dietary source of the latter compound.

Other major algal groups not on the main evolutionary lines have a fatty acid composition compatible with their postulated phyletic position in the Cronquist-Klein scheme. Thus the euglenids were probably derived from the volvocids (presumably via freshwater forms) and have retained the complex fatty acid spectrum of the stem group (Fig. 23). Similarly, the cryptomonads and dinoflagellates, classified together as the division Pyrrophyta and which all possess an essentially similar fatty acid composition, were also probably derived from the volvocids, in this case with a loss of the ω6 family of acids and the 16 carbon atom ω3 polyenoic acids (Fig. 23). The postulated origin of the Phaeophyta or brown algae from the Chrysophyta is also compatible with the fatty acid composition of these two groups (Fig. 23).

While the distribution of polyunsaturated fatty acids among the major taxonomic groups is generally in excellent harmony with the phyletic relationships outlined in Fig. 23, this does not mean that they cannot be accommodated by other phyletic schemes of the classic type, for while alternative classic schemes differ in many important details from that summarized in Fig. 23, they generally share its essential premises, namely that eukaryotes arose from blue-green algae, that red algae are relatively primitive, that the volvocids are the most primitive of the green algae, that the metaphytes were derived from green algae, that the fungi and animal-like protozoa arose from various members of the Chrysophyta, and, finally, that the metazoa evolved from protozoan ancestors.

In contrast, the alternative phylogenetic scheme proposed by Margulis (see Fig. 24) cannot be correlated in any meaningful way with the patterns of polyunsaturated fatty acid biosynthesis found in various eukaryotic organisms. The acceptability of the concepts of Margulis will obviously not be decided by the applicability of a single biochemical characteristic (for a general discussion of the strengths and weaknesses of the endosymbiont theory, see Stanier, 1970). Nevertheless, the fact remains that this phylogenetic scheme is of little value in interpreting this important body of data.

2. Physiological Significance of Polyunsaturated Fatty Acids

a. ω3 Polyenoic Fatty Acids and Photosynthetic Systems. The observed phyletic patterns of polyunsaturated fatty acids, particularly the tendency of the more highly advanced phyletic lines to specialize in the synthesis of only a single type of polyunsaturated acid, pose a basic question for the understanding of the biology of membranes: Are the particular physiological

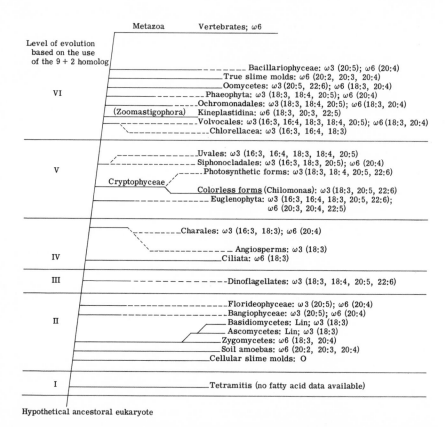

Fig. 24. Radical scheme of the origin and phylogenetic relationships of eukaryotic microorganisms and the distribution of polyunsaturated fatty acids. Adapted from Sagan (1967). Solid lines indicate heterotrophic ancestory; broken lines indicate the acquisition of photosynthetic capacities. The key for fatty acid data is the same as that for Fig. 23. Key to the level of evolution based upon the use of the 9 + 2 homolog: I, used only as a basal body to flagella. II, used as intranuclear divisions centers; flagella absent. III, used both as a basal body to flagella and as intranuclear division centers. IV, used as basal body to flagella and other 9 + 2 homologs permanently differentiated as intranuclear division centers. V, used both as a basal body to flagella and extranuclear division centers. VI, used as a basal body to flagella and other 9 + 2 homologs permanently differentiated into extranuclear division centers.

functions of a given type of cellular membrane dependent on that membrane being composed of a specific kind of lipid containing a specific fatty acid structure? An unequivocal answer is not yet possible, but available data suggest that this may indeed be true. Circumstantial evidence in favor of

TABLE XIII

FATTY ACIDS OF THE MONOGALACTOSYL DIGLYCERIDES OF PHOTOSYNTHETIC TISSUE

	Fatty Acids (wt. %)[b]												
	Saturated		Monoenoic		Dienoic		Unsaturated Polyenoic						
									ω3			ω6	
Source[a]	16:0	18:0	16:1	18:1	16:2	18:2ω6	16:3	16:4	18:3	18:4	20:5	18:3	20:4
Prokaryotic organisms													
Photosynthetic bacteria													
Chloropseudomonas ethylicum (4)	45	0	49	4	0	0	0	0	0	0	0	0	0
Cyanophyta													
Anacystis nidulans (8)	43	4	34	20	0	0	0	0	0	0	0	0	0
Anabaena cylindrica (15)	24		8	3	10	22	3	0	30	0	0	0	0
A. flos-aquae (15)	15		6	2	6	6	13	0	52	0	0	0	0
A. variabilis (8)	27	0	28	12	1	19	0	0	15	0	0	0	0
Spirulina platensis (11)[c]	45	3	2	1	0	6	0	0	5	0	0	43	0
Eukaryotic organisms													
Algae													
Rhodophyta													
Porphyridium cruentum (10)	24[g]	1	1	2	0	4	0	0	0	0	40	0	26
Chrysophyta													
Monodus subterraneus (10)	12[g]	0	16	2	0	2	0	0	0	0	62	0	6
Ochromonas danica (10)[d]	2[g]	0	3	7	0	31	0	0	9	14	0	25	1
Pyrrophyta													
Glenodinium sp. (6)	3	0	0	1	0	3	0	0	4	86	0	0	0
Cryptomonas sp. (2)	1[g]	0	1	5	2[i]	0	0	0	8	74	6	0	0

TABLE XIII (Continued)

FATTY ACIDS OF THE MONOGALACTOSYL DIGLYCERIDES OF PHOTOSYNTHETIC TISSUE

Source[a]	Saturated		Monoenoic		Dienoic		Polyenoic ω3					ω6	
Fatty Acids (wt. %)[b]	16:0	18:0	16:1	18:1	16:2	18:2ω6	16:3	16:4	18:3	18:4	20:5	18:3	20:4
Euglenophyta													
Euglena gracilis (auxotrophic) (5)[e]													
Low light intensity	40	1	2	5[h]	7	7	—[j]	15	23	0	0	0	0
High light intensity	11	1	2	4[h]	5	5	—[j]	32	41	0	0	0	0
Photoheterotrophic	5	1	7	9[h]	15	12	15	8	27	0	0	0	0
Chlorophyta													
Chlamydomonas mundana (9)	8[g]	0	8	10	9	8	27	0	26	0	0	0	0
C. reinhardii (3)[d]	2	3	2	13		6		33	33	0	0	0	0
Chlorella vulgaris (15)	3	0	2	0	18	22	10	0	46	0	0	0	0
(5)	1	0	1	4[h]	36	17	10	0	31	0	0	0	0
Phaeophyta[f]													
Fucus vesiculosus (14)	5							6	13	30	25		0
Metaphyta													
Lycopodiophyta (mosses)[f]													
Hypnum cupressiforme (9)	2	6	0	1	2	4	11	0	48	0	12	0	28
Magnoliophyta (Angiosperms, flowering plants)													
Alfalfa leaf (13)[f]	3	0	0	0	0	2	0	0	95	0	0	0	0
Anchusa leaf (15)[f]	3	0	0	0	0	0	0	0	72	19	0	4	0
Artemisia princeps leaf (12)[f]	2	0	0	0	0	4	0	0	94	0	0	0	0

Castor leaf (9)[f]	6[g]	1	0	0	0	0	0	0	0	91	0	0	0
Holly leaf (9)[f]	1	0	0	0	0	2	2	0	0	97	0	0	6
Myosotis scorpioides (forget-me-not) leaf (7)[f]	3	0	1	0	0	3	3	0	0	42	44	6	0
Runner bean leaf (16)[f]	1	0	0	0	0	2	2	0	0	96	0	0	0
Spinach leaf (5)[f]	1	0	0	2	0	1	1	26	70	0	0	0	0
Spinach leaf (15)[f]	2	0	1	0	0	2	2	21	74	0	0	0	0
Spinach leaf, chloroplast lamella (1)[f]	0	0	0	0	0	2	2	25	72	0	0	0	0

[a] References: (1) C. F. Allen et al. (1967); (2) D. H. Beach et al. (1970); (3) Bloch et al. (1967); (4) Constantopoulos and Bloch (1967a); (5) Constantopoulos and Bloch (1967b); (6) Harrington et al. (1970); (7) Jamieson and Reid (1969); (8) Nichols et al. (1965); (9) Nichols et al. (1966); (10) Nichols and Appleby (1969); (11) Nichols and Wood (1968); (12) Noda and Fujiwara (1967); (13) O'Brien and Benson (1964); (14) Radunz (1968); (15) Safford and Nichols (1970); (16) Sastry and Kates (1964).

[b] See Table III.

[c] This preparation was obtained from algae grown in unialgal culture in which bacterial contamination was reported to be slight.

[d] Isolated from cells grown axenically as photoheterotrophs under constant illumination. In all other cases the lipid was isolated from algae (except where otherwise indicated); grown axenically in essentially mineral media as either photoautotrophs or photoauxotrophs.

[e] Euglena was grown photoheterotrophically in media containing organic carbon sources and also as a photoauxotroph (on a mineral-vitamin medium); in the latter case the intensity of illumination was varied from a low light intensity (120 footcandles) to a high light intensity (610 footcandles).

[f] The biological material from which the MGDG was isolated was not cultured but either grown in greenhouses or collected from the wild.

[g] Also contains small amounts of 14:0 (1–3%).

[h] Shown to be 18:1ω9.

[i] Actually a mixture of 16:2ω7, 16:2ω4, 16:3ω5, and 16:4ω1.

[j] Very small amounts of 16:3ω3 were present and are included in the values for 18:1.

this thesis is particularly strong in the case of photosynthetic membranes. Organisms that carry out the oxygen-evolving type of photosynthesis universally contain a group of glycolipids that are primarily, if not exclusively, localized in their photosynthetic membranes (for a more detailed discussion of this subject, see Chapter 5). One of the most prominent of these lipids is MGDG. The fatty acids of this lipid isolated from a variety of sources have been analyzed in detail, and the results are summarized in Table XIII.

The ω3 polyenoic acids are universal and major components of MGDG isolated from eukaryotic photosynthetic cells or their chloroplasts; the ω6 acids are rarely found in this lipid (Table XIII). (It may well be that individual molecular species of MGDG that contain ω6 acids have functions in the chloroplast membrane that are completely unrelated to that of the more common ω3 acid-containing species.) This preference for ω3 polyenoic acids as components of MGDG is not simply a reflection of the biosynthetic capacities of the algae, since volvocid green algae, euglenids, and brown algae synthesize both types of polyunsaturated fatty acids but incorporate only the ω3 variety into their MGDG (Table XIII). Hence it must be of functional significance. However, it is not yet possible to evaluate what contribution the ω3 acids make to the functioning of MGDG in photosynthetic membranes. Thus it was formerly suggested that an ω3 polyenoic fatty acid containing MGDG was an absolute requirement for the oxygen-evolving type of photosynthesis common to both blue-green algae and eukaryotic algae (Erwin and Bloch, 1963b), but this cannot be true. While ω3 polyenoic acids are components of the MGDG of the photosynthetic membranes of all eukaryotic cells, they are absent from the MGDG isolated from some blue-green algae (Table XIII). In certain blue-green algae the MGDG is devoid of any polyunsaturated fatty acids (Table XIII).

It is also true that the ω3 polyenoic fatty acids of MGDG isolated from eukaryotic algae vary considerably in chain length and degree of unsaturation. The MGDG of green algae and euglenids contains primarily α-linolenic acid and 16:4ω3; MGDG obtained from brown algae and members of Pyrrophyta contain primarily the 18:4ω3 and 20:5ω3 acids (Table XIII). Indeed, 22:6ω3 appears to be the only member of the ω3 family of polyunsaturated fatty acids that is not a significant component of any MGDG isolated to date (Table XIII).

At present the significance, if any, of such variation in chain length and degree of unsaturation of these ω3 polyenoic acids is not clear. There may be a relationship between the utilization of specific fatty acids and the structure of the photosynthetic membrane. Various algal groups differ not only in the type of ω3 acid found in their MGDG but also in the ultra-structure of their chloroplasts, particularly in the arrangement of the thylakoid membranes. In contrast, higher plants, which have the greatest degree of chloroplast structural complexity (Kirk and Tilney-Basset, 1967; Echlin, 1970), exhibit a high degree of specificity in the type of acid found in their MGDG; the latter contains great concentrations of a single ω3 acid, α-linolenic acid, which occupies both the 1 and 2 positions of the glycolipid (Safford and Nichols, 1970). This may be of physiological value to the organisms since it has been suggested that increased complexity of ultrastructure of chloroplast membranes may be correlated with increased efficiency in energy conversion during photosynthesis (Echlin, 1970).

  b. Specificity of Fatty Acids in Phospholipids. The glycolipids, which constitute the bulk of photosynthetic membranes, appear to possess a specificity for ω3 acids. The question is, does there exist a comparable specificity for a given type of fatty acid in the phospholipids, the major components of other membranes such as mitochondria, plasma membranes, etc.?

  No simple answer can be given to this question. Patterns of polyunsaturated fatty acid occurrences in membrane glycerol-phospholipids do exist, but these patterns are not unequivocal, and their possible functional significance in membranes is at present purely speculative. A major difficulty is that we are not here dealing with the lipid structure of a unique component of a particular type of cell membrane as was the case in our discussion of the fatty acid composition of MGDG. Rather we are now concerned with the phospholipid fatty acids of the total cell since our knowledge of the lipid and fatty acid composition of individual membranes other than photosynthetic membranes is extremely limited. Thus in eukaryotic microorganisms data on membranes are available only for two organisms, the yeast *Saccharomyces cerevisiae* and the ciliate *Tetrahymena pyriformis*.

  Broadly speaking, there are four such patterns. One of them is found in phytoflagellates, such as volvocid green algae, euglenids, and freshwater chrysomonads (Erwin and Bloch, 1963b, 1964; Hulanicka *et al.*, 1964; Nichols and Appleby, 1969). These nutritionally versatile species synthesize both ω3 and ω6 polyunsaturated fatty acids. However, their glycerol phospholipids contain largely ω6 acids, and little or no ω3 acids are found with the exception of 20:5ω3. This tendency to employ ω6 polyenoic acids as components of lipids of nonphotosynthetic membranes appears to have been carried to completion in the heterotrophic animallike protozoa and higher animals.

The animallike protozoa and the metazoa, which are presumed to form a single phyletic line derived from phytoflagellates (Fig. 23), exhibit a second pattern. These organisms synthesize only ω6 polyenoic acids, which are primarily components of their glycerol phospholipids (see Erwin and Bloch, 1963a; H. Meyer and Holz, 1966; Korn et al., 1965; and Chapter 3 of this volume).

There is some evidence that the ω6 structure of the polyenoic fatty acids of the membrane phospholipids of the animal line is required for their normal physiological function. This evidence comes exclusively from experiments on vertebrates. Vertebrates are dependent on dietary sources of linoleic acid from which they synthesize their usual ω6 polyenoic acids, primarily arachidonic acid. When deprived of a source of linoleic acid, they synthesize a 20:3ω9 acid (Fig. 22), which replaces the usual ω6 polyenoic acids of their membrane phospholipids (Van Golde, 1968). When this occurs a variety of cellular and systemic pathological changes results, including reported damage to mitochondrial membranes (for reviews, see Aees-Jorgensen, 1961; Alfin-Slater and Aftergood, 1968; Soderhjelm et al., 1970). These changes can be completely reversed in whole animals by supplying exogenous sources of ω6 polyenoic acids, though not by supplying exogenous sources of ω3 polyenoic acids (Mohrhauer and Holman, 1963a,b; Rahm and Holman, 1964). While impressive, these studies are difficult to interpret in terms of fatty acid structure and cell membrane function. The performing of experiments on whole animals as complicated as vertebrates necessitates making a distinction between direct pathological effects of alteration of phospholipid fatty acid composition on cell membranes and indirect systematically mediated effects. Experiments on isolated animal cells in tissue culture should be easier to interpret, but only a few such experiments have been attempted (see Gerschenson et al., 1967). Similarly, experiments on animallike protozoa have not yet been extensive enough to provide sufficient data to define the structural requirements of fatty acids of their membrane phospholipids (see Lees and Korn, 1966; Erwin, 1970).

A third pattern of phospholipid fatty acid composition can be discerned in some photosynthetic green algae (Chlorella, for example), higher plants, and the strictly heterotrophic higher fungi. This group has no requirement for the long-chain, highly unsaturated fatty acids of the ω6 type. Their phospholipids contain only linoleic acid and sometimes the ω3 acid, α-linolenic acid (see Safford and Nichols, 1970; Richardson et al., 1962).

Why do the higher fungi and plant cells fail to employ the ω6 acids while the animallike protists and higher animals specialize in producing them? Again we can only speculate, but one possibility is highly suggestive. All three groups—fungi, higher plants, and animallike protozoa—were presumably derived from the phytoflagellates possessing the first pattern of phospholipid fatty acid composition. These phytoflagellates tend to be nutritionally adaptive forms that lack elaborate cell walls and often display a capacity for phagotrophy. Both the higher fungi and plant cells have become more specialized in their cell envelope, are protected by rigid cell walls, have lost the capacity for phagotrophy, and are hence dependent on dissolved nutrients. The animallike protists have developed in a different direction. Their cell envelope consists of a very flexible plasma membrane, unprotected by rigid layers and possessing highly developed phagocytotic capacities. The properties of such a cell envelope may be at least partially related to the prevalence of $\omega 6$ polyenoic fatty acids in its phospholipid components. The initial successful use of $\omega 6$ acids as exclusive components of the phospholipids of the plasma membrane may have been followed by their employment in the various intracytoplasmic membranes. If so, the absence of photosynthetic membranes would then make it possible to dispense entirely with the synthetic pathways for producing $\omega 3$ acids in animal protists.

A fourth pattern of polyunsaturated fatty acid components of phospholipids has been found among marine organisms. Malins and Wekell (1969) have called attention to the existence of such a marine lipid pattern, and they have postulated that this may represent a physiological adaptation to life in the sea. A major feature of this marine pattern is the tendency of $20:5\omega 3$ and $22:6\omega 3$ to be the major polyunsaturated fatty acids of the phospholipids of both marine algae and marine metazoa. Wagner and Pohl (1966) have suggested that in algae the ability to synthesize such fatty acids was evolved in the marine forms and was subsequently lost upon the adaptation of algae to freshwater environments.

These ideas are intriguing, and the possibility that life in a marine environment might impose unique structural requirements on the molecular components of the membranes of marine organisms cannot be denied. However, at the present time there is no real evidence that the presence of lipids containing primarily $20:5\omega 3$ and $22:6\omega 3$ polyenoic acids in biomembranes has any adaptive value for survival in marine environments.

An alternative explanation might be that the widespread occurrence of $20:5\omega 3$ and $22:5\omega 3$ in the lipids of marine metazoa including vertebrates might simply reflect the prevalence of these polyunsaturated fatty acids in the microalgae that constitute part of the phytoplankton lying at the base of the marine food chain (Ackman et al., 1968; Harrington et al., 1970). The presence of these acids in the microalgae themselves may also not be directly related to their marine environment but may be traits of broad taxonomic groups many of which are primarily marine forms. Thus while some marine algae contain substantial amounts of both these compounds, other predominantly marine groups such as red algae, diatoms, and brown algae generally do not contain significant amounts of the 22:6ω3 acid. Analysis of the fatty acids of both freshwater and marine forms of the same algal group should reveal which of these explanations is correct. So far, data of this sort are meager and inconclusive (Tables V, VI, and VIII).

VI. Note Added in Proof

One species of higher fungi, *Dactylaria ampulliforme* (Deuteromycetes), has been reported to contain small amounts (4%) of γ-linolenic acid. This compound was not detected in the lipids of several strains of a related species, *D. gallopava* [Sumner, J. L. and Evans, H. C. (1971). *Can J. Microbiol. 17*, 7].

Acknowledgment

Preparation of this chapter was supported in part by N.I.H. Grant G.M. 18774.

References

Aaronson, S., and Bensky, B. (1965). *J. Protozool.* **12**, 236.
Aaronson, S., Roze, U., Keane, M., and Zahalsky, A. C. (1969). *J. Protozool.* **16**, 184.
Abbott, B., and Casida, L. E., Jr. (1968). *J. Bacteriol.* **96**, 925.
Ackman, R. G. (1969). *In* "Methods in Enzymology" (J. M. Lowenstein, ed.), Vol. 14, p. 329. Academic Press, New York.
Ackman, R. G., Jangaard, P. M., Hoyle, R. J., and Brockerhoff, H. (1964). *J. Fish. Res. Bd. Can.* **21**, 747.
Ackman, R. G., Tocher, C. S., and McLachlan, J. (1968). *J. Fish. Res. Bd. Can.* **25**, 1603.
Ackman, R. G., Addison, R. F., Hooper, S. N., and Prakash, J. (1970). *J. Fish. Res. Bd. Can.* **27**, 251.
Aees-Jorgensen, E. (1961). *Physiol. Rev.* **41**, 1.
Ailhaud, G. P., Vagelos, P. R., and Goldfine, H. (1967). *J. Biol. Chem.* **242**, 4459.
Akamatsu, Y., and Law, J. H. (1970). *J. Biol. Chem.* **245**, 701.
Alberts, A. W., and Vagelos, P. R. (1966). *J. Biol. Chem.* **241**, 5201.
Alberts, A. W., Majerus, P. W., Talamo, B., and Vagelos, P. R. (1964). *Biochemistry* **3**, 1563.
Alberts, A. W., Majerus, P. W., and Vagelos, P. R. (1965). *Biochemistry* **4**, 2265.
Alfin-Slater, R. B., and Aftergood, L. (1968). *Physiol. Rev.* **48**, 758.
Allen, C. F., Hirayama, O., and Good, P. (1967). *In* "The Biochemistry of Chloroplasts" (T. W. Goodwin, ed.), Vol. 2, p. 195. Academic Press, New York.
Allen, M. B. (1969). *Annu. Rev. Microbiol.* **23**, 29.
Andreasen, A. A., and Stier, T. J. B. (1954). *J. Cell. Physiol.* **43**, 271.

Audette, R. C. S., Baxter, R. M., and Walker, G. C. (1961). *Can. J. Microbiol.* **7**, 282.

Bailey, J. M. (1966). *Biochim. Biophys. Acta.* **125**, 226.

Baraud, J., Demassieux, S., and Maurice, A. (1970). *Rev. Fr. Corps Gras* **17**, 155.

Barron, E. J., and Stumpf, P. K. (1962). *J. Biol. Chem.* **237**, pc613.

Barron, E. J., Squires, C. L., and Stumpf, P. K. (1961). *J. Biol. Chem.* **236**, 2610.

Beach, D. H. (1961). M.S. Thesis, Graduate School of the State University of New York, Syracuse.

Beach, D. H., Harrington, G. W., and Holz, G. G., Jr. (1970). *J. Protozool.* **17**, 501.

Bennett, A. S., and Quackenbush, F. W. (1969). *Arch. Biochem. Biophys.* **130**, 567.

Bentley, R., Lavate, W. V., and Sweeley, C. C. (1964). *Comp. Biochem. Physiol.* **11**, 263.

Bishop, D. G., and Still, J. L. (1963). *J. Lipid Res.* **4**, 87.

Bloch, K. (1962). *Fed. Proc., Fed. Amer. Soc. Exp. Biol.* **21**, 1058.

Bloch, K. (1969). *Accounts Chem. Res.* **2**, 193.

Bloch, K., Baronowsky, P., Goldfine, H., Lennarz, W. J., Light, R., Norris, A. T., and Scheuerbrandt, G. (1961). *Fed. Proc., Fed. Amer. Soc. Exp. Biol.* **20**, 921.

Bloch, K., Constantopoulos, G., Kenyong, C., and Nagai, J. (1967). *In* "The Biochemistry of Chloroplasts" (T. W. Goodwin, ed.), Vol. 2, pp. 197–211. Academic Press, New York.

Bloomfield, D. K., and Bloch, K. (1960). *J. Biol. Chem.* **235**, 337.

Bracco, U., and Muller, H. R. (1969). *Rev. Fr. Corps Gras* **16**, 573.

Brady, R. O. (1958). *Proc. Nat. Acad. Sci. U.S.* **44**, 993.

Brady, R. O. (1960). *J. Biol. Chem.* **235**, 3099.

Brady, R. O., Bradley, R. M., and Trams, E. G. (1960). *J. Biol. Chem.* **235**, 3093.

Bressler, R., and Wakil, S. J. (1961). *J. Biol. Chem.* **236**, 1643.

Brock, T. D. (1967). *Science* **158**, 3804.

Brooks, J. L., and Stumpf, P. K. (1966). *Arch. Biochem. Biophys.* **116**, 108.

Brown, C. M., and Rose, A. H. (1969). *J. Bacteriol.* **99**, 371.

Butterworth, P. H. W., and Bloch, K. (1970). *Eur. J. Biochem.* **12**, 496.

Caldwell, R. S., and Vernberg, F. J. (1970). *Comp. Biochem. Physiol.* **34**, 179.

Chuecas, L., and Riley, J. P. (1969). *J. Mar. Biol. Ass. U.K.* **49**, 97.

Collins, R. P., and Kalnins, K. (1969). *Phyton (Buenos Aires)* **26**, 47.

Constantopoulos, G. (1970). *Plant Physiol.* **45**, 76.

Constantopoulos, G., and Bloch, K. (1967a). *J. Bacteriol.* **93**, 1788.

Constantopoulos, G., and Bloch, K. (1967b). *J. Biol. Chem.* **242**, 3538.

Coots, R. H. (1962). *J. Lipid Res.* **3**, 84.

Cronquist, A. (1971). "Introductory Botany," pp. 125–222. Harper, New York.

Crowfoot, P. D., and Hunt, A. L. (1970). *Biochim. Biophys. Acta* **202**, 550.

Daron, H. H. (1970). *J. Bacteriol.* **101**, 145.

Davidoff, F., and Korn, E. D. (1962a). *Biochem. Biophys. Res. Commun.* **9**, 54.

Davidoff, F., and Korn, E. D. (1962b). *Biochem. Biophys. Res. Commun.* **9**, 328.

Davidoff, F., and Korn, E. D. (1963). *J. Biol. Chem.* **238**, 3210.

Dedyukhina, E. G., and Bekhtereva, M. N. (1969). *Mikrobiologiya* **38**, 775.

Dils, R., and Popják, G. (1962). *Biochem. J.* **83**, 41.

Donaldson, W. E. (1967). *Biochem. Biophys. Res. Commun.* **27**, 681.

Dougherty, E. C. (1955). *Syst. Zool.* **4**, 145.

Dougherty, E. C., and Allen, M. B. (1960). *In* "Comparative Biochemistry of Photoreactive Systems" (M. B. Allen, ed.), pp. 129–144. Academic Press, New York.

Echlin, P. (1970). *Symp. Soc. Gen. Microbiol.* **20**, 221–248.

Ellenbogen, B. B., Aaronson, S., Goldstein, S., and Belsky, M. (1969). *Comp. Biochem. Physiol.* **29**, 805.

Erwin, J. (1971). Unpublished data.

Erwin, J. A. (1970). *Biochim. Biophys. Acta* **202**, 21.

Erwin, J., and Bloch, K. (1962). *Biochem. Biophys. Res. Commun.* **9**, 103.

Erwin, J., and Bloch, K. (1963a). *J. Biol. Chem.* **238**, 1618.

Erwin, J., and Bloch, K. (1963b). *Biochem. Z.* **338**, 496.

Erwin, J., and Bloch, K. (1964). *Science* **143**, 1006.

Erwin, J., Hulanicka, D., and Bloch, K. (1964). *Comp. Biochem. Physiol.* **12**, 191.

Erwin, J. A., Beach, D., and Holz, G. G., Jr. (1966). *Biochim. Biophys. Acta* **125**, 614.

Erwin, J. A. (1968). *In* "The Biology of *Euglena*" (D. E. Buetow, ed.), Vol. 2, p. 133. Academic Press, New York.

Farkas, T., and Herodek, S. (1964). *J. Lipid Res.* **5**, 369.

Farrell, J., and Rose, A. (1967). *Annu. Rev. Microbiol.* **21**, 101.

Foppen, F. H., and Gribanovski-Sassu, O. (1968). *Biochem. J.* **106**, 97.

Forrest, H. S., and Van Baalen, C. (1970). *Annu. Rev. Microbiol.* **24**, 91.

Fulco, A. J. (1965). *Biochim. Biophys. Acta* **106**, 211.

Fulco, A. J. (1967a). *Biochim. Biophys. Acta* **144**, 703.

Fulco, A. J. (1967b). *J. Biol. Chem.* **242**, 3608.

Fulco, A. J. (1969). *J. Biol. Chem.* **244**, 889.

Fulco, A. J. (1970). *J. Biol. Chem.* **245**, 2985.

Fulco, A. J., and Bloch, K. (1964). *J. Biol. Chem.* **239**, 993.

Fulco, A. J., and Mead, J. F. (1959). *J. Biol. Chem.* **234**, 1411.

Fulco, A. J., and Mead, J. F. (1960). *J. Biol. Chem.* **235**, 3379.

Fulco, A. J., Levy, R., and Bloch, K. (1964). *J. Biol. Chem.* **239**, 998.

Gellerman, J. L., and Schlenk, H. (1965). *J. Protozool.* **12**, 178.

Gerschenson, L. E., Mead, J. F., Harary, I., and Haggerty, D. F., Jr. (1967). *Biochim. Biophys. Acta* **131**, 42.

Goldfine, H., and Bloch, K. (1960). *J. Biol. Chem.* **236**, 2596.

Goldman, P., Alberts, A. W., and Vagelos, P. R. (1963). *J. Biol. Chem.* **238**, 1255.

Green, D. E., and Wakil, S. J. (1960). *In* "Lipid Metabolism" (K. Bloch, ed.), pp. 1–164. Wiley, New York.

Greenblatt, C. L., and Wetzel, B. K. (1966). *J. Protozool.* **13**, 521.

Greyer, R. P., Bennett, A., and Rohr, A. (1962). *J. Lipid Res.* **3**, 80.

Gribanovski-Sassu, O., and Foppen, F. H. (1968). *Arch. Microbiol.* **62**, 251.

Gunasekaran, M., Raju, P. K., and Lyda, S. A. (1970). *Phytopathology* **60**, 1027.

Gurr, M. I. (1971). *Lipids* **6**, 266.

Gurr, M. I., and Bloch, K. (1966). *Biochem. J.* **99**, 16c.

Haigh, W. G., Safford, R., and James, A. T. (1969). *Biochem. Biophys. Acta* **176**, 647.

Haines, T. H., Aaronson, S., Gellerman, J. L., and Schlenk, H. (1962). *Nature (London)* **194**, 1282.

Hanson, C. D. (1958). *Syst. Zool.* **7**, 16.

Harlan, W. R., and Wakil, S. J. (1964). *J. Biol. Chem.* **239**, 2489.

Harrington, G. W., and Holz, G. G., Jr. (1968). *Biochim. Biophys. Acta* **164**, 137.

Harrington, G. W., Beach, D. H., Dunham, J. E., and Holz, G. G., Jr. (1970). *J. Protozool.* **17**, 213.

Harris, R. V., and James, A. T. (1965). *Biochim. Biophys. Acta* **196**, 456.

Harris, R. V., Harris, P., and James, A. T. (1965). *Biochim. Biophys. Acta* **106**, 465.

Hartman, L., Hawke, J. C., Morice, I. M., and Shorland, F. B. (1960). *Biochem. J.* **75**, 274.

Hartman, L., Morice, I. M., and Shorland, F. B. (1962). *Biochem. J.* **82**, 76.

Haskins, R. H., Tulloch, A. P., and Micetich, R. G. (1964). *Can. J. Microbiol.* **10**, 187.

Hawke, J. C., and Stumpf, P. K. (1965). *J. Biol. Chem.* **240**, 4746.

Hofmann, K. (1963). "Fatty Acid Metabolism in Microorganisms," p. 38. Wiley, New York.

Holloway, P. W., and Wakil, S. H. (1970). *J. Biol. Chem.* **245**, 1862.

Holloway, P. W., Peluffo, R., and Wakil, S. J. (1963). *Biochem. Biophys. Res. Commun.* **12**, 300.

Holman, R. T., and Rahm, J. J. (1966). *Progr. Chem. Fats Other Lipids* **9**, Part 1, p. 15–90.

Holton, R. W., Blecker, H. H., and Onore, M. (1964). *Phytochemistry* **3**, 595.

Holton, R. W., Blecker, H. H., and Stevens, T. S. (1968). *Science* **160**, 545.

Holz, G. G., Jr. (1969). *Progr. Protozool., Proc. Int. Conf. Protozool., 3rd*, p. 126.

Holz, G. G., Jr., Wagner, B., Erwin, J., and Kessler, D. (1961). *J. Protozool.* **8**, 192.

Holz, G. G., Jr., Erwin, J., Rosenbaum, N., and Aaronson, S. (1962). *Arch. Biochem. Biophys.* **98**, 312.

Holz, G. G., Jr., Rasmussen, L., and Zeuthen, E. (1963). *C. R. Trav. Lab. Carlsberg* **33**, 289.

Honigberg, B. M., Balamuth, N., Bovee, E. C., Corliss, J. O., Gojdics, M., Hall, R. P., Kudo, R. R., Levine, N. D., Loeblich, A. R., Jr., Weiser, J., and Wenrich, D. H. (1964). *J. Protozool.* **11**, 7.

Horning, M. G., Martin, D. B., Karmen, A., and Vagelos, P. R. (1961). *J. Biol. Chem.* **236**, 669.

Howanitz, P. J., and Levy, H. R. (1965). *Biochem. Biophys. Acta* **106**, 433.

Howling, D., Morris, L. J., and James, A. T. (1968). *Biochim. Biophys. Acta* **152**, 224.

Huber, F. M., and Redstone, M. O. (1967). *Can. J. Microbiol.* **13**, 332.

Hughes, D. H. (1962). *Mushroom Sci.* **5**, 540.

Hulanicka, D., Erwin, J., and Bloch, K. (1964). *J. Biol. Chem.* **239**, 2778.

Hutner, S. H., and Holz, G. G., Jr. (1962). *Annu. Rev. Microbiol.* **16**, 189.

Hutner, S. H., and Provasoli, L. (1951). *In* "Biochemistry and Physiology of Protozoa" (A. Lwoff, ed.), Vol. 1, pp. 27–128. Academic Press, New York.

Inkpen, J. A., and Quackenbush, F. W. (1969). *Lipids* **4**, 539.

Ishida, Y., and Mitsuhashi, J. (1970). *J. Jap. Oil. Chem. Soc.* **19**, 93.

Jack, R. C. M. (1966). *J. Bacteriol.* **91**, 2101.

James, A. T. (1963). *Biochim. Biophys. Acta* **70**, 9.

James, A. T. (1968). *In* "Cellular Compartmentalisation and Control of Fatty Acid Metabolism" (F. C. Gran, ed.), pp. 25–38. Academic Press, New York.

James, A. T. (1969). *Biochim. Biophys. Acta* **187**, 13.

James, A. T., Harris, P., and Bezard, T. (1968). *Euro. J. Biochem.* **3**, 318.

Jamieson, G. R., and Reid, E. H. (1969). *Phytochemistry* **8**, 1489.

Johnson, A. R., Pearson, J. A., Shenstone, F. S., and Fogarty, A. C. (1967). *Nature (London)* **214**, 1244.

Jonah, M., and Erwin, J. A. (1971). *Biochim. Biophys. Acta* **231**, 80.

Jones, P. D., Holloway, P. W., Peluffo, R. O., and Wakil, S. J. (1969). *J. Biol. Chem.* **244**, 744.

Kaneda, T. (1966). *Can. J. Microbiol.* **12**, 501.

Kates, M. (1964). *Advan. Lipid Res.* **2**, 17.

Kates, M., and Baxter, R. M. (1962). *Can. J. Biochem. Physiol.* **40**, 1213.

Kates, M., and Hagen, P. O. (1964). *Can. J. Biochem.* **42**, 481.

Kates, M., and Volcani, B. E. (1966). *Biochim. Biophys. Acta* **116**, 264.

Katz, I., and Keeny, M. (1967). *Biochim. Biophys. Acta* **144**, 102.

Kenyon, C. N., and Stanier, R. Y. (1970). *Nature (London)* **227**, 1164.

Kidder, G. W., and Dewey, V. C. (1963). *Biochem. Biophys. Res. Commun.* **12**, 280.

Kirk, J. T. O., and Tilney-Basset, R. A. E. (1967). "The Plastids." pp. 1–50. Freeman, San Francisco, California.

Klein, H. P., Volkmann, C. M., and Leaffer, M. A. (1967a). *J. Bacteriol.* **94**, 61.

Klein, H. P., Volkmann, C., and Weibel, J. (1967b). *J. Bacteriol.* **94**, 475.

Klein, H. P., Volkmann, C. M., and Chao, F. C. (1967c). *J. Bacteriol.* **93**, 1966.

Klein, R. M. (1970). *Ann. N.Y. Acad. Sci.* **175**, 623.

Klein, R. M., and Cronquist, A. (1967). *Quart. Rev. Biol.* **42**, 105.

Klenk, E., Knipprath, W., Eberhagen, D., and Koof, H. P. (1963). *Hoppe-Seyler's Z. Physiol. Chem.* **334**, 44.

Knipprath, N. G., and Mead, J. F. (1968). *Lipids* **3**, 121.

Koch, E. G., and Scherbaum, O. H. (1967). *Z. Allg. Microbiol.* **7**, 349.

Koelensmid, W. A. A. B., DeJongh, H., Van Pelt, J. G., Recourt, J. H., Van Rhee, R., and Rijfkogel, L. J. (1962). Unpublished observations, cited in Shaw (1966).

Korn, E. D. (1963). *J. Biol. Chem.* **238**, 3584.

Korn, E. D. (1964a). *J. Biol. Chem.* **239**, 396.

Korn, E. D. (1964b). *J. Lipid Res.* **5**, 352.

Korn, E. D., and Greenblatt, C. L. (1963). *Science* **142**, 1301.

Korn, E. D., Greenblatt, C. L., and Less, A. M. (1965). *J. Lipid Res.* **6**, 43.

Kubeczka, K. H. (1968). *Arch. Mikrobiol.* **60**, 139.

Larrabee, A. R., McDaniel, E. G., Bakerman, H. A., and Vagelos, P. R. (1965). *Proc. Nat. Acad. Sci. U.S.* **54**, 267.

Law, J., Zalkin, H., and Kaneshiro, T. (1963). *Biochim. Biophys. Acta* **70**, 143.

Leegwater, D. G., Craig, B. M., and Spencer, J. F. T. (1961). *Can. J. Biochem. Physiol.* **39**, 1325.

Lees, A. M., and Korn, E. D. (1966). *Biochemistry* **5**, 1475.

Lennarz, W. J. (1961). *Biochem. Biophys. Res. Commun.* **6**, 112.

Lennarz, W. J., and Bloch, K. (1960). *J. Biol. Chem.* **235**, pc26.

Lennarz, W. J., Light, R. J., and Bloch, K. (1962a). *Proc. Nat. Acad. Sci. U.S.* **48**, 840.

Lennarz, W. J., Scheuerbrandt, G., and Bloch, K. (1962b). *J. Biol. Chem.* **237**, 664.

Levin, E., Lennarz, W. J., and Bloch, K. (1964). *Biochim. Biophys. Acta* **84**, 471.

Lewis, R. W. (1962). *Comp. Biochem. Physiol.* **6**, 75.

Light, R. J., Lennarz, W. J., and Bloch, K. (1962). *J. Biol. Chem.* **237**, 1793.

Lodder, J. (1970). "The Yeasts: A Taxonomic Study." North-Holland Publ., Amsterdam.

Longley, R. P., Rose, A. H., and Knights, B. A. (1968). *Biochem. J.* **108**, 401.

Lynen, F. (1961). *Fed. Proc., Fed. Amer. Soc. Exp. Biol.* **20**, 941.

Lynen, F., Oesterhelt, D., Schweizer, E., and Willecke, K. (1968). *In* "Cellular Compartmentalisation and Control of Fatty Acid Metabolism" (F. C. Gran, ed.), pp. 1–24. Academic Press, New York.

McMahon, V., and Stumpf, P. K. (1964). *Biochim. Biophys. Acta* **84**, 359.

Majerus, P. W., Alberts, A. W., and Vagelos, P. R. (1964). *Proc. Nat. Acad. Sci. U.S.* **51**, 1231.

Majerus, P. W., Alberts, A. W., and Vagelos, P. R. (1965). *J. Biol. Chem.* **240**, 4723.

Malins, D. C., and Wekell, J. C. (1969). *Progr. Chem. Fats Other Lipids* **10**, Part 4, 339–363.

Marcel, Y. L., Christiansen, K., and Holman, R. T. (1968). *Biochim. Biophys. Acta* **164**, 25.

Margulis, L. (1968). *Science* **161**, 1021.

Marr, A. G., and Ingraham, J. L. (1962). *J. Bacteriol.* **84**, 1260.

Marsh, J. B., and James, A. T. (1962). *Biochim. Biophys. Acta* **60**, 320.

Martin, D. B., Horning, M. G., and Vagelos, P. R. (1961). *J. Biol. Chem.* **236**, 663.

Matsumura, S., and Stumpf, P. K. (1968). *Arch. Biochem. Biophys.* **125**, 932.

Maurice, A., and Baraud, J. (1967). *Rev. Fr. Corps Gras* **14**, 713.

Mead, J. F. (1960). *In* "Lipid Metabolism" (K. Bloch, ed.), pp. 41–68. Wiley, New York.

Mead, J. F. (1968). *Progr. Chem. Fats Other Lipids* **9**, Part 2, 161–230.

Merdinger, E., Kohn, P., and McClain, R. C. (1968). *Can. J. Microbiol.* **14**, 1021.

Meyer, F., and Bloch, K. (1963). *Biochim. Biophys. Acta* **77**, 671.

Meyer, H., and Holz, G. G. (1966). *J. Biol. Chem.* **241**, 5000.

Meyer, H., and Meyer, F. (1969). *Biochim. Biophys. Acta* **176**, 202.

Mizuno, M., Shimojima, Y., Iguchi, T., Takeda, I., and Senoh, S. (1966). *Agr. Biol. Chem.* **30**, 506.

Mohrhauer, H., and Holman, R. T. (1963a). *J. Lipid Res.* **4**, 151.

Mohrhauer, H., and Holman, R. T. (1963b). *J. Lipid Res.* **4**, 346.

Molitoris, H. P. (1963). *Arch. Mikrobiol.* **47**, 104.

Morris, L. J. (1967). *Lipids* **3**, 260.

Mudd, J. B., and Stumpf, P. K. (1961). *J. Biol. Chem.* **236**, 2602.

Mumma, R. O., and Bruszewski, T. E. (1970). *Lipids* **5**, 115.

Mumma, R. O., Fergus, C. L., and Sekura, R. D. (1970). *Lipids* **5**, 100.

Nagai, J., and Bloch, K. (1965). *J. Biol. Chem.* **240**, pc3702.

Nagai, J., and Bloch, K. (1966). *J. Biol. Chem.* **241**, pc1925.

Nagai, J., and Bloch, K. (1967). *J. Biol. Chem.* **242**, 357.

Nagai, J., and Bloch, K. (1968). *J. Biol. Chem.* **243**, 4626.

Nichols, B. W. (1965). *Biochim. Biophys. Acta* **106**, 274.

Nichols, B. W. (1968). *Lipids* **3**, 354.

Nichols, B. W., and Appleby, R. S. (1969). *Phytochemistry* **8**, 1907.

Nichols, B. W., and Moorhouse, R. (1969). *Lipids* **4**, 311.

Nichols, B. W., and Wood, B. J. B. (1968). *Lipids* **3**, 46.

Nichols, B. W., Harris, R. V., and James, A. T. (1965). *Biochem. Biophys. Res. Commun.* **20**, 256.

Nichols, B. W., Stubbs, J. M., and James, A. T. (1966). "The Biochemistry of Chloroplasts" (T. W. Goodwin, ed.), Vol. 1, p. 677. Academic Press, New York.

Nichols, B. W., James, A. T., and Breuer, J. (1967). *Biochem. J.* **104**, 486.

Noda, M., and Fujiwara, N. (1967). *Biochim. Biophys. Acta* **137**, 199.

Nugteren, D. H. (1962). *Biochim. Biophys. Acta* **60**, 656.

Nugteren, D. H. (1965). *Biochim. Biophys. Acta* **106**, 280.

O'Brien, J. S., and Benson, A. A. (1964). *J. Lipid Res.* **5**, 434.

Oesterheult, D., Bauer, H., and Lynen, F. (1969). *Proc. Nat. Acad. Sci. U.S.* **63**, 1377.

Oró, J., Tornabene, T. G., Nooner, D. W., and Gelpi, E. (1967). *J. Bacteriol.* **93**, 1811.

Overath, P., and Stumpf, P. K. (1964). *J. Biol. Chem.* **239**, 4103.

Pande, S. V., and Mead, J. F. (1970). *J. Biol. Chem.* **245**, 1856.

Paquot, C., Pham-Quang, L., and Laur, M. H. (1970). *Rev. Fr. Corps Gras* **17**, 547.

Parker, P., Van Baalen, C., and Maurer, L. (1967). *Science* **155**, 707.

Patterson, G. W. (1970). *Lipids* **5**, 597.

Patton, S., Fuller, G., Loeblich, A. R., III, and Benson, A. A. (1966). *Biochim. Biophys. Acta* **116**, 579.

Phillips, G. T., Nixon, J. E., Dorsey, J. A., Butterworth, P. H. W., Chesterton, C. J., and Porter, J. W. (1970). *Arch. Biochem. Biophys.* **138**, 380.

Pohl, P., Wagner, H., and Passig, T. (1968). *Phytochemistry* 7, 1565.

Pollard, W. O., Shorb, M. S., Lund, P. G., and Vasaitis, V. (1964). *Proc. Soc. Exp. Biol. Med.* 116, 539.

Prescott, D. J., Elovson, J., and Vagelos, P. R. (1969). *J. Biol. Chem.* 244, 4517.

Privett, O. S. (1966). *Progr. Chem. Fats Other Lipids* 9, Part I, 91–157.

Pugh, E. L., and Wakil, S. J. (1965). *J. Biol. Chem.* 240, 4727.

Pugh, E. L., Sauer, F., and Wakil, S. J. (1964). *Fed. Proc., Fed. Amer. Soc. Exp. Biol.* 23, 166.

Radunz, A. (1968). *Hoppe-Seyler's Z. Physiol. Chem.* 349, 1091.

Rahm, J. J., and Holman, R. T. (1964). *J. Lipid Res.* 5, 169.

Raju, P. K., and Reiser, R. (1967). *J. Biol. Chem.* 242, 379.

Raju, P. K., and Reiser, R. (1969). *Biochim. Biophys. Acta* 176, 48.

Rasmussen, R. K., and Klein, H. P. (1968a). *J. Bacteriol.* 95, 157.

Rasmussen, R. K., and Klein, H. P. (1968b). *J. Bacteriol.* 95, 727.

Rasmussen, R. K., and Klein, H. P. (1968c). *J. Bacteriol.* 95, 1090.

Reiser, R., and Raju, P. K. (1964). *Biochem. Biophys. Res. Commun.* 17, 8.

Renkonen, O., and Bloch, K. (1969). *J. Biol. Chem.* 244, 4899.

Richardson, T., Tappel, A. L., Smith, L. M., and Houle, C. R. (1962). *J. Lipid Res.* 3, 344.

Rosenbaum, N., Erwin, J., and Holz, G. G. (1965). *Progr. Protozool., Proc. Int. Conf. Protozool., 2nd* pp. 273–274.

Rosenberg, A. (1963a). *Biochemistry* 3, 254.

Rosenberg, A. (1963b). *Biochemistry* 2, 1145.

Rosenberg, A., and Pecker, M. (1964). *Biochemistry* 3, 254.

Rosenberg, A., Pecker, M., and Moschides, E. (1965). *Biochemistry* 4, 680.

Rothstein, M., and Götz, P. (1968). *Arch. Biochem. Biophys.* 126, 131.

Safford, R., and Nichols, B. W. (1970). *Biochim. Biophys. Acta* 210, 57.

Sagan, L. (1967). *J. Theor. Biol.* 14, 225.

Sastry, P. S., and Kates, M. (1964). *Biochemistry* 3, 1271.

Scheuerbrandt, G., and Bloch, K. (1962). *J. Biol. Chem.* 237, 2065.

Scheuerbrandt, G., Goldfine, H., Baronowsky, P. E., and Bloch, K. (1961). *J. Biol. Chem.* 236, pp. 70.

Schiff, J. A., and Epstein, H. T. (1968). *In* "The Biology of *Euglena*" (D. E. Buetow, ed.), Vol. 2, p. 286. Academic Press, New York.

Schlenk, H., and Gellerman, J. L. (1965). *J. Amer. Oil. Chem. Soc.* 42, 504.

Schlenk, H., and Sand, D. M. (1967). *Biochim. Biophys. Acta* 144, 305.

Schneider, H., Gelpi, E., Bennett, E. O., and Oró, J. (1970). *Phytochemistry* 9, 613.

Schroepfer, G. J., and Bloch, K. (1965). *J. Biol. Chem.* 240, 54.

Shaw, R. (1965). *Biochim. Biophys. Acta* 98, 230.

Shaw, R. (1966). *Advan. Lipid Res.* 4, 107.

Shorb, M. S. (1963). *Progr. Protozool., Proc. Int. Conf. Protozool., 1st*, pp 153–158.

Shorb, M. S., Dunlap, B. E., and Pollard, W. O. (1965). *Proc. Soc. Exp. Biol. Med.* 118, 1140.

Shorland, F. B. (1962). *Comp. Biochem.* 3, 1.

Silbert, D. F., Ruch, F., and Vagelos, P. R. (1968). *J. Bacteriol.* 95, 1658.

Silva, P. C. (1962). *In* "Physiology and Biochemistry of Algae" (R. A. Lewin, ed.), p. 827. Academic Press, New York.

Simoni, R. D., Criddle, R. S., and Stumpf, P. K. (1967). *J. Biol. Chem.* 242, 573.

Simmons, R. O., and Quackenbush, F. W. (1954). *J. Amer. Oil. Chem. Sci.* 31, 441.

Smith, G. M. (1950). "The Fresh Water Algae of the United States." McGraw-Hill, New York.

Smith, S., Easter, D. J., and Dils, R. (1966). *Biochim. Biophys. Acta* **125**, 445.

Soderhjelm, L., Wiese, H. F., and Holman, R. T. (1970). *Progr. Chem. Fats Other Lipids* **9**, Part 4, 557–585.

Stanier, R. Y. (1970). *Symp. Soc. Gen. Microbiol.* **20**, 1–38.

Stanier, R. Y., Doudoroff, M., and Adelberg, E. A. (1963). "The Microbial World." Prentice-Hall, Englewood Cliffs, New Jersey.

Stanier, R. Y., Kunisawa, R., Mandel, M., and Cohen-Bazire, G. (1971). *Bacteriol. Rev.* **35**, 171.

Stoffel, W., and Ach, L. L. (1964). *Hoppe-Seyler's Z. Physiol. Chem.* **337**, 123.

Stone, K. J., and Hemming, F. W. (1968). *Biochem. J.* **109**, 877.

Stumpf, P. K., and Boardman, N. K. (1970). *J. Biol. Chem.* **245**, 2579.

Stumpf, P. K., and James, A. T. (1963). *Biochim. Biophys. Acta* **70**, 20.

Stumpf, P. K., Brooks, J., Galliard, T., Hawke, J. C., and Simoni, R. (1967). *In* "The Biochemistry of Chloroplasts" (T. W. Goodwin, ed.), Vol. 2, pp. 213–237. Academic Press, New York.

Sumner, J. L., and Morgan, E. D. (1969). *J. Gen. Microbiol.* **59**, 215.

Sumner, J. L., Morgan, E. D., and Evans, H. C. (1969). *Can. J. Microbiol.* **15**, 515.

Suomalainen, H., and Keranen, A. J. A. (1968). *Chem. Phys. Lipids* **2**, 296.

Talbort, G., and Vining, L. C. (1963). *Can. J. Bot.* **41**, 639.

Thiele, O. W. (1964). *Biochim. Biophys. Acta* **84**, 483.

Thomas, P. J., and Law, J. H. (1966). *J. Biol. Chem.* **241**, 5013.

Tulloch, A. P. (1964). *Can. J. Microbiol.* **10**, 359.

Tulloch, A. P., and Ledingham, G. A. (1964). *Can. J. Microbiol.* **10**, 351.

Tyrrell, D. (1967). *Can. J. Microbiol.* **13**, 755.

Tyrrell, D. (1968). *Lipids* **3**, 368.

Tyrrell, D. (1969). *Can. J. Microbiol.* **15**, 818.

Van Etten, J. L., and Gottlieb, D. (1965). *J. Bacteriol.* **89**, 409.

Van Golde, L. M. G. (1968). *Biochim. Biophys. Acta* **152**, 84.

Wagner, H., and Pohl, P. (1965). *Biochem. Z.* **341**, 476.

Wagner, H., and Pohl, P. (1966). *Phytochemistry* **5**, 903.

Wakil, S. J. (1958). *J. Amer. Chem. Soc.* **80**, 6465.

Wakil, S. J. (1961). *J. Lipid Res.* **2**, 1.

Wakil, S. J. (1964). *In* "Metabolism and Physiological Significance of Lipids" (R. M. C. Dawson and D. N. Rhodes, eds.), pp. 3–28. Wiley, New York.

Wakil, S. J., and Gibson, D. M. (1960). *Biochim. Biophys. Acta* **41**, 122.

Wakil, S. J., Pugh, E. L., and Sauer, F. (1964). *Proc. Nat. Acad. Sci. U.S.* **52**, 106.

Wells, W. N., Schultz, J., and Lynen, F. (1966). *Proc. Nat. Acad. Sci. U.S.* **56**, 633.

White, G. L., and Hawthorne, J. N. (1970). *Biochem. J.* **117**, 203.

Williams, P. M. (1965). *J. Fish. Res. Bd. Can.* **22**, 1107.

Williams, V. R., and McMillan, R. (1961). *Science* **133**, 459.

Wirth, J. C., and Anand, S. R. (1964). *Can. J. Microbiol.* **10**, 23.

Wirth, J. C., Anand, S. R., and Kish, Z. L. (1964). *Can. J. Microbiol.* **10**, 811.

Wood, B. J. B., Nichols, B. W., and James, A. T. (1965). *Biochim. Biophys. Acta* **106**, 261.

Yang, S. F., and Stumpf, P. K. (1965). *Biochim. Biophys. Acta* **98**, 19.

Yokokawa, H. (1969). *J. Jap. Oil Chem. Soc.* **18**, 258.

Yokokawa, H. (1970). *J. Jap. Oil Chem. Soc.* **19**, 97.

Yuan, C., and Bloch, K. (1961). *J. Biol. Chem.* **236**, 1277.

Zalkin, H., Law, J. H., and Goldfine, H. (1963). *J. Biol. Chem.* **238**, 1242.

CHAPTER 3

Phospholipids

D. Mangnall and G. S. Getz

I. Introduction

The involvement of phospholipids in the general metabolism of the cell has been an area of increasing investigation over the past few years. Phospholipids are almost exclusively located in cellular membranes. It is now recognized that membranes not only define the cell boundary and subdivide the cell into discrete compartments, but also serve to regulate the metabolic flux between compartments as well as between the cell and its surrounding environment. Furthermore, they provide structural supports for the spatial organization of complex enzyme systems, such as the electron transport chain of the mitochondrion, and also have a role in the modulation of enzyme activity. Clearly, if one is to gain a complete understanding of how membranes may fulfill these functions it is necessary that the properties of the individual membrane components be studied. It should be borne in mind, however, that the assembly of membranous components into a functioning unit may confer upon the completed membrane properties not found in any of its isolated components.

The unifying feature of the membranes so far examined is that they are composed predominantly of associations of lipid and protein. Some membranes have smaller amounts of carbohydrate associated with them, and it is possible that in some cases nucleic acids are also minor constituents, although this has not yet been fully resolved. The lipids involved are largely phospholipids, with polar glycolipids playing a significant role in some membranes. Sterols, though absent from the membranes of prokaryotes, are present in larger or smaller quantities in all eukaryote membranes. Early analyses of animal membranes showed that phosphatidyl choline, phosphatidyl ethanolamine, and sphingomyelin were the major phospholipid components. Improved separation and analytical tools revealed the protean nature of the phospholipids associated with the various membranes. With these techniques, more discriminating analyses of the lipids of various membranes have indicated that the membranes of different organisms, cells, and subcellular organelles possess unique compositional characteristics. [See, for example, the reviews by Korn (1969) and Getz (1970).] With the advent of radioactive isotopes it became clear that these lipids were not static and inert, as had previously been supposed, but were in a continual state of flux and turnover. This demonstration of the dynamic nature of these molecules stimulated interest in the problem of phospholipid biosynthesis and breakdown. A large number of physicochemical methods were subsequently developed to facilitate the study of the individual phospholipid classes and to examine the turnover of the glycerol, fatty acid, and phosphoryl base moieties. Consequently much is now known about the physical properties of phospholipids and the way in which the phospholipid molecules may be assembled. A general thesis of how the physical attributes of phospholipids contribute to the structure and properties of membranes may be deduced from this information. However, details of how this knowledge relates to the problem of membrane function and turnover are largely undiscovered. Particularly notable is the almost complete absence of information concerning the regulation of the phospholipid metabolic pathways. Clearly, knowledge of this regulation is of the utmost importance to the understanding of the control of the biosynthesis of membrane structures. Also, little is known about how the phospholipid and protein components associate with one another and are assembled into a functioning membrane *in vivo*. However, *in vitro* models employing purified phospholipids and purified proteins (Kimelberg and Lee, 1970; Kimelberg *et al.*, 1970) and studies of the action of phospholipases on isolated membranes (Trump *et al.*, 1970; Glaser *et al.*, 1970) have afforded a means of examining protein-lipid interactions. It is the aim of this chapter to consider what is known about the phospholipids of eukaryotic microorganisms and to try to relate such information to the function and biogenesis of their membranes. In doing so, attention will be directed toward those aspects of phospholipids which are specific to eukaryotic microorganisms rather than the higher forms. However, the number of eukaryotic microorganisms which have been studied with respect to phospholipid metabolism is surprisingly small. Indeed, the comprehensive treatise of Ansell and Hawthorne (1964) ignores the phospholipids of the eukaryotic microbes almost completely. This is largely due to technical difficulties in culturing the organisms on a large scale. For this reason this review will deal mainly with yeasts, *Euglena*, *Tetrahymena*, and the few other organisms which have been critically investigated.

II. Methodology

In general the methods of phospholipid extraction and analysis developed for mammalian systems have proved to be adequate for application to microorganisms. However, yeasts, by virtue of their rigid, tough cell wall, present a special problem in ensuring complete extraction of the phospholipid in an unaltered form.

The chloroform-methanol mixtures usually used as the exclusive extractants with mammalian tissues (Folch *et al.*, 1957) were shown by Letters (1966) to be inadequate for the complete extraction of phospholipids from yeast. He described a procedure using an ethanol treatment followed by chloroform-methanol and acid chloroform-methanol extractions, and, at least for the strain he studied, it can be applied to whole, unbroken cells. Letters' procedure for the complete extraction of lipids from unbroken yeast seems, however, to be strain-dependent. The first neutral solvent treatment did not specifically extract a particular phospholipid class, and the acid solvent did not result in significant lysophosphatide production. Deierkauf and Booij (1968) confirmed the observation of Letters (1966) that extraction of *Saccharomyces cerevisiae* by the chloroform method of Folch *et al.* (1957) resulted in very low lipid yields, while prior breakage of the yeast by shaking for several hours with glass beads gave rise to the variable appearance of lyso derivatives. They made use of neutral solvents to give incomplete but reproducible extraction. Getz *et al.* (1970) have used the Letters (1966) method successfully on mechanically broken yeast without any gross formation of lyso derivatives. Similar neutral solvent-acid solvent techniques have been used on whole cells of *Schizosaccharomyces pombe* (White and Hawthorne, 1970) and broken cell preparations of *Saccharomyces carlsbergensis* (Paltauf and Johnston, 1970). Grinding of lyophilized yeast with glass beads in chloroform-methanol has also been used (Shafai and Lewin, 1968), although no data were presented as to the formation of lyso derivatives. In a comparative study by Nyns *et al.* (1968) of the efficiency of various procedures for the extraction of lipids from the yeast *Candida lipolytica*, treatment of lyophilized cells with cold chloroform-methanol (1:1) was found to give maximal extraction. Other treatments, including cold chloroform-methanol as a 2:1 v/v mixture, were less effective. It should be noted that the di- and triphosphoinositides (which form a significant percentage of the phospholipids of mammalian brain) are not readily extracted by chloroform-methanol mixtures, and Lester and Steiner (1968) extracted *Saccharomyces cerevisiae* protoplasts with an ethanol-ether mixture followed by petroleum ether in order to demonstrate these lipids in this organism.

The other eukaryotic organisms, which are comparatively easy to break by standard homogenization procedures, do not appear to present such extraction problems, and most workers have used chloroform-methanol mixtures. Subcellular fractions are similarly readily extracted. However, even for these membranes it may be appropriate to include an extraction with chloroform-methanol containing base or acid in order to obtain a quantitative recovery of such acidic phospholipids as cardiolipin (Rouser *et al.*, 1963) and polyphosphoinositides (Wells and Dittmer, 1965). The extraction media should be selected with care not only because of their differing extracting properties, but also because some phospholipases are activated by organic solvents (Kates, 1957; Letters and Snell, 1963), which may lead to loss of lipid and increased content of breakdown products.

Letters (1968) has in fact suggested that the initial step (for yeast at least) should be a hot alcohol treatment [80% (v/v) ethanol at 80°C for 15 minutes] to initiate extraction and inactivate such enzymes.

Once extraction has been accomplished, the phospholipids must then be purified, since there is no known method which selectively extracts only phospholipids. Normally this involves a separation from nonlipid contaminants by partitioning against an aqueous salt solution, as with the Folch procedure, or by immobilizing the water-soluble contaminants on Sephadex, as described by Wuthier (1966). Chromatography on silica gel thin-layer chromatography plates or on silicic acid or DEAE-cellulose columns is normally employed to separate the individual phospholipid classes from one another and from the other lipids such as neutral lipids. Quantitation is often achieved by estimation of the phosphorus content of separated phospholipids, although specific tests for the different phospholipid bases are also used. Within a given phospholipid class, subclasses may be distinguished, according to their fatty acid composition, by argentation silica gel thin-layer chromatography (Hill *et al.*, 1968; Holub and Kuksis, 1971). However, surprisingly little application of this technique has been made to microorganisms. For further practical details of purification, separation, and quantitation of various phospholipids, the reader is referred to articles by Radin (1969), Wuthier (1966), Rouser and Fleischer (1967), Rouser *et al.* (1969), Skipski and Barclay (1969), Dittmer and Wells (1969), Getz *et al.* (1970), Letters (1966), Ansell and Hawthorne (1964), and G. A. Thompson and Kapoulas (1969). Procedures for the hydrolysis of phospholipids and the subsequent analysis of the phosphate, glycerol, fatty acid, and base moieties are described by Ansell and Hawthorne (1964) and Dittmer and Wells (1969).

III. Phospholipid Structure

The structure of the most commonly encountered phospholipids is shown in Fig. 1(a)–(j). The simplest phospholipid is phosphatidic acid. Esterification of the phosphate group with a number of alcohols gives rise to the major phospholipid species. Choline, ethanolamine, serine, inositol (or phosphoinositol), and glycerol esterification with phosphatidic acid gives rise to phosphatidyl choline, phosphatidyl ethanolamine, phosphatidyl serine, phosphatidyl inositol (and phosphorylated derivatives), and phosphatidyl glycerol, respectively. Another commonly encountered phospholipid is cardiolipin or diphosphatidyl glycerol. R_1, R_2, etc., represent fatty acid residues so that each structure represents not a single compound

(a)

(b)

(c)

(d)

(e)

(f)

(g)

(h)

(i)

(j)

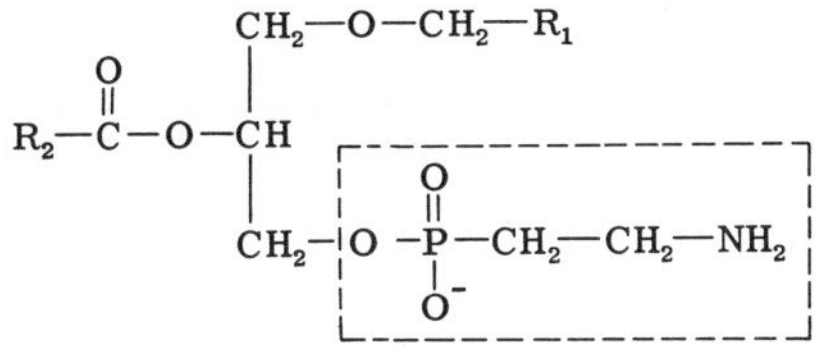

(a)

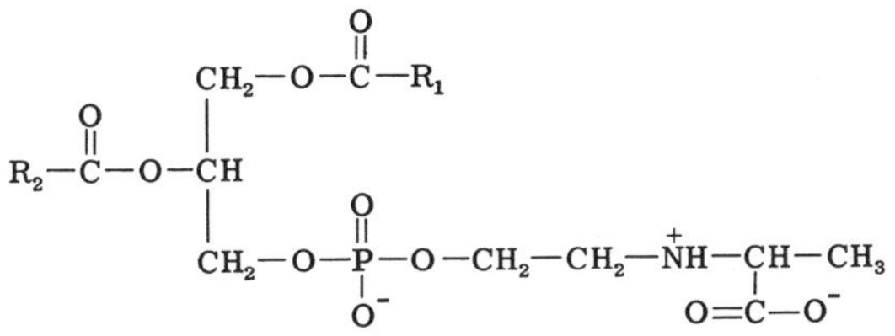

(b)

Fig. 2. Structure of two unusual phospholipids. (a) 2-Aminoethyl phosphonolipid. 2-Aminoethylphosphonic acid, which substitutes for the phosphoryl base alcohol of conventional phosphodiester lipids, is enclosed by the hatched line. This phosphonolipid has been found in *Tetrahymena pyriformis*. Note the saturated ether linkage at C-1. (b) Phosphatidyl-*N*-(2-hydroxyethyl)alanine, so far found only in rumen protozoa.

but rather a whole class of phospholipids, all with the same phosphoryl base but distinguishable by their fatty acid composition. These lipids also exist in much smaller quantities as the lyso derivatives in which there is only one esterified fatty acid. Plasmalogen forms, in which one of the fatty acids is present as a vinyl ether, are found in small quantities. The saturated ether forms may also occur. Figure 1(j) shows another class of phospholipids, the sphingomyelins, composed of phosphoryl choline, a fatty acid, and the long-chain amino alcohol sphingosine. All the phospholipids shown in Fig. 1 have been found in microorganisms as well as in the higher forms.

Fig. 1. Phospholipid structure. (a) Phosphatidic acid, (b) Phosphatidyl choline, (c) Phosphatidyl ethanolamine, (d) Phosphatidyl serine, (e) Phosphatidyl inositol (1-phosphatidyl-myoinositol). With the di- and triphosphoinositides the —OH groups at positions C-4 and C-5 are esterified with phosphate to give 1-phosphatidyl-myoinositol-4-phosphate and 1-phosphatidyl-myoinositol-4,5-diphosphate, respectively, (f) Phosphatidyl glycerol, (g) Diphosphatidyl glycerol (cardiolipin), (h) Plasmalogen or phosphatidal form, (i) Saturated ether form, (j) Sphingomyelin. The portion of the molecule to the left of the phosphate group (as drawn) is a ceramide.

Figure 2 shows two other phospholipids which do not have such a wide distribution. Figure 2(a) shows the phosphonolipid 2-aminoethyl phosphonolipid. This lipid forms a large proportion of the phospholipids of *Tetrahymena pyriformis*. The only other organisms known to possess phosphonolipids are the sea anemone (Rouser *et al.*, 1963) and rumen protozoa (Coleman *et al.*, 1971). The latter organism is the sole source hitherto detected for the phospholipid phosphatidyl-*N*-(2-hydroxyethyl)alanine shown in Fig. 2(b). Phosphatidyl sugars, such as those containing glucosamine, have been found in prokaryotes but will not be dealt with further here, since they have not yet been demonstrated in eukaryotes.

IV. Phospholipid Composition

In animal species it has been shown that for any given tissue or cell type there exists a remarkably, specific pattern of phospholipid composition (the exception being the erythrocytes) irrespective of the species of origin. For example, the phospholipid composition in terms of percent total phospholipid for rat, frog, sheep, mouse, and bovine livers is very similar (Rouser *et al.*, 1968). It would be of interest, therefore, to see if any similar constellation of specific lipid patterns occurs with eukaryotic microorganisms. Unfortunately, the number of detailed phospholipid analyses performed on these microorganisms is too small to allow firm generalizations to be made. Table I shows that although many of the phospholipids of the higher life forms are present in eukaryotic microbes, no obvious conformity to a specific lipid pattern exists, except possibly for the yeasts. The trypanosomes, for example, seem to be unique in their possession of large amounts of sphingomyelin. Sphingomyelin has also been shown to be a constituent of the parasitic amoeba *Entamoeba histolytica* (Sawyer *et al.*, 1967), along with phosphatidyl serine, phosphatidyl inositol, phosphatidyl ethanolamine and phosphatidyl choline, but a quantitative assay was not performed. Generally the bulk of the phospholipid in eukaryotic organisms is phosphatidyl ethanolamine and phosphatidyl choline, although the latter is infrequently found in prokaryotic cells. The prokaryotic pattern of phospholipid composition is encountered in some eukaryotic microorganisms. For example, Merdinger (1969) was unable to identify any choline or inositol in lipid extracts of the yeast-like fungus *Pullularia pullulans*, in which phosphatidyl ethanolamine and phosphatidyl serine were the major phospholipids. Also, in agreement with previous reports (Nichols *et al.*, 1965; Echlin and Morris, 1965), Nichols (1968) was unable to find any phosphatidyl choline, phosphatidyl inositol, or phosphatidyl ethanolamine in extracts of blue-green algae.

It seems that the phospholipid composition of some eukaryotic microorganisms is subject to variation depending on the genetic and nutritional status of the organism. This is exemplified for yeast, in particular.

A. TETRAHYMENA PYRIFORMIS

The lipids of *Tetrahymena pyriformis* are characterized by the presence of a phospholipid in which the phosphoryl base is replaced by 2-aminoethylphosphonic acid (AEP). The latter class of phospholipids has a direct carbon-phosphorus linkage and is known as phosphonolipids. Extracts of *Tetrahymena* contain AEP in a number of forms. It is found as the phosphatidyl ethanolamine analog, in which a high proportion of molecules contain alkyl glyceryl ether linkages (see Fig. 2) (G. A. Thompson, 1967), covalently linked to an insoluble residue and also in the free form (Rosenberg, 1964). *Tetrahymena* also contains a ceramide aminoethyl phosphonate. Two unusual branched-chain sphingosines, tentatively identified as Δ^4-15-methyl C_{16}-sphingosine and Δ^4-17-methyl C_{18}-sphingosine, have been shown to be constituents of this particular ceramide (H. E. Carter and Gaver, 1967). Trace amounts of the corresponding saturated bases as well as other branched-chain sphingosines (methyl C_{17}- and methyl C_{19}-sphingosines) were also found, although it was not indicated that the latter compounds were also associated with a ceramide aminoethyl phosphonate. (β-Phosphono)alanine, another phosphonic acid found free and in insoluble residues of *Tetrahymena*, although not in lipid extracts, is believed to be involved in the biosynthesis of AEP phosphonolipids (Smith and Law, 1970b).

As shown in Table II, the AEP phosphonolipids account for about half of the cilia phospholipid (Nozawa and Thompson, 1971a; Kennedy and Thompson, 1970). It has been suggested (Kennedy and Thompson, 1970) that this confers some degree of stability upon these structures, since the glyceryl ether and carbon-phosphorus bonds are less susceptible to hydrolytic and phospholipolytic attack than the more commonly encountered phosphodiester lipids. The physiological advantage of having such phospholipids in the cilia is not yet understood. It should be noted that the cilia phospholipids account for only 2 to 4% of the total cell phospholipid (Nozawa and Thompson, 1971a; Kennedy and Thompson, 1970) so that the cilia phosphonolipid represents about 1 to 1.5% of the whole cell phospholipid of which about 30% is phosphonolipid. Clearly, most of the phosphonolipid is not associated with the cilia and in fact considerable amounts are found in the membranes of the microsomes and mitochondria and in the pellicle (Table II). Electron micrographs of the pellicle by Nozawa and Thompson (1971a) show a structure composed of three membranes, the outer one of which is continuous with the ciliary membrane.

TABLE I

Phospholipid Composition of some Eukaryotic Microbes as Percent Total Phospholipid

Phospholipid[i]	Dictyostelium discoideum[a,i]	Acanthamoeba castellanii[b]	Tetrahymena pyriformis[c]	Schizosaccharomyces pombe[d]	Saccharomyces cerevisiae[e]	Saccharomyces cerevisiae[f]	Saccharomyces cerevisiae[g]	Trypanosoma lewisi[h]	Trypanosoma vivax[h]	Trypanosoma congolense[h]	Trypanosoma brucei[h]
DPG	8	4.2	8.7	4.4	5.3[m]	2.2					
PC	34	44.5	21.2	44.9	35.2	45.2	42.0	60-65	15-21	53-57	43-47
PE	32	33.2	19.9	19.4	19.1	16.7	22.4	18-25	13-20	8-13	19-24
PS	7	9.5		5.7	8.5	9.3	7.5				
PI		5.8[k]		17.6	21.2	19.8[n]	22.5				
PA					2.6						
PG				Trace	2.5		1.4				
DMPE				Trace			2.2				
LPC	10			0.2				4-10	2-9	2-4	5-8
LPE		0.5	17.6	0.2							
PNE			31.7								

						0-6°	2-6°	5-10°	2-5°
						6-11	53-62	22-24	22-29
Acidic phospholipids	2.3[l]								
Methyl glycerophosphate	1.8		3.9						
Unknown			4.1	1.2					
Sphingomyelin									

[a] Davidoff and Korn (1963).

[b] Ulsamer et al. (1969).

[c] Jonah and Erwin (1971).

[d] White and Hawthorne (1970).

[e] Getz et al. (1970), aerobic stationary phase culture.

[f] Deierkauf and Booij (1968).

[g] Letters (1966), late exponential cultures.

[h] Godfrey (1967).

[i] Abbreviations: DPG, diphosphatidyl glycerol (cardiolipin); PC, phosphatidyl choline; PE, phosphatidyl ethanolamine; PS, phosphatidyl serine; PI, phosphatidyl inositol; PA, phosphatidic acid; PG, phosphatidyl glycerol; DMPE, dimethyl phosphatidyl ethanolamine; LPC, lysophosphatidyl choline; LPE, lysophosphatidyl ethanolamine; PNE, 2-aminoethyl phosphonolipid.

[j] Contains, in addition, 8% of the plasmalogen form of phosphatidyl ethanolamine and possibly small amounts of the plasmalogen form of cardiolipin.

[k] Contains fatty acid aldehyde, inositol, and phosphate but no glycerol.

[l] Contains two unidentified acidic lipids.

[m] Includes phosphatidic acid.

[n] Includes phosphatidyl glycerol phosphate.

[o] This fraction is possibly phosphatidyl inositol but a firm identification was not achieved.

TABLE II

Phospholipid Composition of *Tetrahymena pyriformis*

	Phospholipid (% Total)				
Fraction	Phosphonolipid	DPG[a]	PE[a]	PC[a]	LPC[a] and Trace Compounds
Whole cells	23[b] (29.0,[c] 31.7[d])	5 (8.7[d])	37 (19.9[d])	33 (21.2[d])	2 (17.6[d])
Mitochondria	18 (32.2,[c] 16.1[d])	10 (16.2[d])	35 (19.4[d])	35 (30.2[d])	2 (17.1[d])
Microsomes	23 (29.8[c])	1	34	35	1
Cilia	47 (63.0,[c] 56.2[d])	1 (0[d])	11 (14.7[d])	28 (8.6[d])	1 (20.2[d])

[a] See abbreviations in Table I.

[b] All data from Nozawa and Thompson (1971a) except as otherwise noted.

[c] Data from Kennedy and Thompson (1970).

[d] Data from Jonah and Erwin (1971).

Although subfractionation of the pellicle has not yet been accomplished, calculations based upon the assumption that the outer membrane has the same phosphonolipid composition as the ciliary membrane while the inner membranes are assumed to be similar to the microsomes show the percent phosphonolipid for the pellicle as isolated to be 42%. This is identical to the experimentally observed result. It is possible, therefore, that the high proportion of phosphonolipid is a feature not simply of the cilia but of the whole of the outer membrane.

Phospholipids from the subcellular fractions of *Tetrahymena pyriformis* have also recently been analyzed with respect to fatty acid distribution (Jonah and Erwin, 1971). Mitochondrial fatty acids were more unsaturated than those of whole cells or cilia. In general, mitochondrial phospholipids contained more linoleic and linolenic acid than the corresponding phospholipids of whole cells and cilia. As with higher eukaryotes, the neutral lipids, by contrast, contained only small amounts of polyenoic acids. Assuming that the extraction of cardiolipin from both preparations was complete, the marked difference between whole cell and mitochondrial cardiolipin fatty acids reported by Jonah and Erwin (1971) strongly suggests that, in *Tetrahymena* at least, the cardiolipin may not be solely mitochondrial, as it appears to be in many other eukaryotic cells.

B. Yeast

Changes in the phospholipid composition of the yeast *Saccharomyces cerevisiae* at various stages of growth have been demonstrated by Getz *et al.* (1970) and Gailey and Lester (1968). Log-phase yeast had more phospholipid than stationary-phase yeast and, in particular, more phosphatidyl choline, phosphatidyl ethanolamine, and phosphatidyl inositol, which are the major yeast phospholipids and are found in similar proportions in cytoplasmic respiratory-deficient yeast. The phospholipid composition of log-phase yeast cells was very similar, irrespective of the strain and of the presence or absence of oxygen during growth. Stationary wild-type yeast, grown anaerobically, contained less phosphatidyl choline, phosphatidyl ethanolamine, and cardiolipin and more phosphatidyl inositol than aerobically grown cells. White and Hawthorne (1970), on the other hand, found no notable differences between exponential and stationary cultures of *Schizosaccharomyces pombe*. Jakovcic *et al.* (1971) have examined the cardiolipin content of wild-type and mutant yeasts in relation to mitochondrial development and function. They showed that, as with mammalian systems, the cardiolipin is almost exclusively mitochondrial and that the amount present is related to the degree of mitochondrial development, as measured by electron microscopy and the presence of respiratory enzymes. Changes in the cardiolipin content of *Saccharomyces cerevisiae* have also been demonstrated during the development of respiratory competence following the transfer of glucose-grown cells to a nonfermentable medium (Mangnall and Getz, 1970). During the 400-minute period following the transfer of cells to a growth-sustaining lactate medium, the respiratory capacity of the cells increased more than threefold, and associated with this was a doubling of the cardiolipin content. This is in keeping with the above observation of Jakovcic *et al.* (1971) that the cardiolipin content of yeast is a good measure of the development of the mitochondrial membrane. Cardiolipin production is not dependent on continued mitochondrial protein synthesis, since growth of cells in galactose containing either erythromycin or chloramphenicol, while decreasing the respiratory activity of the yeast, does not decrease the cardiolipin content. Changes in the phospholipid composition of whole yeast due to growth media and yeast type are shown in Table III. Interestingly, it was shown (Jakovcic *et al.*, 1971) that the cytoplasmic petite mutants, which have a deficient respiratory capacity and a decreased cardiolipin content, showed variations in cardiolipin content with changes in the growth conditions in the same way as the wild-type yeast. Thus, the mechanism responsible for the regulation of the mass of mitochondrial membranes by glucose repression still appears to be operative in cells genetically incapable of terminal respiration. Such mutants clearly retain the capacity to manufacture mitochondrial phospholipids. It seems unlikely, therefore, that the loss of respiratory activity is due to a loss of phospholipid components from the

TABLE III

Phospholipid Composition of Log-Phase Wild-Type Yeast ($YF_p{}^+$) and Petite Yeast ($YF_p{}^-$) Grown on Different Carbon Sources (% Total Phospholipid Phosphorus)[a]

Yeast:	$YF_p{}^+$			$YF_p{}^-$	
Carbon Source:	Glucose	Galactose	Lactate	Glucose	Galactose
Phospholipid[b]					
DPG	3.3	7.1	6.6	1.3	3.8
PC	41.6	39.6	47.3	39.8	43.4
PE	21.6	20.5	17.9	21.8	15.9
PS	7.5	5.2	7.0	7.7	4.7
PI	20.3	24.6	16.0	21.8	26.1
PA	0.8	0.5	2.1	1.8	2.3
PG	0.3	0.5	1.2	0.6	0.9
DMPE	1.4	0.7		1.0	
LPC	0.3	1.3	3.0	2.2	1.5
LPE	1.0	0.2		0.2	1.2
Unknown	1.3		1.4	2.4	0.1
PLP[c]/μg/gm of protein	2800		2990	2840	3200

[a] Data reproduced from Table IV of Jakovcic et al. (1971). Used with permission of authors and J. Cell Biol.

[b] Abbreviations: Same as in Table I.

[c] PLP, phospholipid phosphorus.

mitochondrial membranes and an associated loss of activity of the phospholipid-requiring respiratory enzymes. Table IV shows the distribution of phospholipids in the homogenate, mitochondrial, and microsomal fractions of the wild-type and petite yeasts. The mitochondrial pattern is similar to that of several mammalian tissues (Fleischer et al., 1967; Getz et al., 1968) and agrees well with the results of Letters (1966) and Paltauf and Schatz (1969) for yeast. A high content of cardiolipin and a high phosphatidyl ethanolamine to phosphatidyl choline ratio are characteristic of this fraction. The wild-type and petite mitochondrial fractions were very similar in composition, the only difference being a decreased cardiolipin and increased phosphatidyl inositol content in the mutant.

Respiratory-deficient mitochondria isolated from anaerobically grown wild-type cultures (promitochondria) similarly had decreased amounts of cardiolipin and increased amounts of phosphatidyl inositol, although changes in the other phospholipids were reported also (Paltauf and Schatz,

1969). This increase in phosphatidyl inositol was not observed when glucose catabolite repression was employed as the means of decreasing the respiratory activity (Lukins *et al.*, 1968).

The existence of promitochondria was for some time a question of controversy, since electron micrographs frequently failed to show any mitochondria-like structures in anaerobic cells. This was largely a reflection of technical difficulties in the fixation and processing of lipid-poor, anaerobic cells for electron microscopy. It was then shown that even though promitochondria could be demonstrated in the lipid-poor anaerobic cells they were more readily demonstrable under the electron microscope if the growth media were supplemented with ergosterol and unsaturated fatty acids (Swift *et al.*, 1967, 1968; Swift and Wolstenholme, 1969), because anaerobic conditions prevent the formation of unsaturated fatty acids and sterols since the oxidation reactions involved require molecular oxygen. Consequently, osmium tetroxide or potassium permanganate visualization of the promitochondrial membranes, dependent, at least in part, on the re-

TABLE IV

PHOSPHOLIPID COMPOSITION OF SUBCELLULAR FRACTIONS OF WILD-TYPE AND PETITE
YEASTS (% TOTAL PHOSPHOLIPID PHOSPHORUS)[a]

	Homogenate		Mitochondria		Microsomes	
	Wild Type	Petite	Wild Type	Petite	Wild Type	Petite
Phospholipid[b]						
DPG	6.5	4.9	15.6	9.1	1.6	1.2
PC	28.6	37.6	32.5	35.8	40.5	38.7
PE	14.1	14.7	22.0	20.5	12.4	13.7
PS	4.6	9.7	3.3	7.4	7.3	3.9
PI	12.7	23.9	10.7	16.7	27.8	7.2
PA	1.5	2.2	1.0	1.5	0.7	1.3
PG	2.7	Trace	2.0	0.8		0.6
DMPE	1.7	1.1	0.9	Trace	1.4	0.8
LPC	10.8	1.6	4.2	1.9	2.7	14.2
LPE	7.9	Trace	3.6	1.9	2.0	5.4
Unknown	8.5	4.3	4.2	4.7	3.7	13.8
PLP[c]/µg/gm of protein	3285	3316	8030	14,340	5235	3109

[a] Data compiled from Jakovcic *et al.* (1971).

[b] Abbreviations: Same as in Table I.

[c] PLP, phospholipid phosphorus.

action of the stain with the double bonds of unsaturated fatty acid, could not occur. Promitochondria from lipid-supplemented, anaerobically grown cells contained more phospholipid but less ergosterol than aerobic mitochondria. Even less sterol was present in promitochondria from non-supplemented media (Paltauf and Schatz, 1969). Further differences were observed in the fatty acid composition of aerobic mitochondria and promitochondria. Under aerobic conditions oleic and palmitoleic acids accounted for most of the unsaturated fatty acids. Lipid-supplemented promitochondria had a decreased palmitoleic acid and an increased oleic acid content, the bulk of which was probably derived from the Tween 80 (polyoxyethylene sorbitan monooleate) supplement. In the absence of Tween 80 and ergosterol the unsaturated fatty acids were much decreased and short-chain (C_{14} or less) saturated fatty acids were present in unusually large quantities. The unsaturated fatty acids present here were probably derived from the yeast extract of the growth media, and the amounts present seem to be about the minimum necessary for survival, since at harvest (at early stationary phase) 40% of the lipid-deficient cells were already nonviable. Anaerobic growth does not occur in a medium completely free of lipid (Paltauf and Schatz, 1969).

The reduction of the mitochondrial unsaturated fatty acid content associated with anaerobiosis also occurs with glucose repression (Vignais et al., 1970; Lukins et al., 1968).

Letters (1968) found that cells of Saccharomyces carlsbergensis grown anaerobically contained large amounts of both lysophosphatidyl choline (27.2% of the total phospholipid) and lysophosphatidyl ethanolamine (8.5% of the total phospholipid). This was not the case for anaerobically grown Saccharomyces cerevisiae (Getz et al., 1970). It is possible that in Saccharomyces carlsbergensis phospholipases are activated under anaerobic conditions.

The composition of the plasma membrane of Saccharomyces cerevisiae has been investigated using whole yeast (Suomalainen and Nurminen, 1970) or protoplasts (Longley et al., 1968) as starting material for the membrane isolation. Comparison of the results presented in these reports is difficult because different strains were studied using different fractionation techniques. Membranes obtained from protoplasts (Longley et al., 1968) were very similar in phospholipid composition to whole cell extracts and contained only slightly more phosphatidyl ethanolamine, phosphatidyl inositol, and phosphatidyl serine and slightly less phosphatidyl choline. The fatty acid composition of the corresponding phospholipid fractions from each was also similar. Starting with whole yeast, cells were mechanically broken, a cell wall fraction was removed, and, after snail juice digestion, a membrane fraction was obtained which contained less phosphatidyl ethanolamine and phosphatidyl choline and more phosphatidyl inositol and phosphatidyl serine than whole cell extracts. Hunter and Rose (1971) have also reported the presence of cardiolipin in extracts of these membranes. There were no detectable cytochromes and only barely detectable amounts of succinate dehydrogenase (Diamond and Rose, 1971); it was concluded that the preparations were essentially free from any mitochondrial contamination.

A method for the isolation of purified plasma membrane by differential and Urografin density gradient centrifugation, with details of the fine structure, chemical composition, associated enzymes, and subfractionation of this membrane, has also been described by Matile (1970), although the phospholipid composition was not specifically discussed.

Little appears to be known about yeast sphingolipids, although Wagner and Zofcsik (1966a, b) examined the sphingolipid fractions from *Saccharomyces cerevisiae* and from *Candida utilis*. They isolated from this fraction an inositol- and mannose-containing lipid which they termed a "mycoglycolipid." Inositol- and mannose-containing sphingolipids have also been detected in yeast extracts by Trevelyan (1968) and Tanner (1968).

In a more recent investigation of the inositol phospholipids of yeast, S. Steiner and Lester (1972) demonstrated that at least seven inositol-containing phospholipids were present, four of which were stable-to-mild alkaline methanolysis (the other three being the more usually encountered mono-, di-, and triphosphoinositides). One of the alkaline stable phospholipids was a mannosyl di(inositolphosphoryl)ceramide that had been previously characterized by Steiner *et al.* (1969) as the major phosphoinositol containing sphingolipid of bakers yeast. Radioactively labeled 2-deoxyglucose was rapidly incorporated, apparently intact, into some of these alkaline stable inositol phospholipids. So far, these inositol-containing phosphosphingolipids have only been found in organisms possessing cell walls and they may well be involved in cell wall production or cell wall function.

C. EUGLENA GRACILIS

Hulanicka *et al.* (1964) found that *Euglena gracilis* phospholipids represented 74% of the total lipid of etiolated cells and only 21% of green cells. On the basis of the intensity of staining of the various phospholipid spots after thin-layer chromatography, these authors estimated that in green cells phosphatidyl serine predominated, with phosphatidyl ethanolamine and phosphatidyl choline as minor components. In etiolated cells, phosphatidyl ethanolamine and phosphatidyl choline were the principal phospholipids. In a later publication (Erwin, 1968) it was shown that heterotrophic cells contained 55.6% choline phospholipids, 18.7% serine and ethanolamine phospholipids, 5.9% galactolipids, and 20.5% neutral lipids (mainly as triglycerides). In the photoauxotrophic cells the values were, respectively, 4.5, 16.4, 70.4, and 8.2% (mainly diglycerides). On the basis of these data, Erwin has suggested that the choline phospholipids and triglycerides of the heterotrophic cells are replaced by galactolipids and diglycerides in the photoauxotrophs. However, as no details were given of changes in the absolute amounts of lipid present in the two situations, the possibility exists that the decreased percentage of the phospholipid in the green cells is due to an increase in the relative amounts of the other lipids, particularly the galactolipids and diglycerides, rather than an absolute decrease in phospholipid. The phospholipids of this organism were shown to be rich in the polyunsaturated C_{20}-C_{24} fatty acids in both green and heterotrophic cells. On the other hand, the neutral lipids of the etiolated cells (mainly triglycerides) contained largely saturated and monounsaturated fatty acids, suggesting that they functioned as a carbon fuel reserve. The neutral lipids of green cells (mainly diglycerides) contained large amounts of linolenic acid and may serve as a precursor for galactosyl glycerides (Hulanicka et al., 1964; Erwin, 1968). Trans Δ^3-hexadecenoic acid has been shown to be peculiar to green cells, where it seems to be exclusive to phosphatidyl glycerol (Van Deenen and Haverkate, 1966). This seems to be specific to cells in which photosynthesis is of the type found in higher plants. Although such unsaturated fatty acids are clearly associated with the chloroplasts, their participation in the reactions leading to the evolution of oxygen (Nichols et al., 1965; Holton et al., 1964) has been questioned by studies in blue-green algae, where these acids are less prominent. Such fatty acids are also absent from the photosynthetic bacteria (Harris et al., 1965), although in these prokaryotic cells the photosynthetic apparatus is less complex.

A further difference between green and etiolated *Euglena* cells is found in the differential use of an acyl carrier protein (ACP) for fatty acid synthesis. ACP derivatives of fatty acids are produced only by the green cells and are used specifically in the formation of the monogalactosyl diglycerides in the chloroplasts and not for acyl donation to phospholipids or neutral glycerides. Acyl-CoA esters, which are the major activated fatty acid form of the etiolated cells, serve as the acyl donors for all lipids, including the phospholipids of these cells. Further consideration of this topic is beyond the scope of this chapter, and the reader is referred to the papers by Renkonen and Bloch (1969), Nagai and Bloch (1966, 1967), Delo et al. (1971), and Ernst-Fonberg and Bloch (1971).

D. NEUROSPORA CRASSA

Phosphatidyl choline and phosphatidyl ethanolamine form the bulk of the phospholipids in wild-type *Neurospora crassa*, the rest being mainly phosphatidyl serine (Crocken and Nyc, 1964). Mutants deficient in one of the steps involved in methylation of phosphatidyl ethanolamine to phosphatidyl choline showed a decreased phosphatidyl choline content and either traces or significant amounts of monomethyl and dimethyl phosphatidyl ethanolamine when grown in the absence of choline. The fact that the fatty acid composition of monomethyl phosphatidyl ethanolamine, dimethyl phosphatidyl ethanolamine, and phosphatidyl choline from the mutants was very similar suggested that a common phosphatidyl moiety was employed in their biosynthesis (Hall and Nyc, 1962). Phosphatidyl ethanolamine, from which these compounds presumably arise, was not analyzed.

E. ACANTHAMOEBA CASTELLANII

The phospholipids of *Acanthamoeba castellanii* were shown to be composed of 45% phosphatidyl choline, 33% phosphatidyl ethanolamine, 10% phosphatidyl serine, 6% of a phosphoinositide, and 4% cardiolipin. The phosphoinositide component had an unusual composition. It contained fatty acid aldehyde, inositol, and phosphate in the ratio 1.4:0.5:1.1, but no glycerol component was found (Ulsamer *et al.*, 1969).

F. RUMEN PROTOZOA

Kemp and Dawson (1969a, b) have reported the presence of an unusual phospholipid, phosphatidyl-N-(2-hydroxyethyl)alanine (Fig. 2), in lipid extracts of rumen protozoa. They were unable to find this phospholipid in *Tetrahymena* or any source other than rumen protozoa and suggested that it may substitute for phosphatidyl serine, which is absent from these organisms (Dawson and Kemp, 1967). The fatty acid composition of this phospholipid was notable in that it contained relatively large quantities of Δ^{11}-*trans*-octadecenoic acid, which is present in rumen digesta in large amounts as a result of its involvement in the rumenal hydrogenation of linolenic and linoleic acids. Further studies have been made on one of the rumen protozoa (Coleman *et al.*, 1971), which contains, in addition to this unusual phospholipid, phosphatidic acid, phosphatidyl inositol, phosphatidyl choline, phosphatidyl glycerol, phosphatidyl ethanolamine, ethanolamine plasmalogen, and rather substantial amounts of ceramide phosphoryl ethanolamine and diglyceride aminoethyl phosphonate.

G. Dictyostelium Discoideum

Davidoff and Korn (1962, 1963) have studied the phospholipid composition of an aggregateless mutant of the cellular slime mold *Dictyostelium discoideum* (Agg 204). In the first of these studies (Davidoff and Korn, 1962), cells grown in submersion culture on lipid-free, autoclaved *Escherichia coli* in 0.04 M phosphate buffer, pH 6.4, for 48 hours yielded 180 mg of lipid per gram lipid-free dry weight. The neutral lipids represented 40% and the phospholipids 60% of this. The phospholipid composition, determined after chromatography on silicic-acid-impregnated paper, was cardiolipin, 6%; phosphatidyl serine, 4.5%; phosphatidyl ethanolamine (plus some phosphatidyl serine), 39%; phosphatidyl inositol, 14%; and phosphatidyl choline, 37%. A second report (Davidoff and Korn, 1963) carried out in essentially the same way, except that the buffer pH was 6.0 and the cells were grown for up to 72 hours, again quoted a lipid yield of 18% of the lipid-free dry weight with the phospholipids again responsible for 60%. This time, however, there was no phosphatidyl inositol present, but instead the plasmalogen form of phosphatidyl ethanolamine and lysophosphatidyl ethanolamine accounted for 17% of the phospholipid. The other phospholipids were present in about the same proportions as before, although trace amounts of the plasmalogen form of cardiolipin were also possibly present. The authors noted that further growth beyond the stage at which the bacteria were consumed (which presumably represents a stationary phase of growth) did not dramatically alter the phospholipid composition, although they did not comment upon the absence of inositol phospholipid. It is apparent that this organism, in common with many others, exhibits some flexibility in the phospholipid composition, at least with respect to phosphatidyl inositol. Whether this change is associated with the possession of some internal cell structure or reflects changes in the nutritional properties of the growth media is not clear. The phospholipids of this organism contain an overwhelming amount (about 90%) of unsaturated fatty acids, so that the fatty acid at position 1 of the glycerol, which is generally saturated in most situations, must here be largely unsaturated. The structure was further unusual in that 40% of the fatty acid was accounted for by the previously unknown *cis,cis*-5,11-octadecadienoic acid, while *cis,cis*-5,9-hexadecadienoic acid accounted for an additional 10%. The more usual type of unsaturated fatty acid with methylene-interrupted double bonds was not present. The reason for the high proportion of unsaturated acids in the phospholipids of this organism is not clear. Perhaps it is related to the fact that optimal growth appears at 20°C (and is inhibited above 30°C), since the adaptation to a low-temperature environment is frequently accompanied by an increased content of unsaturated fatty acids (Mead, 1961).

V. Phospholipid Biosynthesis

The metabolic. pathways generally involved in the biosynthesis of the major phospholipid classes are outlined in Fig. 3. The demonstration and substantiation of the reactions of these pathways in eukaryotic microorganisms is extremely limited, as will be discussed below. In particular, the reactions leading to the formation of phosphatidyl ethanolamine and phosphatidyl serine have been only barely characterized in the eukaryotic microbes, although a CDP-diglyceride-dependent incorporation of serine into phosphatidyl serine and decarboxylation of phosphatidyl serine to phosphatidyl ethanolamine have recently been reported for *Neurospora*

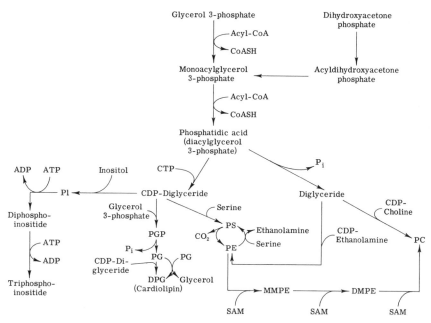

Fig. 3. Pathways for the biosynthesis of diacyl phospholipids. Abbreviations: PI, phosphatidyl inositol; PGP, phosphatidyl glycerophosphate; PG, phosphatidyl glycerol; DPG, diphosphatidyl glycerol (cardiolipin); PS, phosphatidyl serine; PE, phosphatidyl ethanolamine; PC, phosphatidyl choline; SAM, *S*-adenosyl methionine; MMPE, monomethyl phosphatidyl ethanolamine; DMPE, dimethyl phosphatidyl ethanolamine; Pi, inorganic phosphate.

crassa (Sherr and Byk, 1971). M. R. Steiner and Lester (1972) recently also showed that cell-free preparations of *Saccharomyces cerevisiae* could also incorporate serine into phosphatidyl serine by a CDP-diglyceride-dependent mechanism. Evidence for the decarboxylation of the phosphatidyl serine to phosphatidyl ethanolamine as well as for the production of phosphatidyl ethanolamine from CDP-ethanolamine and diglyceride was also presented. The limited amount of detailed information concerning lipid enzymology is even more surprising when compared to knowledge of the control and regulation of the enzymes of carbohydrate and protein metabolism. This is partly due to the fact that the lipid substrates, being generally insoluble in aqueous solutions, are more difficult to handle than water-soluble intermediates. Furthermore, the lipid-metabolizing enzymes are often bound to membrane from which they are not readily isolated and purified. These enzymes frequently form part of a large protein complex, where they are in close association with other enzymes and as such are not readily subject to the detailed definitive kinetic analyses that are possible with the more soluble enzymes of carbohydrate and protein metabolism. Indeed, the enzymic properties of membrane-bound enzymes may be considerably altered upon removal from their usual membrane microenvironment. Such a situation has been graphically explored by Racker and his colleagues in their work on the allotopic properties of the oligomycin-sensitive ATPase of the inner mitochondrial membrane (Racker, 1970).

One of the most useful features of microbial systems is the facility with which nonlethal genetic mutation may be induced and appropriate mutants selected and isolated. This has proved invaluable in the elucidation of metabolic pathways. Mutants lacking one or more enzymes of a biosynthetic sequence generally require supplementation of their growth media with a metabolic product distal to the site of the lesion for growth to occur. Such mutants also, generally, accumulate relatively large quantities of the intermediates immediately proximal to the lesion. In cases where there are two pathways to the same end product, supplementation may not be necessary for growth, although often in such cases the growth is slowed in the absence of the supplement. This approach has not been extensively applied to the study of phospholipid metabolism, since it has been assumed that most intermediates of such metabolic sequences would not permeate the cell membrane as such. Furthermore, the intermediates which would be required for feeding are not readily available and are not easily solubilized in growth media. However, as discussed below, the intracellular accumulation of intermediates by such mutants during growth on nonsupplemented media has proved useful in the elucidation of some of the sequences of phospholipid metabolism.

A. PHOSPHATIDIC ACID

Esterification of glycerol 3-phosphate by long-chain acyl-CoA's by extracts of *Saccharomyces cerevisiae* was first reported by Kuhn and Lynen (1965). Although the ratio of CoA released to glycerol 3-phosphate esterified was 1.5, the reaction product was tentatively identified by thin-layer chromatography as lysophosphatidic acid. In these experiments the fatty acid synthetase activity remained in the supernatant and could be made to release free palmitic acid in the absence of added CoA. Use of such a system for supply of fatty acids to the particulate fatty acyl transferase did not lead to any glycerol 3-phosphate esterification unless free CoA was added. Thus acyl-CoA appeared to be an obligatory intermediate, and the synthetase was not the immediate acyl donor. The reaction had a pH optimum of 7.0, was inhibited by thiol poisons, and exhibited no preference for unsaturated or saturated fatty acids. *Schizosaccharomyces pombe* has a similar acylating system (White and Hawthorne, 1970), with a pH optimum of 7.4, although the activity was shown to vary with different fatty acids ($C_{14:0} > C_{18:1} > C_{16:0} > C_{18:0} > C_{12:0} = C_{10:0} > C_{8:0}$). Under similar conditions *Saccharomyces cerevisiae* showed a similar order of activity. In *Saccharomyces carlsbergensis* glycerol 3-phosphate acylating activity was distributed equally between the mitochondrial and mitochondrial-free supernatant fractions (Johnston and Paltauf, 1970). Evidence has been presented in rat liver for the occurrence of a two-step acylation of glycerol 3-phosphate (Lamb and Fallon, 1970). The formation of monoacyl glycerol 3-phosphate occurs at pH 6.5 and preferentially forms the C_1 isomer with palmitoyl-CoA and the C_2 isomer with oleoyl-CoA. The second acylation that yields phosphatidic acid occurs at a higher pH. This mechanism may account for the specific distribution of fatty acids found in the glycerolipids, and conceivably such a mechanism operates in microbial systems, too. White and Hawthorne (1970) concluded that phosphatidic acid formation by phosphorylation of diglycerides, as has been described for *Escherichia coli* (Pieringer and Kunnes, 1965), did not occur in *Schizosaccharomyces pombe*. The acylation of dihydroxyacetone phosphate, which may also represent a means of regulating the fatty acid distribution, described for mammalian systems by Hajra and Agranoff (1968) and Agranoff and Hajra (1971), has been reported to occur in *Saccharomyces carlsbergensis* (Johnston and Paltauf, 1970), mainly in the mitochondrial fractions. Formation of phosphatidic acid by transfer of acyl groups from an ACP to glycerol phosphate has been proposed for green *Euglena* cells, mainly to explain the glycerol phosphate stimulation of fatty acyl-ACP incorporation into galactolipids (Renkonen and Bloch, 1969). It was suggested that the phosphatidic acid was then dephosphorylated before reaction of the diglyceride with UDP galactose and subsequent galacto-lipid formation.

B. CDP-DIGLYCERIDE

The capacity of a yeast particulate fraction to synthesize the liponucleo-tide CDP-diglyceride from CTP and phosphatidic acid was first reported by Hutchison and Cronan (1968), although the presence of this enzyme in *Escherichia coli* and *Micrococcus cerificans* was already known (J. R. Carter, 1968; McCaman and Finnerty, 1968). The enzyme had a pH optimum of 7.1 and required Mg^{2+} and K^+ ions for maximal activity but was not dependent on added phosphatidic acid. The requirement for this substrate was apparently satisfied by the endogenous content of the particulate fraction. Mangnall and Getz (1971) subsequently characterized this enzyme from a different yeast strain and found it to be associated with the mitochondria. The enzyme had a pH optimum of 6.5, a requirement for Mg^{2+} ions, and, unlike the preparation of Hutchison and Cronan, was dependent on added phosphatidic acid. The activity was stimulated several fold by the nonionic detergent Triton X-100, but sonication had no com-parable effect. KCl, but not NH_4Cl, produced a further stimulation only in the presence of Triton. Plots of the reaction velocity against phos-phatidic acid concentration showed a hyperbolic curve typical of Michaelis-Menten kinetics, and a double reciprocal plot of this data gave a straight line corresponding to an apparent K_m of 0.4 mM for phosphatidic acid. Similar experiments with CTP gave a Lineweaver-Burke plot that showed a sharp, distinct break at about 1 mM CTP. The plot was linear on each side of this point, corresponding to K_m's of 0.11 mM at concentrations below 1 mM CTP and 1.4 mM above 1 mM CTP. This deviation from standard kinetics, together with the probable prominent involvement of CDP-diglyceride in the synthesis of cardiolipin (see Fig. 3), prompted a search for metabolic modifiers of this enzyme which could correlate with the known accumulation of cardiolipin associated with mitochondrial development (vide supra). ATP, ADP, and AMP, either singly or in combination over the range 0 to 5 mM, were without effect, suggesting that this enzyme is not responsive to the energy charge of the adenylate system. Similarly, the glycolytic intermediates glucose 6-phosphate, fructose 1,6-diphosphate, and glycerol 3-phosphate and the pyridine nucleotides NAD and NADP were without any modulating effect. However, small amounts of NADH, in the presence of cyanide, showed a slight but re-producible stimulation (20% at 1 mM). NADPH produced increasing inhibition up to 2 mM (50% inhibition). Increasing citrate concentrations, on the other hand, gave increasing stimulation (50% at 5 mM), while malate, another Krebs cycle intermediate, was without effect. Phosphatidyl inositol and phosphatidyl serine, potential end products of CDP-diglyceride metabolism (see Fig. 3), produced a 40 to 50% stimulation of activity up to 5 mM, as did phosphatidyl ethanolamine. However, cardiolipin, the other metabolic end product, over the same range caused a 50% inhibition of activity. The activity was shown to be associated with both the outer and inner mitochondrial membranes (Mangnall and Getz, 1972). The presence of lipid-metabolizing enzymes on the outer membrane of the mitochondrion is not a new phenomenon, but several phospholipid-synthesizing enzymes have recently been localized to yeast and mammalian liver inner mitochondrial membranes (see also Hostetler and Van Den Bosch, 1972). Furthermore, the specific activity of this enzyme and the capacity to respond to both citrate stimulation and cardiolipin inhibition have been shown to undergo changes during mitochondrial development following transfer of yeast from a high-glucose-containing medium to a lactate medium. These studies suggest that this enzyme may play an important role in the biosynthesis of cardiolipin.

Steiner, M. R. and Lester (1972) have recently presented evidence suggesting that for *Saccharomyces cerevisiae* CDP-diglyceride may be a precursor, either directly or indirectly, of all the major phospholipids. If such is indeed the case the mechanism by which the production of this intermediate is controlled is of particular importance. Their data also suggest that CDP-diglyceride may be produced from an endogenous phospholipid other than phosphatidic acid.

C. PHOSPHATIDYL INOSITOL

Mutants have proved useful in studies of phosphatidyl inositol metabolism. Studies with inositol-deficient yeast (*Saccharomyces carlsbergensis*) showed that growth in the absence of inositol resulted in an accumulation of intracellular lipids, mainly triglycerides, while phospholipids, especially phosphatidyl inositol, decreased in content. The nuclear vacuole was also shown to disappear with inositol deficiency and to reappear with subsequent inositol provision (Shafai and Lewin, 1968; Challinor and Daniels, 1955; Lewin, 1965).

A later study (Paltauf and Johnston, 1970) showed that the production of excess triglycerides occurred with deficient cells grown on glucose or ethanol, while lactate-grown cells did not show any of the inositol deficiency characteristics. It was suggested that inositol biosynthesis may be more efficient under conditions of lactate growth, since the phosphatidyl inositol content was the same as that of the supplemented cells. Electron micrographs of mitochondria from deficient cells showed no morphological abnormality. Further studies (Johnston and Paltauf, 1970) showed that the triglyceride accumulation was not due to a decreased fatty acid oxidative capacity of the deficient cell, since fatty acid oxidation occurred at similar rates in both deficient and supplemented cells. However, this yeast had only a limited capacity to oxidize fatty acids, and it was suggested that the triglyceride served as a fatty acid pool for phospholipid biosynthesis. The possibility that fatty acyl-CoA utilization might be altered by inositol deficiency was also investigated. However, the biosynthetic pathway for triglycerides could not be established in this yeast. The rate of hydrolysis of phosphatidic acid was very low and the usual pathway of phosphatidic acid → diglyceride → triglyceride was not demonstrable. Addition of potential acyl acceptors (1,2-diglyceride, 1- or 2-monoglyceride, CDP-diglyceride, phosphatidic acid, acyl dihydroxyacetone phosphate) did not stimulate the *in vitro* production of triglycerides. Acylation of *sn*-glycerol 3-phosphate and of dihydroxyacetone phosphate occurred in both supplemented and deficient cells. However, CDP-diglyceride formation from phosphatidic acid and CTP was consistently higher in depleted cells, suggesting the involvement of this liponucleotide in phosphatidyl inositol formation in this yeast. Fatty acid synthesis was also more active in depleted cells. It was concluded that triglyceride accumulation represents a storage of excess fatty acids in a nontoxic form, although the mechanism by which inositol depletion brings this about is still unclear. Of interest is the observation by these authors that phosphatidyl inositol of depleted cells turns over more rapidly than other phospholipids of depleted or inositol-supplemented cultures; 82% of the ^{32}P in prelabeled phosphatidyl inositol of depleted cells was lost within a single doubling time. Diglycerides derived from phosphatidyl inositol breakdown may contribute to the triglyceride accumulation. It is perhaps worth noting that the depleted cells which had decreased amounts of phosphatidyl inositol also showed an increased phosphatidyl serine content. This may be a reflection of the continued production of CDP-diglyceride in these cells.

Phosphatidyl inositol metabolism in another yeast, the fission yeast *Schizosaccharomyces pombe*, which has an absolute requirement for inositol, has been reported by White and Hawthorne (1970). Interestingly, addition of CDP-diglyceride produced a marked inhibition of the rate of phosphatidyl inositol formation from inositol by dialyzed homogenates. Phosphatidyl ethanolamine and, to a lesser degree, phosphatidic acid were stimulatory. Both CTP and CMP were stimulatory in assays using dialyzed homogenates but were inhibitory with nondialyzed preparations, even when phosphatidic acid was added. Thus the CDP-diglyceride pathway of phosphatidyl inositol synthesis, which is operative in mammalian systems

(Paulus and Kennedy, 1960), does not appear to function here, and all attempts to show the existence of such a pathway in this yeast have failed. Neither was there any evidence for exchange reactions of the type phosphatidic acid + inositol $\rightleftharpoons$ phosphatidyl inositol + H_2O or phosphatidyl ethanolamine + inositol $\rightleftharpoons$ phosphatidyl inositol + ethanolamine, which might have led to phosphatidyl inositol synthesis. Hydrolysis of phosphatidyl inositol to phosphatidic acid and inositol was reported, presumably as a result of phospholipase activity, but there was no indication that reversal of this reaction could account for phosphatidyl inositol formation. The inability to demonstrate the pathway of phosphatidyl inositol biosynthesis is surprising when it is considered that phosphatidyl inositol represents about 20% of the total phospholipids of this organism.

More recently a CDP-diglyceride-dependent incorporation of inositol into phosphatidyl inositol by cell-free extracts of *Saccharomyces cerevisiae* has been reported (M. R. Steiner and Lester, 1972).

Saccharomyces cerevisiae, prelabeled with ³H-inositol and ³²P-phosphoric acid and then transferred to nonlabeled growth media, showed a subsequent loss of ³H and ³²P from phosphatidyl inositol with a half-life of 1.3 cell doublings. This was accompanied by increases in the radioactivity of mannosyl di(inositol phosphoryl)ceramide and by accumulation in the medium of glycerophosphoryl inositol (the deacylated form of phosphatidyl inositol) (Angus and Lester, 1971).

Di- and triphosphoinositides have been reported to occur in *Saccharomyces cerevisiae* (Lester and Steiner, 1968), but their metabolism in microorganisms has thus far not been studied. In mammalian systems these phospholipids are particularly abundant in brain tissues. It would be interesting to know if inclusion of relatively large quantities of these phospholipids confer upon a membrane any special characteristics, particularly with respect to its permeability and ion transport properties. The intracellular location of the di- and triphosphoinositides has not been reported, but it is clearly of interest to determine if an association with a particular cellular membrane exists.

D. PHOSPHATIDYL CHOLINE

As shown in Fig. 3, two pathways for the biosynthesis of phosphatidyl choline exist. One of these involves the successive addition of methyl groups to phosphatidyl ethanolamine, while the second pathway operates by transfer of phosphoryl choline from CDP-choline to a diglyceride acceptor.

1. Methylation Pathway

Mutants of *Neurospora crassa* have been employed in a study of the formation of phosphatidyl choline (Scarborough and Nyc, 1967a, b). The methylating system, which was shown to be microsomal, utilized S-adenosyl methionine as a methyl group donor and had the requirement for phosphatidyl ethanolamine satisfied by the endogenous microsomal content of this phospholipid (since addition of exogenous phosphatidyl ethanolamine caused no stimulation of activity). In the wild-type cells about half of the monomethyl phosphatidyl ethanolamine produced was converted further to dimethyl phosphatidyl ethanolamine and phosphatidyl choline. One of the mutants, strain 47904, produced monomethyl phosphatidyl ethanolamine at almost normal rates but could metabolize it further only at a very slow rate. Microsomes from a second mutant, strain 34486, had a much reduced rate of production of monomethyl phosphatidyl ethanolamine (possibly as a result of an increased K_m for S-adenosyl methionine), but once formed, it was rapidly converted to dimethyl phosphatidyl ethanolamine and phosphatidyl choline. Mixtures of microsomes from both mutants produced appreciable amounts of all the methylated forms. These experiments suggest that at least two enzymes are involved in the trimethylation of phosphatidyl ethanolamine to phosphatidyl choline. One enzyme transforms phosphatidyl ethanolamine to monomethyl phosphatidyl ethanolamine and is distinguishable from the enzyme(s) responsible for the further methylations. However, it was shown that the enzymes involved in the methylations of monomethyl and dimethyl phosphatidyl ethanolamine exhibited similar pH optima and similar heat denaturation characteristics for both methylations (Scarborough and Nyc, 1967b). Furthermore, it was not possible to separate the two methylating activities by ammonium sulfate precipitation, and the ratio of activities remained the same in the fractions so obtained. This is consistent with, but does not prove, the earlier postulation of these authors (Scarborough and Nyc, 1967a) that one enzyme catalyzes the methylation of both monomethyl phosphatidyl ethanolamine and dimethyl phosphatidyl ethanolamine.

Evidence for a similar methylation pathway in *Saccharomyces cerevisiae* has been presented by Waechter et al. (1969), who suggested that the activity of this pathway was dependent on the level of free choline in the growth medium. This was confirmed by Steiner and Lester (1970), who showed that the activity was associated with a particulate fraction obtained after 5.4×10^6 g min centrifugation. Preparations from cells grown either in the presence or absence of choline supplements had the same K_m for S-adenosyl methionine and similar responses to pH, but the V_{max} for the choline-deficient cell preparation was 14 times that of the extract from choline-grown cells. This difference was not explicable on the basis of a different content of phosphatidyl ethanolamine in the cells grown under the two conditions and seems to represent a repression of enzyme synthesis in the presence of choline. Addition of choline to the *in vitro* assay system had no effect on the methylating capacity of the extracts. Similarly, *in vivo* addition of choline to cells grown with or without choline supplementation was without immediate effect upon the methylation activity of extracts derived from these cells. The involvement of monomethyl phosphatidyl ethanolamine and dimethyl phosphatidyl ethanolamine as intermediates was shown by *in vitro* assays. With increasing time the amount of mono- and dimethyl phosphatidyl ethanolamine increased and eventually reached a steady-state level, although phosphatidyl choline continued to increase. As would be expected, the monomethylated derivative reached a steady-state concentration before the dimethylated form. However, at this time, information concerning the number of enzymes involved and their cellular localization and details of how medium choline causes a modification of their activity are still largely unresolved. In *Tetrahymena pyriformis* the methylating activity was again microsomal and could utilize exogenous monomethyl phosphatidyl ethanolamine as substrate but not phosphatidyl ethanolamine or dimethyl phosphatidyl ethanolamine. In this respect the *Tetrahymena* system differs from that of *Neurospora* where both the mono- and dimethyl phosphatidyl ethanolamine act as substrate for the same enzyme. Growth on methyl ^{14}C-methionine caused labeling of phosphatidyl choline but methylation of 2-aminoethylphosphonic acid did not occur.

Microsomal methylation of phosphatidyl ethanolamine has also been shown to occur in the protozoan *Ochromonas malhamensis* (Lust and Daniel, 1964, 1966). Also, the operation of this pathway in *Euglena* has been suggested by Tipton and Swords (1966). They were unable to demonstrate any activity of the CDP-choline diglyceride pathway in this organism.

2. CDP-Choline Diglyceride Pathway

Neurospora can utilize choline for phosphatidyl choline synthesis (Crocken and Nyc, 1964). The enzymic basis for this observation was not studied, although the fact that mutant 34486 did not accumulate detectable quantities of mono- and dimethyl phosphatidyl ethanolamine during growth on choline suggests that some pathway other than that involving the methylation of phosphatidyl ethanolamine is employed for the formation of lecithin under these conditions.

In a very recent publication, Sherr and Byk (1971) reported that during growth of *Neurospora crassa* 47904 on either [Me-14C] or [1,2-14C]choline about 90% of the incorporated counts appeared in phosphatidyl choline, with less than 1.5% appearing as phosphatidyl ethanolamine. These authors were unable to demonstrate any reaction between CDP-choline and diglyceride or any exchange reaction between phosphatidyl ethanolamine and choline of the type reported between serine and phosphatidyl ethanolamine (Borkenhagen *et al.*, 1961).

Saccharomyces cerevisiae may also utilize the CDP-choline diglyceride pathway for phosphatidyl choline biosynthesis, since Ostrow (1971) has demonstrated the presence of a CDP-choline diglyceride phosphoryl choline transferase in mitochondrial and microsomal fractions of this yeast. The mitochondrial activity is associated with the outer membrane of the organelle. The specific activity of the outer mitochondrial enzyme was comparable to that of the microsomal membranes, suggesting that contamination by microsomes could not account for the mitochondrial activity although the microsomal activity could be due to contamination by outer mitochondrial membrane fragments. Thus yeast mitochondria may be capable of synthesizing their own lecithin. However, recently, Angus and Lester (1971) have demonstrated that during growth of *Saccharomyces cerevisiae* after prelabeling with ^{32}P-phosphoric acid and ^{3}H-myoinositol the specific activity of ^{32}P of phosphatidyl serine and phosphatidyl ethanolamine decreased and was matched by an increase in phosphatidyl choline ^{32}P specific activity, in a manner consistent with a precursor-product relationship. Although the composition of the growth medium was not recorded in this report, one presumes that it was low in choline. This suggests that under these growth conditions choline is derived from phosphatidyl ethanolamine and that the CDP-choline diglyceride pathway is preserved solely for the production of phosphatidyl choline in mitochondria.

Such a situation may also occur in *Tetrahymena pyriformis*, where the CDP-choline diglyceride pathway is mitochondrial and the methylating pathway is microsomal (Smith and Law, 1970a). The studies of phosphatidyl choline biosynthesis in *Tetrahymena pyriformis* by Smith and Law were prompted by the suggestion of G. A. Thompson (1969) that phosphatidyl choline was involved in the formation of the carbon-phosphorus bond of the phosphonolipids. They showed this organism to have the capacity to utilize both pathways of synthesis. CDP-choline diglyceride phosphoryl choline transferase was found to be mitochondrial, and higher activities were observed with Mn^{2+} than Mg^{2+} ions, while Ca^{2+} or Hg^{2+} were inhibitory.

E. PHOSPHONOLIPIDS AND GLYCERYL ETHERS

The presence in *Tetrahymena pyriformis* of a class of phosphonolipids having 2-aminoethylphosphonic acid (AEP) as the phosphoryl base has resulted in extensive use of this organism as a model for phosphonolipid studies. Liang and Rosenberg (1966) isolated the phosphonolipid along with phosphatidyl ethanolamine in a 1:13 ratio and showed that extracts of *Tetrahymena* could form a CMP–AEP complex from CTP and AEP by a reaction analogous to the formation of CDP-choline or CDP-ethanolamine. A phosphonate containing glycerophosphatide was formed during *in vitro* incubation of the CMP–AEP complex with diglyceride and *Tetrahymena* cell-free preparations. However, previous work showing that during growth in the presence of ^{32}P-phosphoric acid the phosphonolipid was more rapidly and more highly labeled than the free AEP (Rosenberg, 1964) suggested that this pathway represented either a salvage mechanism for AEP or simply reflected a lack of specificity for bases in the normal phospholipid synthesizing system. This phosphonic acid analog of phosphatidyl ethanolamine has also been characterized by G. A. Thompson (1967) and shown to have a significant glyceryl ether content. Growth in radioactive glucose or glycerol caused only a minimal labeling of lipids, while ethanolamine was shown to moderately label both ethanolamine and choline phospholipids. No labeled AEP was produced (see also Liang and Rosenberg, 1968; Coleman *et al.*, 1971), and there was no support for the suggestion of Segal that AEP was derived from lipid-bound ethanolamine (Segal, 1965). ^{14}C-Acetate and ^{14}C-palmitate, on the other hand, were rapidly incorporated into the phosphonolipid. Incorporation of the glyceryl ether, ^{3}H-chimyl alcohol, occurred at the same rate as palmitate uptake for the choline phospholipid but, surprisingly, was more slowly incorporated into the ethanolamine lipid fractions, which included the AEP phospholipids. It was suggested that phosphoenol pyruvate might serve as the precursor of AEP (Warren, 1968; Trebst and Geike, 1967). To satisfy the previously observed labeling pattern, Liang and Rosenberg (1968) postulated the formation of a phosphatidyl enol pyruvate which would subsequently undergo transamination and decarboxylation. Growth on ^{14}C-phosphoalanine (the aminophosphonic analog of serine) likewise gave rise to labeled phospholipids containing AEP (Smith and Law, 1970b). Labeled phosphoalanine was not found as such in the phospholipids. It was suggested that phosphoalanine is a precursor of AEP and that the conversion occurs at the level of the free phosphoalanine rather than at the phospholipid level. However, a rapid decarboxylation of the phospholipid form, in a manner similar to that involved in the conversion of phosphatidyl serine to phosphatidyl ethanolamine, could not be excluded. In such a situation, the failure to demonstrate any phospholipid-containing phosphoalanine means that the reaction would have to be irreversible and very rapid.

Tetrahymena also contains large quantities of glyceryl ethers and has been employed for the study of the biosynthesis of these compounds. Kapoulas and Thompson (1969) showed that homogenates of *Tetrahymena* and a 20,000 *g* supernatant would take up 70% of added ^{14}C-cetyl alcohol or ^{14}C-palmitate into lipids in 60 minutes. This process required supplementation with CoASH, ATP, Mg^{2+}, NADH, and glycerol 3-phosphate. Two to five percent of the palmitate was incorporated into free glyceryl ethers or alkali-stable phospholipids, while there was a 10% incorporation of cetyl alcohol into these products. Substitution of glyceraldehyde 3-phosphate for glycerol 3-phosphate increased the cetyl alcohol incorporation into alkali-stable lipids. The 100,000 *g* supernatant was the most active fraction in the formation of alkali-stable lipids, although it was less effective than microsomal preparations for the synthesis of the usual phospholipids. Pyridine nucleotides were not required for alcohol uptake by this fraction, but addition of NADPH markedly stimulated total glyceryl ether synthesis. Most of the glyceryl ethers occurred in choline, ethanolamine, and aminoethylphospholipids. There was no evidence for any accumulation of a proposed ketone derivative of glyceryl ethers, as had been found by Snyder *et al.* (1969) in a system utilizing a mouse microsomal preparation. These observations, however, do suggest that the pathway for the biosynthesis of glyceryl ethers from long-chain alcohol and dihydroxyacetone phosphate recently established for such animal tissues as cancer cells and preputial glands (Wykle and Snyder, 1970) may also be operative in the formation of these compounds in *Tetrahymena pyriformis*. Further studies showed that the mode of utilization of the added glyceryl ethers by *Tetrahymena* was dependent on the glyceryl ether concentration of the growth media (Kapoulas *et al.*, 1969). At low concentrations, incorporation of the intact molecule into phospholipids occurred to a large extent, and there was only limited degradation and conversion to fatty acid. In the presence of large amounts of substrate, however, considerable ether cleavage occurred. This was apparently not due to any induced enzyme synthesis but rather represented extensive involvement of an otherwise minor pathway in the presence of a large substrate excess. However, growth for long periods in high chimyl alcohol concentrations resulted in an abnormal phospholipid composition in which diacylphospholipids were replaced by monoether-monoacyl analogs. A model was proposed in which uptake by the oral apparatus and uptake by membrane phagocytosis or pinocytosis gave rise to two distinct chimyl alcohol pools in the ciliate. One pool was considered to provide substrate for the phospholipid if the pool size was small, but would go increasingly to triglyceride as the pool size increased and the capacity for phospholipid synthesis was extended. After utilization of the substrate of this first pool it was postulated that the cells employed triglyceride and the substrate of the second pool for phospholipid biosynthesis.

Nozawa and Thompson (1971b) have shown that radioactive palmitic acid was incorporated into the lipids of different membrane fractions of logarithmically growing cultures of *Tetrahymena pyriformis* at widely differing rates. Within 1 minute the microsomes and postmicrosomal supernatant achieved a high level of specific radioactivity, while such a level was found in the membranes enveloping the cilia only after several hours incubation. Radioactive acetate or hexadecylglycerol produced the same kind of labeling pattern. For all the fractions, incorporation of ^{14}C-palmitate into phosphonolipid was slower than into other major phospholipids.

F. PHOSPHATIDYL-*N*-(2-HYDROXYETHYL)ALANINE

The biosynthesis of phosphatidyl-*N*-(2-hydroxyethyl)alanine has been examined by Coleman *et al.* (1971) using cultures of the rumen protozoan *Entodinium caudatum*. Labeling experiments using ^{32}P inorganic phosphate, 2-^{14}C-ethanolamine and [32]P- and ^{14}C-labeled phosphatidyl ethanolamine showed that the direct precursor of the *N*-(2-hydroxyethyl)alanine containing phospholipid was probably intact phosphatidyl ethanolamine. When the source of radioactivity was ^{14}C-starch, ^{14}C-lactate, or ^{14}C-pyruvate, the phosphatidyl-*N*-(2-hydroxyethyl)alanine was more heavily labeled than other phospholipids. The latter observations suggested that the *N*-(1-carboxyethyl) group, which is linked to the phosphatidyl ethanolamine amino group, was probably derived from a three-carbon glycolytic intermediate. It seems most unlikely therefore that this phospholipid is derived from a phosphorylated derivative of *N*-(2-hydroxyethyl)alanine via a cytidine nucleotide-mediated pathway similar to that utilizing CDP-choline for lecithin synthesis but rather that the *N*-(1-carboxyethyl) grouping is transferred directly onto preexisting phosphatidyl ethanolamine molecules. These studies of the protozoan *Entodinium caudatum* also indicated that radioactive phosphatidyl ethanolamine was taken up by these organisms and catabolized, forming glycerophosphorylethanolamine followed by glycerophosphate and inorganic phosphorus.

VI. Metabolic Studies

A. EFFECTS OF UNSATURATED FATTY ACIDS AND TEMPERATURE OF GROWTH

The effect of hyperthermia has been extensively studied in *Tetrahymena pyriformis*. Constant exposure to an elevated temperature results in morphological abnormalities and is normally lethal for this organism (Rosenbaum *et al.*, 1966). Rosenbaum *et al.* (1966) showed that continued growth at an elevated temperature could be maintained by supplementation of the growth medium with phospholipids. These results suggested that elevated temperatures disrupted the normal membrane structure and function. This interpretation was supported by the observed release of 2-aminoethylphosphonic acid from *Tetrahymena* at elevated temperatures (Chou and Scherbaum, 1965, 1966). Levy *et al.* (1969) presented data to show that release of hydrolytic enzymes from lysosomes was not a primary cause of the defects observed here. Although release of the hydrolytic enzymes might follow destruction of the lysosomal membrane, it was concluded that a temperature-induced lesion in phospholipid biosynthesis was the most likely explanation for these observations. In studies of the phospholipid-maintained growth at 40°C, Erwin (1970) showed that only Asolectin, a crude soybean phospholipid extract with a high unsaturated fatty acid content, would allow serial transfer at 40°C. For successful transfer during growth on synthetic phospholipids poor in unsaturated fatty acids, it was necessary that the temperature be briefly lowered at each transfer. Also, while the cells cultured in synthetic phospholipids always had an abnormal morphology, the cells transferred to fresh Asolectin media produced a population of normal-looking cells after 24 hours, and abnormalities were not apparent until the cultures were 48 hours old. Fatty acid analyses suggested that during growth on synthetic phospholipids these were incorporated intact into the ciliate membranes. Asolectin, on the other hand, was first hydrolyzed, and the liberated unsaturated fatty acids were apparently used for the production of new membrane lipids. Saturated and monounsaturated fatty acids were shown to be ineffective in supporting growth at high temperature (Rosenbaum *et al.*, 1966), whereas purified ciliate fatty acids (which contain polyunsaturated fatty acids) did support growth at the high temperatures and produced morphologically normal cells. Fatty acid toxicity is a major problem in this type of experiment, and Asolectin was superior to free fatty acids in supplementing growth probably because it represented a less toxic form. A similar situation is found in *Paramecium aurelia*, which requires oleic acid for growth but grows better on oleic acid containing phospholipids, glycerides, and Tweens than on the free fatty acid (Soldo and Van Wagtendonk, 1967). It thus appears that the temperature-induced lesion is one involving unsaturated fatty acid biosynthesis or oxidation and turnover. This system provides a useful model for the study of the relationship between the degree of unsaturation and the functioning of biological membranes.

Unsaturated fatty acid mutants of *Saccharomyces cerevisiae* also afford a means of studying the involvement of unsaturated fatty acids in membrane function (Keith *et al.*, 1969; Wisnieski *et al.*, 1970). Unsaturated fatty acid depletion has been shown to lead to a dependence on fermentable substrates for energy supply (Proudlock *et al.*, 1969). Although the cells conta ned cytochromes and respired normally, the mitochondria lacked respiratory control and had only minimal capacity for phosphorylation of ADP. The mechanisms involved are as yet unclear. Growth of wild-type yeast in an anaerobic environment without unsaturated fatty acid and ergosterol supplementation gave rise to mitochondria which lacked the mitochondrial ribosomal RNA present in cells grown with lipid supplement (Forrester *et al.*, 1971). This has led to the suggestion that the mitochondrial ribosome is closely associated with, or may be an integral part of, the mitochondrial membrane. Levin *et al.* (1971) have used such mutants in a study of the unsaturated fatty acid requirements of yeast for growth and development of respiratory capacity. At 28°C the yeast could grow on both *cis* and *trans* unsaturated fatty acids with good development of respiratory competence after transfer from high glucose to low glucose media. Below 23°C, however, only the *cis* isomer supported growth and allowed derepression from high glucose media. These experiments stress the physiological importance of unsaturated fatty acids in membrane biology and emphasize the need for a greater understanding of their metabolism.

B. Phospholipids and Transport

The involvement of phospholipids in the active transport of arsenate into the yeast *Saccharomyces carlsbergensis* has been investigated by Cerbón (1969, 1970). Arsenic enters by a constitutive system normally used for phosphate uptake, but unlike phosphate causes continual inactivation of the transport mechanism so that its uptake eventually ceases. Experiments with radioactive (^{74}As) arsenate led to the formation of an arsenic-lipid complex during active transport, in which phosphatidyl inositol had a predominant involvement. It was suggested that this represented an intermediate, or part of the transport system. Analysis of the phospholipid composition of normal cells and of arsenate-adapted cells (in which arsenate inflow was much decreased) showed phosphatidyl choline, phosphatidyl ethanolamine, and phosphatidyl inositol as the major phospholipids, and the only difference was a 50 to 100% increase in the proportion of inositides in the adapted cells. Exposure of cells to arsenate for 30 to 60 minutes caused inactivation of the transport system and was associated with a 30 to 50% increase in the inositide level. A decreased rate of lipid biosynthesis was observed in cells exposed to arsenate, and there was no detectable turnover of phosphatidyl choline and phosphatidyl ethanolamine and only slow turnover of phosphatidyl inositol. The low rate of synthesis and the increased inositide content might alter the membrane such as to inactivate the arsenate-active transport system in the exposed cells. Alternatively, it was proposed that arsenic may combine only with newly made inositides, perhaps by substituting for the phosphate group, and that a normal rate of inositide biosynthesis was necessary for normal rates of arsenate uptake.

A role for lecithin in the process of lysine transport into *Neurospora crassa* has been suggested by the work of Sherr (1969), who showed a decreased capacity for lysine uptake by deficient mutants. Addition of phosphatidyl choline to the growth media restored the capacity for lysine uptake.

Bianchi and Turian (1967) examined the lipid changes of *Neurospora* during germination. They concluded that lipids do not play an important role as an energy source in this process and that the sugar trehalose is the major energy supply.

During encystment of *Acanthamoeba castellanii* there was an increase in the concentration of neutral lipid and a decrease in the phospholipid content without any major change in the total lipid amount (Bowers and Korn, 1969).

The production of extracellular lipids by eukaryotic cells has largely been ignored, although the topic of yeast extracellular lipid has been reviewed by Stodola *et al.* (1967).

B. Synchronously Growing Cultures

The growth of microbial cells in batch or continuous cultures does not lend itself to studies of the temporal events that occur during the division of a single cell and the growth and subsequent division of the daughter cells. A number of methods have been developed for producing cultures in which all or most of the cells divide and grow in synchrony, at least for two or three cell cycles. Such a synchronized population of cells clearly offers a valuable tool for studying the biochemical events of the cell cycle. Much information concerning the timing of the biosynthesis of macromolecules and the relation of such events to the cell genome has been derived from studies of synchronously dividing yeast cultures (Tauro *et al.*, 1969). It has been shown, for example, that the number of periods of α-glucosidase synthesis is proportional to the number of structural genes for this enzyme in *Saccharomyces cerevisiae* and that each of the structural genes is expressed at a characteristic time during the cell cycle (Tauro and Halvorson, 1966). In studies of phospholipid synthesis, King (1971) has shown that synchronized cultures of *Saccharomyces cerevisiae* grown on 1% galactose exhibit a stepwise increment in phospholipid content. The synthesis of the phospholipid was coincident with the periods of budding in the cell cycle. The accumulation of radioactivity from galactose-1-^{3}H or glycerol-2-^{3}H in the total phospholipid also occurred in a more or less stepwise fashion, although the periods during which no accumulation occurred were relatively short. Separation of phosphatidyl choline, phosphatidyl ethanolamine, and cardiolipin by thin-layer chromatography and subsequent determination of their tritium content also revealed that for these three phospholipids accumulation of labeled precursors occurred in a stepwise manner. The periods of greatest cumulative increments of radioactivity in these phospholipids generally corresponded to the period when oxygen consumption increased. Pulse labeling experiments using glycerol-2-^{3}H showed a cyclic pattern of incorporation into the individual phospholipid classes, indicating variations in the rate of labeling during the cell cycle. Although some variation in the timing of the period of maximum rate of incorporation was evident, this period was generally associated with a period of budding and daughter cell growth. Particularly worthy of note from these results is the variation in the rate of synthesis of the mitochondrial phospholipid cardiolipin. Williamson and Moustacchi (1971), using the same synchronously growing yeast strain, showed that the mitochondrial DNA did not exhibit any stepwise or oscillatory pattern of synthesis but appeared to be manufactured more or less continuously throughout the cell cycle. Thus it appears that the potential availability of an increase of the mitochondrial gene dosage is not tightly linked to a proportional increase in cardiolipin. This is in accord with the above-referenced data, indicating that mitochondrial protein synthesis is not necessary for the increments in the cardiolipin content of yeast associated with a lack of glucose repression.

Also studied in these experiments were the changes in activity of the enzyme CDP-choline diglyceride phosphorylcholine transferase, which represents the terminal step in one of the biosynthetic pathways for phosphatidyl choline. There was an increase in specific activity of this enzyme in yeast homogenates obtained at each period of budding. The elevated activity was not maintained, however, but decreased at the end of the budding cycle. This enzyme thus showed a cyclic variation in activity during the cell cycle. In contrast, succinate cytochrome c reductase and cytochrome oxidase exhibited stepwise increments in activity. The oscillatory pattern of activity of CDP-choline diglyceride phosphorylcholine transferase resembles that of DNA polymerase (Eckstein *et al.*, 1967). In contrast to most of the enzymes studied in synchronized yeast, these two enzymes are anabolic rather than catabolic. Whether such oscillatory patterns of activity are characteristic of the phospholipid-synthesizing and other anabolic enzymes remains to be established. Also worthy of further investigation is the manner by which such oscillations in enzymic activity are produced.

VII. Phospholipid Breakdown

The breakdown of phospholipids occurs via a series of phospholipases specific for the hydrolysis of particular linkages in the molecule (see Fig. 4). These enzymes are widely distributed. *Penicillium notatum* extracts were found by Fairbairn (1948) to contain high lysophospholipase (phospholipase B) activity, capable of acting on phosphatidyl ethanolamine and phosphatidyl choline. Bangham and Dawson (1959, 1960) and Dawson (1958a, b) showed that phospholipase A was also present but required the addition of phosphatidyl inositol, dicetyl phosphate, phosphatidic acid, cardiolipin, or some similar anionic amphipath for maximal activity. Kates *et al.* (1965) considered that such agents facilitated the initial hydrolysis of the substrate particles. Dawson and Hauser (1967) considered that the molecular packing arrangement of the substrate molecules was also a critical feature in determining the hydrolytic rate.

Hoffmann–Ostenof *et al.* (1961) reported on the finding of phospholipase C activity in yeast. Harrison and Trevelyan later showed (1963) that the

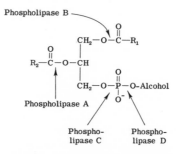

Fig. 4. Site of action of the phospholipases. Recently, the enzymes catalyzing hydrolysis of the acyl ester linkages have been designated as A_1 and A_2 and B_1 and B_2. The A_1 and A_2 enzymes utilize the diacyl compound, hydrolyzing the bonds at R_1 and R_2, respectively, while the B_1 and B_2 enzymes cause hydrolysis of the lysoderivatives at R_1 and R_2, respectively.

TABLE V

EFFECT OF INITIAL MEDIA GLUCOSE CONCENTRATION ON THE
PHOSPHATIDYL CHOLINE AND PHOSPHATIDYL INOSITOL CONTENT
OF YEAST PRETREATED WITH ACETONE PRIOR TO LIPID
EXTRACTION[a]

Glucose in Media (%)	Amount (% Total Lipid P)			
	PC[b]		PI[b]	
	Control	Acetone-treated	Control	Acetone-treated
1.0	45	45	22	25
2.5	40	31	28	31
5.0	39	19.4	28	41
10.0	36	1.8	27	80

[a] Data from Letters (1968).
[b] See Abbreviations in Table I.

breakdown of phospholipids, particularly phosphatidyl ethanolamine and phosphatidyl choline, which occurred on drying yeast cells was related to the internal nitrogen content, and they attributed this breakdown to phospholipase C activity. Kokke *et al.* (1963) showed that in aqueous yeast extracts, phosphatidyl choline was broken down to yield lysophosphatidyl choline, which was then converted to phosphatidyl choline and glycerolphosphoryl choline by enzymic transacylation, involving transfer of a fatty acid residue from one lysomolecule to another. However, in contradistinction to the results of Trevelyan, no phospholipase C was detected in the experiments of Van Den Bosch *et al.* (1967). Letters (1968) has investigated the relationship between enzymic breakdown of yeast phospholipids and membrane function. Acetone was found to activate yeast phospholipid breakdown in a manner that was dependent on the carbohydrate content of the growth medium. In cells grown on increasing concentrations of glucose and subjected to a 30-minute treatment with aqueous acetone before normal extraction, the phosphatidyl inositol content was increased and the phosphatidyl choline content was decreased with increasing glucose content as compared to the control, non-acetone-treated cells (see Table V). The phospholipids of cells grown anaerobically on 2.25% glucose underwent significant breakdown with acetone treatment. Methanol treatment had a similar effect. The nature of the breakdown products of

phosphatidyl choline during partial and complete degradation was investigated. If phospholipase C were responsible, as was suggested by
Harrison and Trevelyan (1963), then there should be no increase in free
fatty acids. Considerable release of free fatty acid was observed, however,
supporting the suggestion of Kokke *et al.* Furthermore, a lysophosphatidyl
choline, containing fatty acid only in the C-1 position, could be obtained
by employing conditions in which acetone-stimulated phosphatidyl choline
breakdown was incomplete. It has been proposed that this organic solvent
stimulation of yeast phospholipases might have a physiological role in the
disposal of alcohol, a major metabolic product accumulating during growth
in high glucose. Intracellular alcohol is toxic to the yeast and can be dis-
posed of by release into the medium. It was suggested that accumulated
alcohol may stimulate the phospholipases and modify the protoplast
membrane in such a way as to facilitate its disposal.

Abdulla and Davison (1965) have demonstrated an inhibition of yeast
succinate dehydrogenase by phospholipase A. This was prevented if the
phospholipase was inhibited with EDTA. In the course of these studies the
formation of a phospholipid with succinate esterified at the C-2 position
was suggested. In this system the succinate of a chemically synthesized
succinylphosphatidyl choline underwent oxidation which was not inhibited
by phospholipase A. The site of inhibition by phospholipase A was therefore
considered to be at the level of formation of the succinyl phospholipid.
Letters (1968) suggested that these observations could explain the decrease
in succinate dehydrogenase activity observed by Schatz (1963) after
transfer of yeast from an aerobic to an anaerobic environment, since his
results indicated that phospholipase activity was probably increased under
anaerobic conditions and might well interfere with the succinate dehydrogenase in the manner described by Abdulla and Davison (1965). It is
surprising that no later work on this postulation has been reported.

VIII. Phospholipids and Membrane Function and Biogenesis

In considering the importance of phospholipids to membrane function
and synthesis, it should be kept in mind that the phospholipids may play a
role beyond that of being purely structural elements and may, for example,
act as modifiers of enzymic activity. We have chosen therefore to consider
the structural and nonstructural roles separately.

A. STRUCTURAL ROLES

The major structural roles of membranes are currently considered to
be (1) subdivision of the cell into discrete compartments and (2) formation

of a structural framework upon which enzyme systems may be assembled in a spatially ordered way. The phospholipid molecule is ideally suited to a role in membrane formation by virtue of the possession of both a long-chain hydrophobic region (which is responsible for the insolubility of these molecules in aqueous solutions, a clear prerequisite for the separation of aqueous phases) and a hydrophilic, charged, polar grouping. When dispersed in aqueous solutions, these molecules spontaneously form micelles or bilayers in which the hydrophobic regions are oriented toward one another and the polar groups are presented to the aqueous phase. Such bilayers have been extensively studied since they exhibit a number of the properties of membranes, such as free permeability to water, relative impermeability to simple cations, and high electrical capacitance. (See, for example, the review by T. E. Thompson and Henn, 1970.) Furthermore, such a bilayer arrangement may occur *in vivo* in biological membranes, as was proposed by Danielli and Davson in 1935. Such a model, in which a phospholipid bilayer is sandwiched between two protein layers, was originally proposed for all membranes. It is now recognized that such is probably not the case (Benson, 1964). While the membranes of myelin and the nuclear envelope may conform to the Danielli–Davson model, mitochondrial and chloroplast membranes probably have a more complex arrangement. For further details of proposed models, the reader is referred to such reviews as those of Korn (1966, 1968, 1969), Stoeckenius and Engelman (1969), Kavanau (1966), and Vanderkooi and Green (1970).

B. NONSTRUCTURAL FUNCTIONS

Much of the work carried out on nonstructural functions of phospholipids has involved mammalian rather than microbial eukaryotic systems, but similar nonfunctional roles are, in all probability, encountered also in the microbial systems. Phospholipids (in particular, cardiolipin) have been shown to restore the succinate cytochrome c reductase activity of lipid-depleted mitochondria and cytochrome oxidase complexes, possibly due to an involvement in the bonding of cytochrome c (Fleischer et al., 1962; Tzagoloff and MacLennan, 1965). The phospholipids were shown to have different effects on different respiratory complexes (Bruni and Racker, 1968), indicating that the phospholipid distribution is important to the optimal functioning of the mitochondrial respiratory process. This is further exemplified by studies on the purified components comprising and regulating the ATPase activity of the inner mitochondrial membrane. Phospholipids have been shown to be involved in the conferment of the oligomycin (or rutamycin) sensitivity upon the ATPase and also protect the enzyme against heat inactivation (Racker, 1967; Kagawa and Racker, 1966a, b; Bulos and Racker, 1968a, b). Phospholipids are required for the maximal activity of many other membrane-bound enzymes such as the calcium transport ATPase of sarcoplasmic reticulum (Martonosi et al., 1968), glucose 6-phosphatase (Duttera et al., 1968), β-hydroxybutyrate dehydrogenase (Sekuzu et al., 1963), NADH cytochrome c reductase (Jones and Wakil, 1967), and stearyl-CoA desaturase (Jones et al., 1969). The most precise information on the involvement of phospholipids in cell wall or membrane biosynthetic reactions is exemplified by the special role of phosphatidyl ethanolamine in the formation of the outer core of the Salmonella lipopolysaccharide, so elegantly studied by Rothfield and his colleagues (Rothfield and Romeo, 1971), and of undecaprenyl phosphate as a cofactor in the synthesis of peptidoglycan of Staphylococcus (Getz, 1970; Osborn, 1969). The special role of phosphatidyl glycerol in transport reactions of E. coli is also noteworthy in this connection (see review by Kaback, 1970). The precise nature of the interaction(s) remains to be elucidated. The physiological function of such interactions as the differential effects of particular phospholipids with various enzymes can currently only be speculated upon in very general terms, except in a very few cases. A full comprehension of such events may not be forthcoming until the molecular architecture of the membrane, particularly with respect to the relative positions of the phospholipids, has been established.

C. MEMBRANE BIOGENESIS

Since phospholipids are clearly important both structurally and functionally to the properties of membranes, they have, potentially at least, value as markers of membrane biogenesis and degradation. This is particularly true for lipids with a more or less specific membrane association, such as that of cardiolipin with the membranes of the mitochondrion. Specific associations of this kind are, however, the exception rather than the rule. Also, the capacity for exchange of lipids between membranes presents a further complicating factor. Nonetheless, the use of lipids as membrane markers has yielded, and will no doubt continue to yield, useful information concerning membrane biogenesis and turnover. In this connection the eukaryotic microbes have been particularly useful in studies of mitochondrial biogenesis. In 1963, Luck used choline-requiring mutants of Neurospora crassa to examine the question of whether mitochondria arose de novo or by division of preexisting mitochondria. After growth of the mutant on ^{3}H-choline to label the membranes, cells were transferred to nonradioactive choline media and allowed to grow for a further three generations. Radioautography of purified mitochondria, isolated from these cells by density gradient centrifugation, showed all the mitochondria to be labeled. Such a result provided strong evidence for the notion of growth and division of preexisting mitochondria rather than *de novo* synthesis; however, it did not furnish definitive proof for such a postulate since the possible exchange of labeled phospholipid between old and new mitochondria could not be ruled out. Subsequent studies (Luck, 1965a, b) showed that the content of choline and amino acids in the culture media regulated the mitochondrial phospholipid composition. Growth of the mutant in low choline media produced cells containing relatively dense mitochondria that were poor in phospholipid. The transfer of such cells to a medium rich in choline produced progressive changes in the mitochondria, which became steadily richer in phospholipid and consequently progressively less dense. The observation of this gradual change offered further support for the conclusion that mitochondria arise by growth and division of preexisting mitochondria.

Another eukaryotic microorganism which has been extensively studied with respect to mitochondrial biogenesis is the yeast *Saccharomyces cerevisiae*, since this particular organism has the ready capacity to modulate the mitochondrial activity, depending on the genetic and environmental condition. Growth under conditions of anaerobiosis or aerobic growth in the presence of large, repressive quantities of glucose leads to the production of cells with no or little mitochondrial respiratory activity. Growth in a nonfermentable substrate, such as ethanol, lactate, or glycerol, or a weakly repressive hexose, such as galactose, produces cells with a much greater respiratory competence (Slonimski, 1953; Yotsuyanagi, 1962; Ephrussi *et al.*, 1956; Utter *et al.*, 1967). Jakovcic *et al.* (1971) have shown that the cardiolipin content of such cells (greater than 90% of which is associated with the mitochondria) is correlated with, and is a good indicator of, the state of development of the mitochondrial membrane. The value of cardiolipin as a mitochondrial marker is clearly seen in the case of cytoplasmic mitochondrial mutants (petites), which, though deficient in any mitochondrial respiratory function, were found to respond to changes in media glucose content by changes in cardiolipin in a similar fashion to wild-type cells. The small though significant fluctuations in cardiolipin concentration relative to the rather large variations in respiratory activity in wild-type yeast grown under different conditions suggests that cardiolipin is an important component of the "scaffold" membrane of the inner mitochondrion, into which respiratory elements may be separately and independently inserted. In other words, the data on cardiolipin content of anaerobic and glucose-repressed yeast support the proposition that in yeast there exist mitochondrial structures (promitochondria) (Criddle and Schatz, 1969) which do not contain a complete and functional complement of respiratory and phosphorylative components but which can be rendered fully functional by the later addition of those not present.

Amoebae have also been used to study phospholipid and membrane formation. For example, Chlapowski and Band (1971a) examined the labeling pattern of *Acanthamoeba palestinensis* membranes obtained by using either ^{14}C-choline or ^{3}H-glycerol as membrane precursors. They observed that the turnover rate of ^{14}C-choline labeled phosphatidyl choline varied with the concentration of nonradioactive choline present in the growth medium. Also, ^{14}C-choline could be incorporated into phosphatidyl choline by cell-free microsomal suspensions, suggesting that a choline exchange reaction occurs both in cell free systems and the intact cell. Although the nature of this exchange reaction was not specifically examined, it is clear that, for this particular organism of least, radioactive choline is not a suitable marker with which to examine membrane synthesis or turnover. This is particularly so for chase experiments involving the subsequent addition of large amounts of nonlabeled choline. By contrast, ^{3}H-glycerol was found to be a much more specific membrane label. Although neutral lipids accounted for about 33% of the total ^{3}H-glycerol incorporated, these were in the form of lipid droplets. The ^{3}H-glycerol in the membranes was almost entirely in the phospholipid.

In a subsequent study (Chlapowski and Band, 1971b) in which ^{3}H-glycerol was used as a membrane precursor for *Acanthamoeba palestinensis* the major sites of phospholipid synthesis appeared to be the rough endoplasmic reticulum and the nuclear membrane. It was hypothesized that newly made phospholipids were then transported to the Golgi membranes and then to a structure which the authors termed collapsed vesicles which seemed to act as plasma membrane precursors. It was also suggested that collapsed vesicles could be formed from food vacuoles so that these membranes were then returned to the cell surface.

In addition to these experiments, the accumulating evidence for factors catalysing the *in vitro* exchange of phospholipids between different cellular membranes (McMurray and Dawson, 1969; Wirtz and Zilversmit, 1968) emphasizes the need to choose the precursor to be used in phospholipid turnover studies with considerable care. Indeed, if all of the exchanges demonstrable *in vitro* occur *in vivo* then unless accurate assessment of the effects of such reactions can be made the correct interpretation of the results may be very difficult.

The very rapid formation of a limiting membrane structure following injury of *Amoeba proteus* has been investigated electron-microscopically

(Szubinska, 1971). Amoebae which had been speared with fine glass micro-needles and then rapidly fixed (35–45 secs) in the presence of ruthenium violet were found to be almost surrounded by an extra trilaminar structure. This trilaminar structure strikingly resembled the plasma membrane, from which it was separated by less than one micron. The formation of this structure was calcium-dependent and apparently involved small, darkly staining, dense droplets which were observed throughout the interior of the undamaged control cells. The formation of this "new membrane" structure was considered to be a protection mechanism for this type of free-living cell and to provide a rapid repair mechanism in case of damage to the delicate plasma membrane. It was suggested that the dense droplets represent some form of membrane precursor which can rapidly form a new membrane if the cell is injured. However, the nature of the dense droplets and of the process by which the trilaminar structure is formed so far remain speculative. The isolation and study of these droplets is of obvious interest in view of the possibility that they represent a form of membrane precursor. Clearly, however, it would be necessary to carry out any isolation procedures under conditions where new membrane would not form spontaneously, perhaps by using chelating agents to provide a calcium-free system. Alternatively, the use of an isolation medium which closely resembles the cell sap may allow for the purification of these droplets in an unaltered form.

One other potential approach to the study of the role of phospholipids in membranes, and one which has been little exploited, is that of the control of phospholipid biosynthesis. Such studies seem especially important since they are almost certainly relevant to the question of how cells regulate and control the amount of particular membrane they contain. Questions such as how does a cell recognize the need for more membrane, how is a require-ment for more membrane manifest within the cell at the molecular level, how does the cell "know" when sufficient membrane has been made, and how does the cell stop the production of excess membrane are of obvious importance. Very few studies concerning the control of phospholipid metabolism have appeared in the literature, and as yet such questions are largely unanswered. As has been mentioned earlier, Mangnall and Getz (1971) have shown that the formation by yeast mitochondria of CDP-diglyceride, a presumed intermediate in the synthesis of cardiolipin, can be modulated by a number of metabolic intermediates. In particular, citrate has a stimulatory effect on the enzyme, while cardiolipin inhibits CDP-diglyceride production. Furthermore, the capacity to respond to these modifiers changes under physiological conditions, that is, during the derepression of glucose-grown cells and development of mitochondrial respiratory function following the transfer of such cells to a lactate media

(Mangnall and Getz, 1972). The possibility exists, therefore, that the signal for cardiolipin biosynthesis to begin is an increase in the mitochondrial citrate content and that the synthesis is turned off when a certain cardio-lipin-to-enzyme ratio is reached. Such a scheme is undoubtedly far too simple, but nevertheless may represent a small step toward a further understanding of how the regulation of phospholipid metabolism may relate to the needs of the cell for membrane biosynthesis. The purification and characterization of the enzymes involved in the synthesis of specific membrane phospholipids and, more importantly, an understanding of the regulatory features of such enzymes will undoubtedly lead to a much greater comprehension of the problem of membrane assembly. The complex regulations which are responsible for the maintenance of a fairly constant and characteristic membrane phospholipid composition and phospholipid-to-protein ratio must be at least partially accounted for on the basis of the properties and regulations of the enzymes catalyzing phospholipid biosynthesis and degradation. The intricate pattern of synthesis of various phospholipids through the cell cycle, revealed by studies of synchronized yeast, suggests that there is a great deal to be learned in this area.

Acknowledgments

Work of the authors has been supported by grants from USPHS, #G.M. 13048; The the American Cancer Society, VC-67B; and the Chicago Heart Association, A 71-36 and F 70-5.

References

Abdulla, Y. H., and Davison, A. N. (1965). *Biochem. J.* **96,** 100.
Agranoff, B. W., and Hajra, A. K. (1971). *Proc. Nat. Acad. Sci. U.S.* **68,** 411.
Angus, W. W., and Lester, R. L. (1971). *Fed. Proc., Fed. Amer. Soc. Exp. Biol.,* **30,** 1244.
Ansell, G. B., and Hawthorne, J. N. (1964). "Phospholipids," *Biochim. Biophys. Acta* Vol. 3, Elsevier Publishing Company, New York.
Bangham, A. D., and Dawson, R. M. C. (1959). *Biochem. J.* **72,** 486.
Bangham, A. D., and Dawson, R. M. C. (1960). *Biochem. J.* **75,** 133.
Benson, A. A. (1964). *Annu. Rev. Plant Physiol.* **15,** 1.
Bianchi, D. E., and Turian, G. (1967). *Nature (London)* **214,** 1344.
Borkenhagen, L. F., Kennedy, E. P., and Fielding, L. (1961). *J. Biol. Chem.* **236,** PC 28.
Bowers, B., and Korn, E. D. (1969). *J. Cell Biol.* **41,** 786.
Bruni, A., and Racker, E. (1968). *J. Biol. Chem.* **243,** 962.
Bulos, B., and Racker, E. (1968a). *J. Biol. Chem.* **243,** 3891.
Bulos, B., and Racker, E. (1968b). *J. Biol. Chem.* **243,** 3901.
Carter, H. E., and Gaver, R. C. (1967). *Biochem. Biophys. Res. Commun.* **29,** 886.

Carter, J. R., Jr. (1968). *J. Lipid Res.* **9,** 748.

Cerbón, J. (1969). *J. Bacteriol.* **97,** 658.

Cerbón, J. (1970). *J. Bacteriol.* **102,** 97.

Challinor, S. W., and Daniels, N. W. R. (1955). *Nature (London)* **176,** 1267.

Chlapowski, F. J., and Band, R. N. (1971a). *J. Cell Biol.* **50,** 625.

Chlapowski, F. J., and Band, R. N. (1971b). *J. Cell Biol.* **50,** 634.

Chou, S. C., and Sherbaum, O. H. (1965). *Exp. Cell Res.* **40,** 217.

Chou, S. C., and Sherbaum, O. H. (1966). *Exp. Cell Res.* **45,** 31.

Coleman, G. S., Kemp, P., and Dawson, R. M. C. (1971). *Biochem. J.* **123,** 97.

Criddle, R. S., and Schatz, G. (1969). *Biochemistry* **8,** 322.

Crocken, B. J., and Nyc, J. F. (1964). *J. Biol. Chem.* **239,** 1727.

Danielli, J. F., and Davson, H. (1935). *J. Cell. Comp. Physiol.* **5,** 495.

Davidoff, F., and Korn, E. D. (1962). *Biochem. Biophys. Res. Commun.* **9,** 54.

Davidoff, F., and Korn, E. D. (1963). *J. Biol. Chem.* **238,** 3199.

Dawson, R. M. C. (1958a). *Biochem. J.* **68,** 352.

Dawson, R. M. C. (1958b). *Biochem. J.* **70,** 559.

Dawson, R. M. C., and Hauser, H. (1967). *Biochim. Biophys. Acta* **137,** 518.

Dawson, R. M. C., and Kemp, P. (1967). *Biochem. J.* **105,** 837.

Deierkauf, F. A., and Booij, H. L. (1968). *Biochim. Biophys. Acta* **150,** 214.

Delo, J., Ernst-Fonberg, M. L., and Bloch, K. (1971). *Arch. Biochem. Biophys.* **143,** 384.

Diamond, R. J., and Rose, A. H. (1971). Unpublished observations (cited by Hunter and Rose, 1971, p. 211).

Dittmer, J. C., and Wells, M. A. (1969). *In* "Methods in Enzymology" (J. M. Lowenstein, ed.), Vol. 14, p. 482. Academic Press, New York.

Duttera, S. M., Byrne, W. L., and Ganoza, M. C. (1968). *J. Biol. Chem.* **243,** 2216.

Echlin, P., and Morris, I. (1965). *Biol. Rev.* **40,** 143.

Eckstein, H., Paduch, V., and Hilz, H. (1967). *Eur. J. Biochem.* **3,** 224.

Ephrussi, B., Slonimski, P. P., Yotsuyanagi, Y., and Tavlitzki, J. (1956). *C. R. Trav. Lab. Carlsberg* **26,** 87.

Ernst-Fonberg, M. L., and Bloch, K. (1971). *Arch. Biochem. Biophys.* **143,** 392.

Erwin, J. A. (1968). *In* "The Biology of Euglena" (D. E. Buetow, ed.), Vol. 2, p. 133. Academic Press, New York.

Erwin, J. A. (1970). *Biochim. Biophys. Acta* **202,** 21.

Fairbairn, D. (1948). *J. Biol. Chem.* **173,** 705.

Fleischer, S., Brierley, G., Klouwen, H., and Slautterback, D. B. (1962). *J. Biol. Chem.* **237,** 3264.

Fleischer, S., Fleischer, B., Rouser, G., Casu, A., and Kritchevsky, G. (1967). *J. Lipid Res.* **8,** 170.

Folch, J., Lees, M., and Sloane-Stanley, G. H. (1957). *J. Biol. Chem.* **226,** 497.

Forrester, I. T., Watson, K., and Linnane, A. W. (1971). *Biochem. Biophys. Res. Commun.* **43,** 409.

Gailey, F. B., and Lester, R. L. (1968). *Fed. Proc., Fed. Amer. Soc. Exp. Biol.* **27,** 458.

Getz, G. S. (1970). *Advan. Lipid Res.* **8,** 175.

Getz, G. S., Bartley, W., Lurie, D., and Notton, B. M. (1968). *Biochim. Biophys. Acta* **152,** 325.

Getz, G. S., Jakovcic, S., Heywood, J., Frank, J., and Rabinowitz, M. (1970). *Biochim. Biophys. Acta* **218,** 441.

Glaser, M., Simkins, H., Singer, S. J., Sheetz, M., and Chan, S. I. (1970). *Proc. Nat. Acad. Sci. U.S.* **65,** 721.

Godfrey, D. G. (1967). *Exp. Parasitol.* **20,** 106.

Hajra, A. K., and Agranoff, B. W. (1968). *Fed. Proc., Fed. Amer. Soc. Exp. Biol.* **27,** 458.

Hall, M. O., and Nyc, J. F. (1962). *Biochim. Biophys. Acta* **56,** 370.

Harris, R. V., Wood, B. J. B., and James, A. T. (1965). *Biochem. J.* **94,** 22P.

Harrison, J. S., and Trevelyan, W. E. (1963). *Nature (London)* **200,** 1189.

Hill, E. E., Husbands, D. R., and Lands, W. E. M. (1968). *J. Biol. Chem.* **243,** 4440.

Hoffmann-Ostenhof, O., Geyer-Fenzl, M., and Wagner, E. (1961). *In* "The Enzymes of Lipid Metabolism" (P. Desnuelle, ed.), p. 39. Pergamon, Oxford.

Holton, R. W., Blecker, H. H., and Onore, M. (1964). *Phytochemistry* **3,** 595.

Holub, B. J., and Kuksis, A. (1971). *J. Lipid Res.* **12,** 510.

Hostetler, K. Y., and Van Den Bosch, H. (1972). *Biochim. Biophys. Acta* **260,** 380.

Hulanicka, D., Erwin, J., and Bloch, K. (1964). *J. Biol. Chem.* **239,** 2778.

Hunter, K., and Rose, A. H. (1971). *In* "The Yeasts" (A. H. Rose and J. S. Harrison, eds.), Vol. 2, p. 211. Academic Press, New York.

Hutchison, H. T., and Cronan, J. E., Jr. (1968). *Biochim. Biophys. Acta* **164,** 606.

Jakovcic, S., Getz, G. S., Rabinowitz, M., Jakob, H., and Swift, H. (1971). *J. Cell Biol.* **48,** 490.

Johnston, J. M., and Paltauf, F. (1970). *Biochim. Biophys. Acta* **218,** 431.

Jonah, M., and Erwin, J. A. (1971). *Biochim. Biophys. Acta* **231,** 80.

Jones, P. D., and Wakil, S. J. (1967). *J. Biol. Chem.* **242,** 5267.

Jones, P. D., Holloway, P. W., Peluffo, R. O., and Wakil, S. J. (1969). *J. Biol. Chem.* **244,** 744.

Kaback, H. R. (1970). *Annu. Rev. Biochem.* **39,** 561.

Kapoulas, V. M., and Thompson, G. A., Jr. (1969). *Biochim. Biophys. Acta* **187,** 594.

Kapoulas, V. M., Thompson, G. A., Jr., and Hanahan, D. J. (1969). *Biochim. Biophys. Acta* **176,** 237.

Kates, M. (1957). *Can. J. Biochem.* **35,** 127.

Kates, M., Madeley, J. R., and Beare, J. L. (1965). *Biochim. Biophys. Acta* **106,** 630.

Kavanau, J. L. (1966). *Fed. Proc., Fed. Amer. Soc. Exp. Biol.* **25,** 1096.

Kagawa, Y., and Racker, E. (1966a). *J. Biol. Chem.* **241,** 2461.

Kagawa, Y., and Racker, E. (1966b). *J. Biol. Chem.* **241,** 2467.

Keith, A. D., Resnick, M. R., and Haley, A. B. (1969). *J. Bacteriol.* **98,** 415.

Kemp, P., and Dawson, R. M. C. (1969a). *Biochim. Biophys. Acta* **176,** 678.

Kemp, P., and Dawson, R. M. C. (1969b). *Biochem. J.* **113,** 555.

Kennedy, K. E., and Thompson, G. A., Jr. (1970). *Science* **168,** 989.

Kimelberg, H. K., and Lee, C. P. (1970). *J. Membrane Biol.* **2,** 252.

Kimelberg, H. K., Lee, C. P., Claude, A., and Mrena, E. (1970). *J. Membrane Biol.* **2,** 235.

King, L. (1971). Ph.D. Thesis, University of Chicago, Chicago.

Kokke, R., Hooghwinkel, G. J. M., Booij, H. L., Van Den Bosch, H., Zelles, L., Mulder, E., and Van Deenen, L. L. M. (1963). *Biochim. Biophys. Acta* **70,** 351.

Korn, E. D. (1966). *Science* **153,** 1491.

Korn, E. D. (1968). *J. Gen. Physiol.* **52,** 257.

Korn, E. D. (1969). *Annu. Rev. Biochem.* **38,** 263.

Kuhn, N. J., and Lynen, F. (1965). *Biochem. J.* **94,** 240.

Lamb, G. R., and Fallon, H. J. (1970). *J. Biol. Chem.* **245,** 3075.

Lester, R. L., and Steiner, M. R. (1968). *J. Biol. Chem.* **243,** 4889.

Letters, R. (1966). *Biochim. Biophys. Acta* **116,** 489.

Letters, R. (1968). *Bull. Soc. Chim. Biol.* **50**, 1385.

Letters, R., and Snell, B. K. (1963). *J. Chem. Soc., London* p. 5127.

Levin, B., Bartzis, H., and Getz, G. S. (1971). Unpublished study.

Levy, M. R., Gollon, C. E., and Elliot, A. M. (1969). *Exp. Cell Res.* **55**, 295.

Lewin, L. M. (1965). *J. Gen. Microbiol.* **41**, 215.

Liang, C. R., and Rosenberg, H. (1966). *Biochim. Biophys. Acta* **125**, 548.

Liang, C. R., and Rosenberg, H. (1968). *Biochim. Biophys. Acta* **156**, 437.

Longley, R. P., Rose, A. H., and Knights, B. A. (1968). *Biochem. J.* **108**, 401.

Luck, D. J. L. (1963). *J. Cell Biol.* **16**, 483.

Luck, D. J. L. (1965a). *J. Cell Biol.* **24**, 445.

Luck, D. J. L. (1965b). *J. Cell Biol.* **24**, 461.

Lukins, H. B., Jollow, D., Wallace, P. G., and Linnane, A. W. (1968). *Aust. J. Exp. Biol. Med. Sci.* **46**, 651.

Lust, G., and Daniel, L. J. (1964). *Arch. Biochem. Biophys.* **108**, 414.

Lust, G., and Daniel, L. J. (1966). *Arch. Biochem. Biophys.* **113**, 603.

McCaman, R. E., and Finnerty, W. R. (1968). *J. Biol. Chem.* **243**, 5074.

McMurray, W. C., and Dawson, R. M. C. (1969). *Biochem. J.* **112**, 91.

Mangnall, D., and Getz, G. S. (1971). *Fed. Proc., Fed. Amer. Soc. Exp. Biol.* **30**, 1226.

Mangnall, D., and Getz, G. S. (1970). Unpublished data.

Mangnall, D., and Getz, G. S. (1972). In preparation.

Martonosi, A., Donley, J., and Halpin, R. A. (1968). *J. Biol. Chem.* **243**, 61.

Matile, Ph. (1970). *FEBS Symp.* **20**, 39.

Mead, J. F. (1961). *Fed. Proc., Fed. Amer. Soc. Exp. Biol.* **20**, 952.

Merdinger, E. (1969). *J. Bacteriol.* **98**, 1021.

Nagai, J., and Bloch, K. (1966). *J. Biol. Chem.* **241**, 1925.

Nagai, J., and Bloch, K. (1967). *J. Biol. Chem.* **242**, 357.

Nichols, B. W. (1968). *Lipids* **3**, 354.

Nichols, B. W., Harris, R. V., and James, A. T. (1965). *Biochem. Biophys. Res. Commun.* **20**, 256.

Nozawa, Y., and Thompson, G. A., Jr. (1971a). *J. Cell Biol.* **49**, 712.

Nozawa, Y., and Thompson, G. A., Jr. (1971b). *J. Cell Biol.* **49**, 722.

Nyns, E. J., Chiang, N., and Wiaux, A. L. (1968). *Antonie von Leeuwenhoek; J. Microbiol. Serol.* **34**, 197.

Osborn, M. J. (1969). *Annu. Rev. Biochem.* **38**, 501.

Ostrow, D. (1971). *Fed. Proc., Fed. Amer. Soc. Exp. Biol.* **30**, 1226.

Paltauf, F., and Johnston, J. M. (1970). *Biochim. Biophys. Acta* **218**, 424.

Paltauf, F., and Schatz, G. (1969). *Biochemistry* **8**, 335.

Paulus, H., and Kennedy, E. P. (1960). *J. Biol. Chem.* **235**, 1303.

Pieringer, R. A., and Kunnes, R. S. (1965). *J. Biol. Chem.* **240**, 2833.

Proudlock, J. W., Haslam, J. M., and Linnane, A. W. (1969). *Biochem. Biophys. Res. Commun.* **43**, 409.

Racker, E. (1967). *Fed. Proc., Fed. Amer. Soc. Exp. Biol.* **26**, 1335.

Racker, E. (1970). *In* "Membranes of Mitochondria and Chloroplasts" (E. Racker, ed.), p. 127. Van Nostrand-Reinhold, Princeton, New Jersey.

Radin, N. S. (1969). *In* "Methods in Enzymology" (J. M. Lowenstein, ed.), Vol. 14, p. 245. Academic Press, New York.

Renkonen, O., and Bloch, K. (1969). *J. Biol. Chem.* **244**, 4899.

Rosenbaum, N., Erwin, J., Beach, D., and Holz, G. G., Jr. (1966). *J. Protozool.* **13**, 535.

Rosenberg, H. (1964). *Nature (London)* **203**, 299.

Rothfield, L., and Romeo, D. (1971). *Bacteriol. Rev.* **35,** 14.

Rouser, G., and Fleischer, S. (1967). *In* "Methods in Enzymology" (R. W. Estabrook and M. E. Pullman, eds.), Vol. 10, p. 385. Academic Press, New York.

Rouser, G., Kritchevsky, G., Heller, D., and Lieber, E. (1963). *J. Amer. Oil Chem. Soc.* **40,** 425.

Rouser, G., Nelson, G. L., Fleischer, S., and Simon, G. (1968). *In* "Biological Membranes" (D. Chapman, ed.), p. 5. Academic Press, New York.

Rouser, G., Kritchevsky, G., Yamamoto, A., Simon, G., Galli, C., and Bauman, A. J. (1969). *In* "Methods in Enzymology" (J. M. Lowenstein, ed.), Vol. 14, p. 272. Academic Press, New York.

Sawyer, M. K., Bischoff, J. M., Guidry, M. A., and Reeves, R. E. (1967). *Exp. Parasitol.* **20,** 295.

Scarborough, G. A., and Nyc, J. F. (1967a). *J. Biol. Chem.* **242,** 238.

Scarborough, G. A., and Nyc, J. F. (1967b). *Biochim. Biophys. Acta* **146,** 111.

Schatz, G. (1963). *Biochem. Biophys. Res. Commun.* **12,** 448.

Segal, W. (1965). *Nature (London)* **208,** 1284.

Sekuzu, I., Jurtshuk, P., Jr., and Green, D. E. (1963). *J. Biol. Chem.* **238,** 975.

Shafai, T., and Lewin, L. M. (1968). *Biochim. Biophys. Acta* **152,** 787.

Sherr, S. I. (1969). *Bacteriol. Proc.* p. 120.

Sherr, S. I., and Byk, C. (1971). *Biochim. Biophys. Acta* **239,** 243.

Skipski, V., and Barclay, M. (1969). *In* "Methods in Enzymology" (J. M. Lowenstein, ed.), Vol. 14, p. 530. Academic Press, New York.

Slonimski, P. P. (1953). "Formation des enzymes respiratoires chez la levure." Masson, Paris.

Smith, J. D., and Law, J. H. (1970a). *Biochim. Biophys. Acta* **202,** 141.

Smith, J. D., and Law, J. H. (1970b). *Biochemistry* **9,** 2152.

Snyder, F., Wykle, R. L., and Malone, B. (1969). *Biochem. Biophys. Res. Commun.* **34,** 315.

Soldo, A. T., and Van Wagtendonk, J. (1967). *J. Protozool.* **14,** 596.

Steiner, M. R., and Lester, R. L. (1970). *Biochemistry* **9,** 63.

Steiner, M. R., and Lester, R. L. (1972). *Biochim. Biophys. Acta* **260,** 222.

Steiner, S., and Lester, R. L. (1972). *J. Bacteriol.* **109,** 81.

Steiner, S., Smith, S., Waechter, C. J., and Lester, R. L. (1969). *Proc. Nat. Acad. Sci. US* **64,** 1042.

Stodola, F. H., Deinema, M. H., and Spencer, J. F. T. (1967). *Bacteriol. Rev.* **31,** 194.

Stoeckenius, W., and Engelman, D. M. (1969). *J. Cell Biol.* **42,** 613.

Suomalainen, H., and Nurminen, T. (1970). *Chem. Phys. Lipids* **4,** 247.

Swift, H., and Wolstenholme, D. R. (1969). *In* "Handbook of Molecular Cytology" (A. Lima-De-Faria, ed.), p. 972. North-Holland Publ., Amsterdam.

Swift, H., Rabinowitz, M., and Getz, G. S. (1967). *J. Cell Biol.* **35,** 131A.

Swift, H., Rabinowitz, M., and Getz, G. S. (1968). *In* "Biochemical Aspects of the Biogenesis of Mitochondria" (E. C. Slater *et al.,* eds.), p. 3. Adriatica Editrice, Bari.

Szubinska, B. (1971). *J. Cell Biol.* **49,** 747.

Tanner, W. (1968). *Arch. Mickrobiol.* **64,** 158.

Tauro, P., and Halvorson, H. O. (1966). *J. Bacteriol.* **92,** 652.

Tauro, P., Schweizer, E., Epstein, R., and Halvorson, H. O. (1969). *In* "The Cell Cycle" (G. M. Padilla, G. L. Whitson, and I. L. Cameron, eds.), p. 101. Academic Press, New York.

Thompson, G. A., Jr. (1967). *Biochemistry* **6**, 2015.

Thompson, G. A., Jr. (1969). *Biochim. Biophys. Acta* **176**, 330.

Thompson, G. A., Jr., and Kapoulas, V. M. (1969). *In* "Methods in Enzymology" (J. M. Lowenstein, ed.), Vol. 14, p. 668. Academic Press, New York.

Thompson, T. E., and Henn, F. A. (1970). *In* "Membranes of Mitochondria and Chloroplasts" (E. Racker, ed.), p. 1. Van Nostrand-Reinhold, Princeton, New Jersey.

Tipton, C. L., and Swords, M. D. (1966). *J. Protozool.* **13**, 469.

Trebst, H., and Geike, F. (1967). *Z. Naturforsch. B* **22**, 989.

Trevelyan, W. E. (1968). *J. Inst. Brew., London* **74**, 365.

Trump, B. F., Duttera, S. M., Byrne, W. L., and Arstila, A. U. (1970). *Proc. Nat. Acad. Sci. U.S.* **66**, 433.

Tzagoloff, A., and MacLennan, D. H. (1965). *Biochim. Biophys. Acta* **99**, 476.

Ulsamer, A. G., Smith, F. R., and Korn, E. D. (1969). *J. Cell Biol.* **43**, 105.

Utter, M. F., Duell, E. A., and Bernofsky, C. (1967). *In* "Aspects of Yeast Metabolism" (A. K. Mills, and H. Krebs, eds.), p. 197. Paris, Philadelphia, Pennsylvania.

Van Deenen, L. L. M., and Haverkate, F. (1966). *In* "Biochemistry of the Chloroplasts" (T. W. Goodwin, ed.), Vol. 1, p. 117. Academic Press, New York.

Van Den Bosch, H., Van Der Elzen, H. M., and Van Deenen, L. L. M. (1967). *Lipids* **2**, 279.

Vanderkooi, G., and Green, D. E. (1970). *Proc. Nat. Acad. Sci. U.S.* **66**, 615.

Vignais, P. M., Nachbaur, J., Huet, J., and Vignais, P. V. (1970). *Biochem. J.* **116**, 42P.

Waechter, C. J., Steiner, M. R., and Lester, R. L. (1969). *J. Biol. Chem.* **244**, 3419.

Wagner, H., and Zofcsik, W. (1966a). *Biochem. Z.* **346**, 333.

Wagner, H., and Zofcsik, W. (1966b). *Biochem. Z.* **346**, 343.

Warren, W. A. (1968). *Biochim. Biophys. Acta* **156**, 340.

Wells, M. A., and Dittmer, J. C. (1965). *Biochemistry* **4**, 2459.

White, G. L., and Hawthorne, J. N. (1970). *Biochem. J.* **117**, 203.

Williamson, D. H., and Moustacchi, E. (1971). *Biochem. Biophys. Res. Commun.* **42**, 195.

Wirtz, K. W. A., and Zilversmit, D. B. (1968). *J. Biol. Chem.* **243**, 3596.

Wisnieski, B. J., Keith, A. D., and Resnick, M. R. (1970). *J. Bacteriol.* **101**, 160.

Wuthier, R. (1966). *J. Lipid Res.* **7**, 558.

Wykle, R. L., and Snyder, F. (1970). *J. Biol. Chem.* **245**, 3047.

Yotsuyanagi, Y. (1962). *J. Ultrastruct. Res.* **7**, 121.

CHAPTER 4

Sulfolipids and Halosulfolipids

Thomas H. Haines

I. Introduction

For one hundred years sulfolipids have been known to be associated with mammalian systems, especially brain (Thudichum, 1874). Their occurrence in eukaryotic microorganisms was first described, however, only ten years ago (Benson *et al.*, 1959a), and since this first report they have been reported in all photosynthetic plants (including algae and plankton), protozoa, fungi, bacteria, insects, and invertebrates. Like the phospholipids with which they are generally associated, the sulfolipids have a wide variety of structures only one of which includes glycerol in its formulation. Also like the phospholipids, the sulfur atom is found as both a sulfate ester $(RCH_2OSO_3^-)$ and a sulfonate $(RCH_2SO_3^-)$. Although the former has been found in a variety of chemical forms, the latter has been described only in the glycolipid 6-deoxyhexose-6-sulfonate. In the case of the chloroplast sulfolipid, this hexose is glucose.

All membrane preparations that have been analyzed with respect to their lipid composition have been found to contain phospholipids. In contrast, each sulfolipid has been reported as one component of a particular membrane. This uniqueness has provided a tool for exploring the biosynthesis and metabolism of one membrane of an organism in the presence of the other membranes.

For the most part the current status of sulfolipids in eukaryotic microorganisms is that of establishing the structures and localizing the substances within the organelles of the cell. Very little has been done on intermediary metabolism, largely because the structures themselves are so new. Although the phospholipids are generally present in higher concentrations than sulfolipids in tissues, it is nonetheless surprising that the structures of most of the phosphatides had been well established before any of the sulfolipids (except for brain sulfatide) were even discovered. There are two major reasons for this, and since these reasons aid in our understanding of the sulfolipids, we shall explore them.

The first and most significant reason for the delay in the discovery of new sulfolipids was the absence of a reliable and sensitive colorimetric test for sulfate comparable to the molybdate test for phosphate. Until the introduction of radioisotopes, the only reliable means for the determination of sulfate was the gravimetric barium sulfate method. The appearance of radioisotopes and chromatography, especially thin-layer chromatography, set the stage for the rapid discovery of sulfolipids.

In addition to the difficulties of analysis and isolation, the physical properties of the sulfolipids have retarded their discovery. The sulfolipids are more polar than the phospholipids or glycolipids. Although sulfatides are usually extracted from tissue by chloroform-methanol mixtures, they are usually sufficiently water soluble to remain in the aqueous phase in an ether-water system. When there is an interfacial fluff, it often contains a high proportion of these substances. Furthermore, in the procedure of Folch *et al.* (1957), the sulfolipids generally are completely extracted into the polar phase. Many investigators working on eukaryotic lipids use the Folch procedure and *discard this phase.*

The recent recognition of the significance of membranes and the emergence of polar lipids as key constituents of membranes have sparked considerable interest in the sulfolipids, as they have the other lipids in this volume. All the work on the sulfolipids of the eukaryotes has been motivated by this consideration, coupled with the emergence of a sensitive colorimetric assay for sulfolipids (Kean, 1968). There is every reason to indicate that this trend will continue.

A review by Goldberg (1961) has been preempted by radical changes in our knowledge of the sulfolipids. Even the structure of cerebroside sulfate has since been corrected (Yamakawa *et al.*, 1962; Stoffyn and Stoffyn, 1963). The plant sulfolipid was reviewed by Benson (1963). The chemistry of all the known sulfolipids was reviewed by Haines (1971). A major portion of the latter review is devoted to hydrolysis, analysis (including physical methods), and synthesis of sulfate esters and sulfonic acids.

This chapter will contain an extensive discussion of the two sulfolipid types that have been well described: the plant sulfonolipid (6-sulfo-O-α-quinovosyl-(1 → 1)-glycerol diglyceride) and the halosulfatides (polyhalo derivatives of 1,14-docosane disulfate and 1,15-tetracosane disulfate). Mention will be made of other eukaryotic sulfolipids that have been discovered but not yet characterized. The two eukaryotic sulfolipids that have been well described have turned out to be radically different from any previously known lipids, and the characterization of the reported but uncharacterized eukaryotic sulfolipids may well turn out to be as interesting.

II. Nomenclature

The term *sulfolipid* was first used from the earliest reports of Thudichum (1874) to describe the sulfur-containing lipid in brain. This substance, cerebroside sulfuric acid, was first characterized by Blix (1933). Because the material is a sulfate ester of galactose, it has also been referred to as sulfatide. Sulfolipids (including the plant sulfolipid) have been discovered that are *not* sulfatides, as the sulfur is in the form of a sulfonic acid (C—S bond) and not a sulfate ester (C—O—S bonds). Daniel *et al.* (1961) have suggested that the term sulfolipid be used to denote a sulfonate lipid. Baer and Stanacev (1964), on the other hand, have suggested that the term *phosphonolipid* be used to denote phospholipids in which the phosphorus occurs as a phosphoric acid. Furthermore, the term sulfolipid has been used for many years in the literature for cerebroside sulfate and is continuing to be so used (Stoffyn, 1966). Nonetheless, some sharpening of nomenclature is obviously necessary, as sulfolipids of diverse structure are appearing each year.

The following nomenclature has been suggested (Haines, 1971) to clarify the situation:

1. The term *sulfolipid:* to denote any sulfur-containing lipid.

2. The term *sulfatide:* to identify a sulfolipid in which the sulfur occurs as a sulfate ester.

3. The term *sulfonolipid:* (pronounced sŭl-fŏń-ō-lĭ-pĭd) in reference to sulfolipids that contain sulfur in the sulfonic acid form.

4. The term *thiolipid:* to denote sulfolipids with sulfur in the reduced form (Daniel *et al.*, 1961).

Although no lipid that contains the sulfoxide or sulfone states of sulfur has been reported, it is clear that *sulfoxolipid* and *sulfonolipid* can be used in reference to such compounds should they be identified.

This distinguishing nomenclature is based not only upon the obvious differences of chemical structure but upon the more significant differences of biochemistry. The formation of sulfatides is clearly through an entirely different biosynthetic route than that of the sulfonolipids (as well as the thiolipids), and their metabolic behavior should be radically different. These terms will be used throughout this chapter.

III. The Plant Sulfonolipid

The plant sulfonolipid was the first sulfolipids to be reported in eukaryotic microorganisms. Since its discovery over ten years ago by Benson and co-workers (1959a, 1960; Lepage *et al.*, 1961; Daniel *et al.*, 1961; Shibuya and Benson, 1961; Miyano and Benson, 1962a,b), this sulfolipid has been found in all green plants. There is now little doubt that the substance is primarily a chloroplast substituent in green plants, localized in the lamellae. Several lines of evidence suggest that its participation in photosynthesis is not merely as a structural component of chloroplast membrane, and this evidence will be discussed in Section III, E.

The sulfonolipid is not restricted to the higher plants and green algae, as it has been reported in red algae (Benson and Shibuya, 1962; Radunz, 1969), blue-green algae, brown algae, and purple bacteria (Radunz, 1969). Neither its function nor its localization is clear in these organisms. The sulfonolipid constitutes 14 to 18% of the lipids in the red, brown, and blue algae but only 4 to 5% of the lipids of isolated chloroplasts of green plants and algae. This suggests that it may be a component of membranes other than those associated with photosynthesis in the red, brown, and blue-green algae. It should also be noted that its fatty acid composition in these organisms is radically different from that of green plants (Radunz, 1969), even after considering the seasonal changes to which the fatty acid composition of sulfonolipid is subjected in higher plants (Klopfenstein and Shigley, 1967).

Because the plant sulfonolipid is concentrated in the chloroplast membrane of higher plants and green algae, it presents a unique tool to the researcher on membrane structure and function. Its structure has implications to the structure of the membrane, its presence in the membrane represents a tool for assaying one membrane in the presence of others, and control of its biosynthetic route would offer another handle for manipulating the membrane's biogenesis.

The plant sulfonolipid has not been reviewed since 1963 (Benson, 1963), although its chemistry has been reviewed in some detail by Haines (1971).

A. ISOLATION

The first structural studies of Benson et al. (1959a) did not require the isolation of substantial amounts of material, as the studies were conducted on ^{14}C-labeled material isolated by paper chromatography. Even subsequent stereochemical work (Miyano and Benson, 1962a,b) relied upon small quantities of radioactive material identified by cocrystallization with synthesized enantiomer. Lepage et al. (1961) conducted the first isolation of visible amounts by ion exchange chromatography. They note that fresh or dried alfalfa or clover are suitable sources, and most subsequent isolations have used these plants. They are normally poor sources for sulfonolipid. The highest reported concentration is that of a species of alfalfa that yielded 24% of its lipid as sulfonolipid when grown at 30°C (Kuiper, 1970). High concentrations of the sulfonolipid occur in red algae (14.9% of the lipid), brown algae (18.3%), and blue algae (13.9%) (Radunz, 1969). A good procedure suitable for the isolation of a reasonable quantity of rather pure sulfonolipid is that of O'Brien and Benson (1964). These authors use a rather cumbersome and elaborate procedure involving chromatography of the lipid extract of alfalfa or Chlorella pyrenoidosa on three successive columns. Nonetheless, their product is pure, and it represents 99% of the ^{35}S-sulfolipid placed on the first column. One gram of alfalfa lipid extract yielded 30 mg of sulfonolipid, whereas the same quantity of Chlorella lipid yielded only 19 mg. The authors' description of the procedures for preparing the supports and packing the columns is clear and easy to follow. The columns are Florisil, DEAE-cellulose, and silicic acid essentially that of Rouser et al. (1961, 1967)—and these original references should be read before conducting the isolation.

O'Brien and Benson (1964) also report that ^{35}S-sulfonolipid is distributed after complete equilibration in a chloroform-water system so that 64% of the radioactivity is in the chloroform phase and 36% in the aqueous phase. In a benzene-water system, 98% of the sulfonolipid remains in the aqueous system. Along with phosphatidyl inositol, the sulfonolipid appears to be the most polar lipid on chromatograms (Kates, 1960; Wintermans, 1960; Mumma and Benson, 1961). These data suggest that the Folch *et al.* (1957) extraction procedure might be effective for the preparation of a crude batch of sulfolipid, particularly if a chloroplast preparation rather than whole leaf is used as the source. It is suggested here that repeated washes of the lower phase with the *upper phase* of the Folch procedure would permit isolation of the sulfonolipid in the lower phase.

Several other column procedures have been reported in the literature. Use has been made of Florisil (Russell and Bailey, 1966) DEAE-cellulose (Nichols and James, 1964; Roughan and Batt, 1968; Allen *et al.*, 1966), and ECTEOLA-cellulose (Klopfenstein and Shigley, 1966).

Preparative thin-layer chromatography has proved useful for obtaining small amounts of pure sulfonolipid, which is particularly useful for doing fatty acid analyses of single spots on chromatograms (Pohl *et al.*, 1970; Klopfenstein and Shigley, 1966; O'Brien *et al.*, 1964). The method of Pohl *et al.* (1970) permits a fatty acid analysis of the sulfonolipid after a single thin-layer chromatogram of crude plant lipids. Sodium methoxide is the esterifying reagent.

B. STRUCTURAL STUDIES

The structure of the sulfonolipid was evolved by the Benson group, who discovered the substance in green plants (Benson *et al.*, 1960). An elegant combination of isotopes, paper chromatography, and chemistry allowed Benson *et al.* (1959a) to identify the substance as a glycerol lipid containing a hexose-6-sulfonate in glycosidic linkage to the glycerol. At first the sulfonolipid had one fatty acid esterified to the glycerol. It was later found by Yagi and Benson (1962) that this was produced during the isolation by an extremely active lipase from the natural diacyl sulfonolipid.

Using ion exchange resin chromatography, Lepage *et al.* (1961) isolated a sufficient quantity of the deacylated sulfonolipid to allow physical studies on the material. The glyceryl sulfoglycoside exhibited a molecular rotation, $(M)_D^{25}$, of $+31,000$ degrees, characteristic of alkyl-α-D-glucopyranosides (Daniel *et al.*, 1961). Further evidence for an α-glycoside was obtained from the nuclear magnetic resonance absorption of an anomeric equatorial proton. The rotational shift in Cupra B of -370 degrees indicated three adjacent equatorial hydroxyls, typical of glucosides (Lepage *et al.*, 1961). The complete structure of the glycerylsulfoquinovoside in-cluding the L configuration of the glycerol moiety was confirmed by an X-ray crystallographic analysis of its rubidium salt by Okaya (1964). The structure of the plant sulfonolipid is thus that shown in formula (I).

Chloroplast
sulfonolipid

(I)

Synthesis of the deacylated sulfonolipid with the correct configuration was achieved by Miyano and Benson (1962b). Its infrared spectrum is available (Haines, 1971), as is that of the intact sulfonolipid (Radunz, 1969). Several salts of sulfoquinovose have been prepared by Helferich and Ost (1963) and Lehmann and Benson (1964a,b).

The fatty acids were first identified as palmitic (43%) and α-linolenic (47%) in alfalfa (O'Brien and Benson, 1964). Subsequent work has shown that the condition of the source affects the fatty acid composition radically (Klopfenstein and Shifley, 1967). This will be discussed in more detail in Section III,E.

C. ANALYSIS

The analytical method used in the initial discovery of the sulfonolipid was paper chromatography of ^{35}S-labeled plant tissue. Chromatography on silicic-acid-impregnated paper in chloroform-methanol, 9:1 ($R_F = 0.4$), was used by O'Brien and Benson (1964). Thin-layer chromatography on silica gel using chloroform-methanol-acetic acid, 65:25:10 ($R_F = 0.7$) (Klopfenstein and Shigley, 1966), or acetone-benzene-water, 91:30:8 ($R_F = 0.4$) (Pohl *et al.*, 1970), has been useful. The latter system separates all the plant lipids remarkably well in a single chromatogram. Anion ex-change paper has also been used (Mumma and Benson, 1961).

Anthrone has become a unique assay reagent for the sulfonolipid. Wintermans (1960) and subsequently Weenink (1963) have described a characteristic absorption peak at 592 nm for the sulfonolipid. Galactose and other plant hexoses generally yield adducts with anthrone in sulfonic acid that peak at 625 nm. Application of this method can be used with either the whole lipids or the deacylated lipids. The reaction was used by Isono and Nagai (1966; Nagai and Isono, 1965) in their efforts to characterize the sea urchin sulfonolipid. The blue color of this anthrone reaction was used by Russell (1966; Russell and Bailey, 1966) to routinely assay samples for sulfonolipid. However, Russell's procedure was apparently erratic in the hands of his fellow New Zealanders Roughan and Batt (1968), who used the phenol-sulfuric acid reagent of Dubois et al. (1956), which doubles the sensitivity and is quantitative for the sulfonolipid (as well as galacto lipids) while it is still adsorbed to silica. The phenol-sulfuric acid reagent does not distinguish the sulfonolipid from other glycolipids.

A procedure used for the assay of the sulfatides of Ochromonas danica by Haines (1965) and elegantly developed for cerebroside sulfate by Kean (1968) has not been applied to the sulfonolipid. The procedure involves mixing the sample with a cationic dye such as azure A and extracting directly in a colorimeter tube with chloroform. It can be conducted on crude samples with only the very slightest contamination from other anionic lipids.

D. Biosynthesis

Little is known about the biosynthesis of the plant sulfonolipid. Most attempts to obtain information on this problem have led to different speculations with a paucity of data to back them up. At least three routes have been put forward in the literature for the biosynthesis of 6-sulfoquinovose.

The first proposal was that of Zill and Cheniae (1962), who suggested in a review that 3'-phosphoadenosine-5'-phosphosulfate (PAPS) transfers a sulfonyl group to a carbon atom of an acceptor molecule. They suggest that the acceptor may be a lipid or a nucleotide-bound precursor.

Benson (1963), in his review of the sulfonolipid, combined this suggestion with one by Kittredge et al. (1962) for the phospholipids and proposed that pyruvate was an appropriate carbanion acceptor. He pointed out that the occurrence of sulfolactaldehyde, sulfolactate, and sulfopropanediol in Chlorella, albeit in small amounts (Shibuya and Benson, 1961), supports this contention. He suggested that the biosynthesis might occur by a "sulfoglycolytic sequence" from sulfopyruvate. This suggestion was further expanded by Davies et al. (1966), who made the proposal in Fig. 1.

A second pathway for the biosynthesis of 6-sulfoquinovose was proposed by Lehmann and Benson (1964a). These authors (Lehmann and Benson,

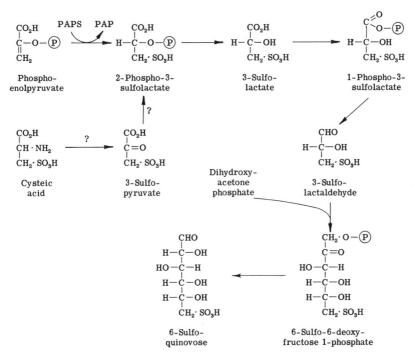

Fig. 1. Pathway for the biosynthesis of 6-sulfoquinovose suggested by Davies *et al.* (1966).

1964b) have synthesized 6-sulfoquinovose via the methyl glucoseenide (the double bond between C-5 and C-6 of a glucoside). Their suggestion was based upon the observation that sulfite adds to methyl-α-D-glucoseenide to form the 6-sulfoquinovoside in 5 minutes at room temperature in aqueous solution at pH 6.4 to 7.0. The addition of sulfite to the glucoseenide is a free radical reaction.

A third pathway for the biosynthesis of 6-sulfoquinovose was put forward by Hodgson *et al.* (1971). It is analogous to the displacement of *O*-acetyl or *O*-succinyl groups by thiols or H_2S (Kaplan and Flavin, 1966; Kredich and Tompkins, 1966; Weibers and Garner, 1967; Giovanelli and Mudd, 1968; Kerr and Flavin, 1968). The synthesis of sulfonic acids by displacement with sulfite is, of course, as familiar to the synthetic chemist as the addition of sulfite to double bonds (Haines, 1971).

It is easy to see how many varied proposals are available in the literature. When one looks for hard data on the biosynthesis, however, all such data are available in only two publications. Nissen and Benson (1964) established that 3-^{14}C-cysteine is not incorporated into the sulfonolipid in *Chlorella*. Davies *et al.* (1966) confirmed this finding for *Euglena gracilis*. They further found that cysteic acid, labeled on either C-3 or the sulfur atom, was incorporated to the same extent into the sulfonolipid. In addition, cysteic acid, but not cysteine, inhibited the incorporation of $^{35}SO_4^{2-}$ into sulfonolipid by *Euglena*, whereas cysteine *but not cysteic acid* inhibited the uptake of $^{35}S_4^{2-}$ by the cells. These data implicate cysteic acid as an intermediate in the biosynthesis of 6-sulfoquinovose.

Davies *et al.* (1966) have also found that molybdate (3.0 mM) inhibits the incorporation of $^{35}SO_4^{2-}$ into sulfonolipid in *Euglena* but has no effect on its uptake by the cells. These data suggest that PAPS is involved in the biosynthesis of the sulfolipid. It is on the basis of the above that these investigators proposed the pathway described in Fig. 1.

In recent years it has become increasingly clear that a specific and active transport process is involved in the uptake of sulfate by microbes (Pardee, 1966, 1967; Dreyfus, 1964; Ellis, 1964; Kylin, 1964, 1966, 1967; Hodgson *et al.*, 1971). This transport system precedes the formation of PAPS and is independent of it. Thus the uptake of sulfate by *Penicillium chrysogenum* is suppressed by inhibitors of energy metabolism (2,4-dinitrophenol or azide), but sulfate reduction is unaffected (Yamamoto and Segel, 1966), as is that of *Chlorella* (Wedding and Black, 1960) and *Euglena gracilis* (Abraham and Bachhawat, 1965). All the evidence in the literature to date indicates that the first step in the metabolism of sulfate is its activation to PAPS, as described by Wilson and Bandurski (1958). These authors had shown that molybdate inhibited the formation of PAPS by the enzyme ATP sulfurylase (EC 2.7.7.4). Inhibition of sulfate activation by molybdate *in vivo* has been demonstrated in many systems, notably *Escherichia coli* and *Bacillus subtilis* (Pasternak, 1962), rat brain (Pritchard, 1966), and salivary gland (Pritchard, 1967a, b). It is therefore not easy to interpret the molybdate inhibition data of Davies *et al.* (1966). Their findings do not necessarily imply the participation of PAPS directly in the biosynthesis of sulfonolipid.

Recently, Hodgson *et al.* (1971) have isolated *Chlorella* mutants that are unable to reduce sulfate, apparently because of a genetic lesion involving the formation of PAPS. Other sulfur metabolism mutants isolated by these investigators suggest that PAPS is involved in the reduction of sulfate by *Chlorella*. The pathway for the reduction of sulfate by *Salmonella typhimurium* is now becoming clear, and PAPS is definitely involved. In this bacterium PAPS reductase has been identified as the product of the cys H gene (Kredich, 1971). This enzyme reduces PAPS to sulfite (Noriko *et al.*, 1971). Since PAPS is involved in the formation of sulfite, inhibition of ATP-sulfurylase by molybdate would inhibit the biosynthesis of the sulfonolipid regardless of which of the proposed pathways is correct.

Also consistent with the proposal in Fig. 1 was the finding of sulfolacetaldehyde, and sulfolactate in *Chlorella* by Benson and Shibuya (1961). A report (Wickberg, 1957) that the red alga *Polysiphonia fastigiata* contains cysteinolic sulfolactone (2-L-amino-3-hydroxy-1-propane sulfonic acid lactone) (II) is of some interest in view of the high concentration of the sulfonolipid in red algae (Benson and Shibuya, 1962; Radunz, 1969).

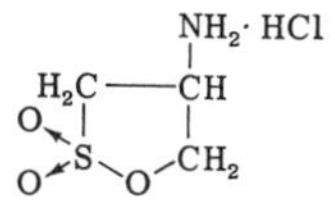

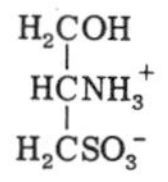

Cysteinolic acid (III) has also been identified in brown and green algae (Ito, 1963) and in the freshwater diatom *Navicula pelliculosa* (Busby, 1966), along with sulfopropanediol (IV).

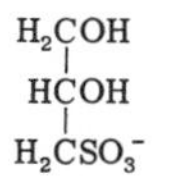

A summary of this discussion on the biosynthesis of 6-sulfoquinovose is illustrated in Fig. 2. The pathway includes the suggestion of a nucleophyllic displacement by sulfite on *O*-acetyl serine as proposed by Hodgson

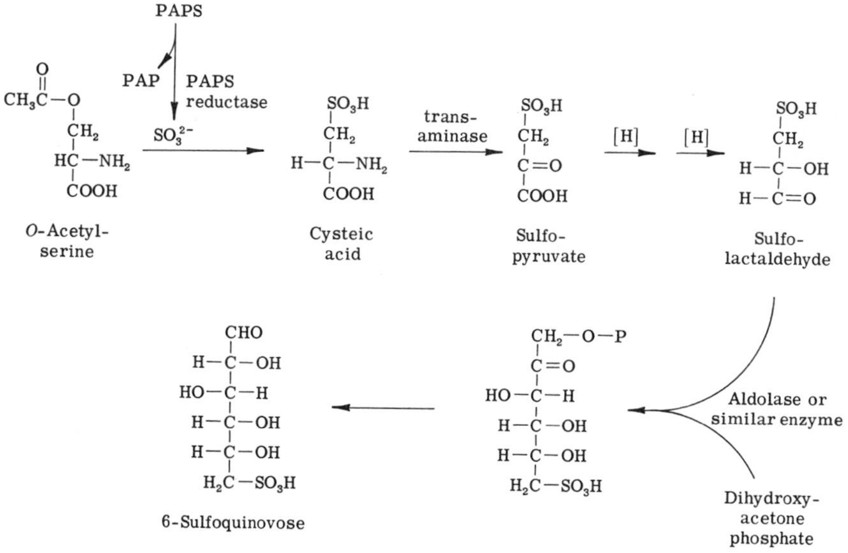

Fig. 2. Proposed biosynthesis of 6-sulfoquinovose.

et al. (1971). The choice of *O*-acetyl serine as the precursor of cysteic acid is based upon its availability as an intermediate in the biosynthesis of cystein (Kredich, 1971). Davies *et al.* (1966) have demonstrated the participation of cysteic acid in the biosynthetic route. A secondary role is proposed for PAPS. The pathway is otherwise similar to that of Fig. 1.

Other studies on the biosynthesis of plant sulfonolipid relate to the participation of nucleoside derivatives of 6-sulfoquinovose as participants in the formation of the glycosidic bond with glycerol. Shibuya *et al.* (1963) identified a nucleoside diphosphosulfoquinovoside among the ^{35}S-labeled components of plant extracts. It was suggested that this compound was the activated intermediate for sulfoquinovosyl glyceride.

A study of the ability of plants to incorporate ^{75}Se-selenate into sulfonolipid was made by Nissen and Benson (1964). Although APSe was identified, no evidence could be found for the formation of either PAPSe or selenolipid.

E. IN CHLOROPLAST MEMBRANE

In the very first publications reporting the sulfonolipid in plants it was recognized as an important component of photosynthetic microorganisms and higher plants. Table I shows a list of the microorganisms and plants in which the sulfonolipid has been identified to date. Virtually every higher plant and photosynthetic microorganism that has been investigated has been shown to contain the material. It is therefore tempting to speculate that the sulfolipid plays a role in photosynthesis. Whether or not this is the case, it has now become very clear that the sulfonolipid is a structural component of the chloroplast membrane in higher plants and green algae.

The occurrence of sulfonolipid in many of the organisms listed in Table I has been confirmed in several laboratories. An outstanding exception has been the photosynthetic bacteria where conflicting information is available. Thus Wood *et al.* (1965) in a survey of five photosynthetic bacteria found that only *Rhodopseudomonas spheroides* contained sulfonolipid and that it was absent from *R. capsulata*, *R. palustris*, *R. gelatinosa*, and *Rhodospirillum rubrum*. On the other hand, Benson *et al.* (1959a), using a sample of ^{35}S-labeled *R. rubrum* provided by John Ormerod, reported sulfonolipid in this bacterium. This discrepancy might be explained by the very small amounts of sulfonolipid found in the photosynthetic bacteria. Park and Berger (1967) estimated that the sulfonolipid was only 0.01% of the dry weight of *Rhodomicrobium vannieli*. This is approximately the value found for *R. spheroides* by Radunz (1969). It would appear that ^{35}S-labeling would have been a more sensitive analytical method than that used by Wood

TABLE I

ORGANISMS IN WHICH SULFONOLIPID HAS BEEN IDENTIFIED

Class	Species	% Total Lipid	Reference
Higher plants	Barley (*Hordeum vulgare*)		Benson *et al.* (1959a)
	New Zealand spinch (*Tetragonia expansa*)		Benson *et al.* (1959b)
	Chive sp.		Benson *et al.* (1959a)
	Coleus sp.		Benson *et al.* (1959a)
	Sweet clover sp.		Benson *et al.* (1959b)
	Tomato sp.		Benson *et al.* (1959b)
	Alfalfa sp.		Lepage *et al.* (1961)
	Runner bean (*Phascolus multiflorus*)		Kates (1960)
	Spinach (*Spinacia oleracea*)		Wintermans (1960)
	Sugarbeet (*Beet vulgaris*)		Wintermans (1960)
	Elder (*Sambucus nigra*)		Wintermans (1960)
	Paul's scarlet rose		Davies *et al.* (1965)
	Maize (corn) (*Zea mays*)		Davies *et al.* (1965)
	Fern (*Dryopteris filix-mas*)	4.0	Radunz (1968)
	(*Antirrhinum majus*)	5.2	Radunz (1969)
	Wheat (*Triticum aestivum*)		Nissen and Benson (1964)
	Sunflower (*Heliantus annus*)		Nissen and Benson (1964)
	Red clover (*Trifoleum pratanse* L.)		Russell and Bailey (1966)
	Moss (*Sphagnum fimbriatum*)		Collier and Kennedy (1963)
	Rye grass sp.		Collier and Kennedy (1963)
	Selaginella Kraussiana A. Br. (a Pteridophyta)		Collier and Kennedy (1963)
	Broad bean (*Vicia faba*) fluffy pericarp		Collier and Kennedy (1963)
	Kalanchoe crenata tissue culture		Thomas and Stobart (1970)
	Sunflower (*Helianthus annus*)		Nissen and Benson (1964)
	Wheat (*Triticum aestivum*)		Nissen and Benson (1964)
Chlorophyceae (green algae)	*Chlorella pyrenoidosa*		Benson *et al.* (1959a)
	Scenedesmus D₃		Benson *et al.* (1959a)
	Chlorella vulgaris		Nichols (1965)
	Chlorella prototecoides		Shibuya and Hase (1965)
	Ulva lactuca var. *latissima* D.C.		Collier and Kennedy (1963)
	Cladophora sp.		Collier and Kennedy (1963)
	Enteromorpha compressa Grev.		Collier and Kennedy (1963)
	Scenedesmus obliquus		Yagi and Benson (1962)
	Chlorella ellipsoidea		Miyachi and Miyachi (1966)

TABLE I   (Continued)

ORGANISMS IN WHICH SULFONOLIPID HAS BEEN IDENTIFIED

Class	Species	% Total Lipid	Reference
Phaeophyceae			
(brown algae)	*Fucus serratus*		Collier and Kennedy (1963)
	Fucus vesiculosus	18.3	Radunz (1969)
Rhodophyceae	*Rhodymenia palmata* Grev.		Collier and Kennedy (1963)
(red algae)	*Plumaria elegans* Schmitz		Collier and Kennedy (1963)
	Antithamniou plumula Thuret		Collier and Kennedy (1963)
	Gigartina stellata Batt.		Collier and Kennedy (1963)
	Dumontia incrassata Lamour		Collier and Kennedy (1963)
	Ceramium rubrum C. A. Agardh		Collier and Kennedy (1963)
	Batrachospermum moniliforme	14.9	Radunz (1969)
Cyanophyceae	*Rivularia atra* Roth.		Collier and Kennedy (1963)
(blue-green algae)	*Oscillatoria chalybea*	13.9	Radunz (1969)
Athiorhodacae	*Rhodopseudomonas spheroides*	2.6	Radunz (1969)
(photosynthetic			Wood *et al.* (1965)
bacteria)			
	Rhodospirillum rubrum		Benson *et al.* (1959a)
	Rhodomicrobium vannielli	0.01	Park and Berger (1967)
Phytoflagellates	*Euglena gracilis*		Rosenberg (1963)
	Ochromonas danica		Miyachi *et al.* (1966)
	Chlamydomonas reinhardii		Ohad *et al.* (1967)
Hemoflagellate	*Crithidia fasciculata*		

et al. (1965), and it would appear that the other *Rhodopseudomonas* species in their study should be reinvestigated.

In addition to its identification in photosynthetic organisms, the sulfonolipid has also been specifically associated with the photosynthetic process and/or membrane. Thus in gross analyses its concentration is highest in those plant tissues associated with photosynthesis (leaves) and lowest in nonphotosynthetic tissues (root, stem, and seed), although it is generally identified in plant tissue in at least trace amounts.

The first study of the variation of sulfonolipid concentration with chloroplast "concentration" (greening) was that of Rosenberg and Pecker (1964). These authors (see also Rosenberg and Gouax, 1967; Helmy *et al.*, 1967) were able to show a direct correlation between the appearance of chlorophyll and that of the galactosyl glycerides and sulfonolipid in *Euglena gracilis*. A dark-grown culture of the organism was exposed to light, and changes in the lipid composition were noted.

Kennedy and Collier (1963; Collier and Kennedy, 1963) have reported three sulfolipids in green plants, green algae, brown algae, and blue-green algae. In contrast, red algae and the fluffy pericarp of the broad bean *Vicia faba* contained only one sulfolipid. Unfortunately, these investigators did not positively identify the sulfur in each of their spots on the paper chromatograms of extracts but did a sulfur analysis on the mixture of three isolated from brown algae. Furthermore, the analysis was low (found, 3.68%; theoretical, 5.5%) based upon sulfonolipid structure. Their method of detection was based upon staining chromatograms with ionic dyes. In this system, the sulfolipids were characteristically pink, apparently because of their acidity. Many lipids were tested in their system (Kennedy and Collier, 1963) but not phosphatidyl glycerol. According to analyses in many other laboratories, this material should have appeared in their chromatograms. Since this very acidic lipid would likely stain as a sulfolipid in their system, it is probable that one of their spots was phosphatidyl glycerol. Additionally, Shibuya and Benson (1961) have shown that an unusually active lipase in photosynthetic tissue converts the sulfolipid to lysosulfolipid. It would therefore appear that the "three plant sulfolipids" in the chromatograms of Collier and Kennedy (1963) are phosphatidyl glycerol, sulfonolipid, and lysosulfonolipid. One implication of this interpretation of their results is that nonphotosynthetic plant tissue (fluffy pericarp of broad bean) and red algae contain the sulfonolipid but not phosphatidyl glycerol and do not contain an active sulfolipase. In addition, Collier and Kennedy (1963) reported a different sulfolipid that they found in two fungi, which will be discussed in Section V.

A striking observation in the phytoflagellate *Ochromonas danica* was made by Miyachi *et al.* (1966), who were surprised by the absence of sulfonolipid in the early reports of Haines (Haines and Block, 1962; Haines 1965) describing the alkyl disulfates in this microbe. The cells of these investigations contained chloroplasts but not the sulfonolipid. The cells had been grown under constant light in a sucrose medium. Benson's group repeated these analyses and obtained similar results. They then cultured the organism autotrophically with the result that the sulfonolipid appeared on chromatograms of lipids, although its concentration was still low (approximately one-fifteenth that of the alkyl disulfates). These experiments imply that the sulfonolipid plays a very direct role in photosynthesis, since it was the utilization of photosynthesis as the principal food supply that enhanced the sulfonolipid concentration. The organism contained large amounts of chloroplast membrane in both heterotrophic and autotraphic cultures.

A study of alfalfa leaf sulfonolipid composition by Klopfenstein and

Shigley (1967) showed that the sulfonolipid concentration varied seasonally. Higher concentrations are noted in spring, and the level drops gradually throughout the summer. They also noted changes in the sulfonolipid's fatty acid composition throughout the season. During the period of most active photosynthesis, the linolenic acid composition is highest. Palmitic acid increases and linolenic acid decreases with age.

The involvement of the sulfonolipid in photosynthesis might well be explained by its fatty acids. Erwin and Bloch (1963, 1964) had proposed that linolenic acid is involved in photosynthesis on the basis of the analytical data then available. Although subsequent studies have not yet demonstrated a direct connection, it has been shown that of the major chloroplast lipids—monogalactosyl diglyceride, digalactosyl diglyceride, phosphatidyl glycerol, and the sulfonolipid—both galacto lipids contain almost exclusively linolenic acid (Weenink, 1962; Sastry and Kates, 1963; Benson, 1963) and the sulfonolipid is approximately 50 # palmitic acid and up to 50% linolenic acid (O'Brien and Benson, 1964; Klopfenstein and Shigley, 1966; Radunz, 1969), although the relative concentrations are variable under different growth conditions. Linolenic acid concentration in *Chorella* has been correlated with photosynthetic oxygen production (Appleman *et al.*, 1966). Recently, Brand *et al.* (1971) were able to show an absolute requirement for polyunsaturated fatty acid glycerides (with linolenic most active) in photosystem I (spinach) that had been extracted with heptane. Thus triglycerides containing polyunsaturated fatty acids but *not* plastoquinones, vitamin K, β-carotene, or α-tocopherolquinone restored photosynthetic activity. Ohad *et al.* (1967) have also reported a lower level of sulfonolipid in dark-grown *Chlamydomonas reinhardii* (which has lost nearly all its chloroplasts). They also found that greening is accompanied by a net synthesis of chloroplast membrane.

An extensive study on the greening of callus (tissue culture) of *Kalanchoe crenata* was recently conducted by Thomas and Stobart (1970). The molar ratios of various lipids to chlorophyll were examined through each of seven generations of cells. It took seven generations for the cultures to achieve the full green. Sulfonolipid was found to appear in lipid extracts in the third generation—*prior to the appearance of chlorophyll*. Although dark-grown cells contained mono- and digalactosyl diglycerides, no sulfolipid could be detected in these cells. Furthermore, although the galacto lipids never reached a constant molar ratio with respect to chlorophyll, the sulfonolipid rapidly achieved a chlorophyll-to-sulfonolipid ratio of 4.4 and remained at that level through greening. The appearance of the sulfonolipid prior to the appearance of chlorophyll was consistent with the earlier data of Rosenberg and Pecker (1964), and the constant molar ratio of chlorophyll to sulfonolipid was in remarkable agreement with Lichenthaler and Park (1963). These data suggest the sulfonolipid may be involved in orienting the chlorophyll molecules in the membrane— whether or not participation in photosynthesis occurs.

Shibuya and Hase (1965) have also studied the destruction of the chloroplast membrane by bleaching *Chlorella protothecoides*. In addition to a decrease in sulfonolipid concentration during bleaching, they noted a large increase in 6-sulfoquinovosyl glycerol, implying that a sulfolipase is involved in the destruction of the chloroplast.

In a recent study of the lipid composition of chloroplast grana and stroma lamellae Allen and Park (1971) have found that these two membranes have very similar composition. These data are consistent with the model chloroplast membrane proposed by Weier and Benson (1967). This latter summary of the status of the chloroplast membrane problem places the sulfonolipid and the galacto lipids in the same role in the membrane.

IV. The Sulfolipids and Halosulfolipids of *Ochromonas*

In 1962, Haines and Block identified some ^{35}S-labeled lipids in extracts of the phytoflagellate *Ochromonas danica*. These substances, which have turned out to be strange membrane components, dominated the ^{35}S-labeled compounds in the cell. The substances were soon identified as sulfate esters that were present in amounts greater than most phospholipids or glycolipids in the cells. Elovson and Vagelos (1969) found them to constitute 3% of the dry weight of the cell. Although early reports indicated that these sulfatides are widespread in microbes and algae (Haines, 1965), some doubt is raised by recent attempts to label the halosulfatides in a wide variety of microbes with [36]Cl (Emanuel *et al.*, 1972). Of 12 microbes screened, only in *O. danica* and *O. malhamensis* were halosulfatides identified in the organisms. There is considerable evidence in the literature, however, that microbes excrete lipoid sulfate esters (Haines, 1965; Mumma and Gahagan, 1964; Roberts *et al.*, 1957). The nature of these sulfatides has not been described, but several were found to cochromatograph with the sulfatides herein described. Mumma and Gahagan (1964) had originally reported the ^{35}S-labeling of sulfatides excreted by higher plants. These substances were later found to be absent from axenic culture of the plants and probably produced by contaminating *algae* (Mumma, 1967).

That these halosulfatides are present in membrane is implied by their large concentration as polar lipids. Evidence obtained from fragmented cells is consistent with this notion.

A. ISOLATION

The early studies of Haines and Block (1962) and later of Haines (1965) were conducted on labeled material and did not require the isolation of any significant amounts of material. Sufficient information was obtained, however, to permit a large-scale isolation of material by Mayers and Haines (1967). At first, cells were extracted with chloroform-methanol, 2:1, and the extract was saponified with 0.2N KOH at 37°C for 45 minutes. The latter procedure, which saponifies the phospholipids and glycolipids completely, does not attack the sulfatides. After removal of the nonsaponifiable fraction by ether extraction followed by acidification and further ether extraction to obtain the fatty acids, the neutralized aqueous solution is extracted with n-butanol. Butanol has several advantages in this procedure: (1) It is the most polar alcohol that is not miscible with water, (2) It inhibits foaming during flash evaporation, and (3) it forms an azeotrope with water so that upon evaporation the sample is dry (several triturations with hexane is generally used to remove the last traces of butanol).

Initial attempts at isolating a pure sample of sulfatide for structural work were thwarted by the tenacious appearance of protein in the preparation. This was overcome by the digestion of the crude mixture with the

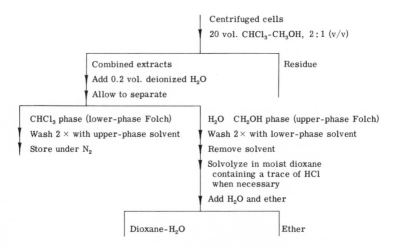

Fig. 3. Folch extraction of halodiols.

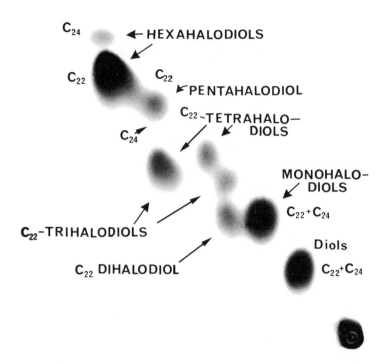

Fig. 4. Autoradiogram of a two-dimensional thin-layer chromatogram of halodiols after 1-¹⁴C-laurate was incubated with *Ochromonas danica*.

proteolytic enzyme pronase (Haines, 1965; Mayers and Haines, 1967). Once the sulfatides were identified as disulfates, the requirements for pure sulfatides diminished, and crude preparations were hydrolyzed or solvolyzed so that structural work could be conducted on the diols that were obtained (Mayers *et al.*, 1969; Elovson and Vagelos, 1969; Haines *et al.*, 1969).

A simpler extraction procedure is that of Folch *et al.* (1957; Lees *et al.*, 1959); see Fig. 3. Although this procedure gives a crude product, it is quick and the *upper phase* contains approximately 90% of the sulfatides that are extracted from the cells by the chloroform-methanol. Unfortunately less than 60% of the sulfatide is extracted by the solvents; the remainder may be released from the fat-extracted residue by digestion with pronase (Haines, 1965).

The sulfatides are only poorly separated into their constituent components by thin-layer or column chromatography as sulfate esters. In addition to the streaking usually associated with acids of this type, the chromatogram is further complicated by the fact that different salts of the sulfatides chromatograph to different positions. Since the mixture consists of halogenated alkyl diol disulfates, hydrolysis or solvolysis produces a quantitative yield of the halogenated alkyl diols, which may then be separated on thin-layer, column, or gas-liquid chromatography. The mixture of diols is quite complicated, however, and two-dimensional thin-layer chromatography (Mooney et al., 1972) has proved quite useful as a method for separating the diols into their components (Fig. 4).

In the early identification work, Mayers and Haines (1967) had separated the nonhalogenated diol by repeated crystallization of the mixture from hexane. Of the entire mixture of diols shown in Fig. 4, only the nonhalogenated diols are insoluble in hexane; these may readily be separated from the halogenated diols by recrystallization.

B. STRUCTURAL STUDIES

The structures of all the sulfatides in the mixture have not yet evolved, but the three principal components have been identified, as have several of the minor ones, and the pattern is rather clear. The first in the mixture was identified by Mayers and Haines (1967) as 1,14-docosanediol-1,14-disulfate. This was demonstrated by analysis and by the identification of primary and secondary sulfate in the infrared spectrum. Removal of the two sulfate groups was effected by hydrolysis or by solvolysis in dioxane, which was found to leave the orientation of a secondary C—O bond undisturbed (Mayers et al., 1969). The resulting diol was identified by mass spectrometry as 1,14-docosanediol. This was confirmed by synthesis of 1,14-docosanediol and comparison of the infrared spectra of the two substances. Rotation of the diol established the configuration of the secondary hydroxyl as S, and since solvolysis of the secondary sulfate did not disturb the C—O bond, the original sulfatide was 1-(S)-14-docosanediol-1-(S)-14-disulfate. The diol resulting from solvolysis of this disulfate is the major diol shown in Fig. 4 and labeled "diols."

Elovson and Vagelos (1969) confirmed the above data and, using a gasliquid chromatograph-mass spectrometer hookup, were able to identify a second diol, which Mayers and Haines (1967) could identify only as a tetracosane diol from its retention time on gas-liquid chromatography. The substance turned out to be 1,15-tetracosanediol. The ratio of the C_{22} to the C_{24} diols is about 8:1, as shown by a gas-liquid chromatogram (Haines, 1971). These diols are shown in Fig. 4 near the origin of the chromatogram.

The spot above these substances has been characterized by Haines et al. (1969) as threo-(R)-13-chloro-1-(R)-14-docosanediol derived from

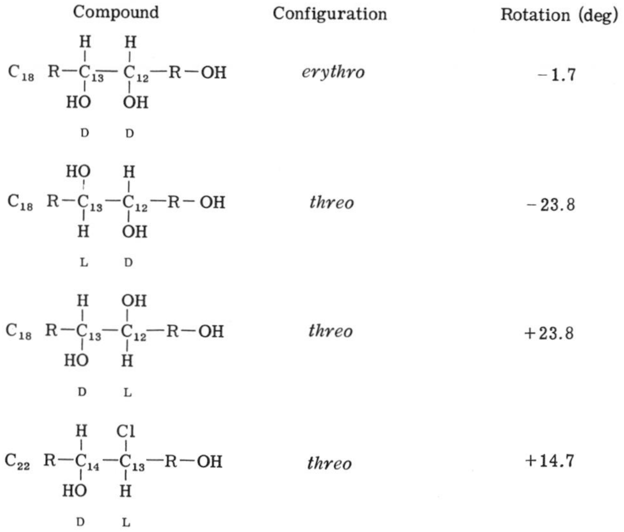

Compound	Configuration	Rotation (deg)

$$C_{18} \quad R-\underset{\underset{HO}{|}}{\overset{\overset{H}{|}}{C_{13}}}-\underset{\underset{OH}{|}}{\overset{\overset{H}{|}}{C_{12}}}-R-OH$$

erythro   −1.7

D   D

$$C_{18} \quad R-\underset{\underset{H}{|}}{\overset{\overset{HO}{|}}{C_{13}}}-\underset{\underset{OH}{|}}{\overset{\overset{H}{|}}{C_{12}}}-R-OH$$

threo   −23.8

L   D

$$C_{18} \quad R-\underset{\underset{HO}{|}}{\overset{\overset{H}{|}}{C_{13}}}-\underset{\underset{H}{|}}{\overset{\overset{OH}{|}}{C_{12}}}-R-OH$$

threo   +23.8

D   L

$$C_{22} \quad R-\underset{\underset{HO}{|}}{\overset{\overset{H}{|}}{C_{14}}}-\underset{\underset{H}{|}}{\overset{\overset{Cl}{|}}{C_{13}}}-R-OH$$

threo   +14.7

D   L

Fig. 5. Rotations of some dihydroxy and chlorohydroxy long-chain alcohols. The glycol data are those of Morris and Wharry (1966). The chlorohydrin data were obtained by Haines *et al.* (1969). Other chlorohydrins in the literature have rotations similar to glycols of analagous configuration.

the corresponding disulfate. The diol was identified by its mass spectrum, infrared spectrum, rotation, analysis, and nuclear magnetic resonance spectrum. The configuration of the chlorohydrin was determined by rotation with comparison to those of Morris and Wharry (1966); see Fig. 5. The configuration of the chlorohydrin was confirmed by conversion to the *cis*-epoxide and comparison in thin-layer chromatography to authentic *cis*- and *trans*-epoxides and *threo*- and *erythro*-chlorohydrins. The identification of the 13-chloro was also confirmed by a mass spectrum of the silyl derivative obtained by Elovson and Vagelos (1969). These investigators also identified a small amount of 14-chloro-1,15-tetracosane-diol by the gas-liquid chromatography-mass spectrometry combination.

Two dichloroalkane diols have been identified in the mixture to date: 11,15-dichloro-1,14-docosanediol by Elovson and Vagelos (1969) and 2,2-dichloro-1,14-docosanediol by Pousada *et al.* (1972b). They have been identified by mass spectral evidence only, but include derivatization in each case.

The trichloro and tetrachloro derivatives have not yet been identified, although one can tell from the mass spectra how many chloro groups are

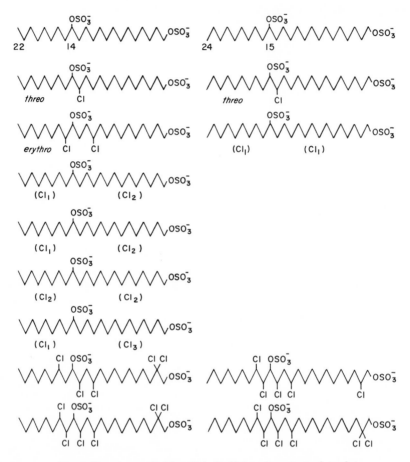

Fig. 6. Structures of chlorodiol disulfates characterized to date.

distal or proximal to the secondary hydroxyl. These are summarized in Fig. 6.

The pentachloro docosanediol was identified by Pousada *et al.*, (1972b) as 2,2,11,13,15-pentachloro-1,14-docosanediol by mass spectral evidence coupled with a nuclear magnetic resonance spectrum. In a like manner these investigators identified 2,12,14,16,17-pentachloro-1,15-tetracosanediol and 2,2,12,14,16,17-hexachloro-1,15-tetracosanediol.

In an elegant combination of degradative chemistry, ^{36}Cl radiocounting, and mass spectrometry, Elovson and Vagelos (1970) identified the major hexachlorodocosanediol as 2,2,11,13,15,16-hexachloro-1,14-docosanediol.

It appears that the two latter hexachloro compounds are the end products of a series of chlorinating enzymes. The structures of all the known compounds are shown in Fig. 6.

In addition to the chlorosulfatides, Pousada et al. (1972a) have isolated a series of bromosulfatides of similar structure. In this series only the threo-(R)-13-bromo-1-(R)-14-docosanediol-1-(R)-14-disulfate has been characterized. It was identified by its mass spectrum, nuclear magnetic resonance spectrum, rotation, and analysis. The 2,2,11,13,15,16-hexabromo-1,14, docosanediol was also obtained from these cultures, as were several of the tri- and tetrabromo intermediates. These were identified by their relative positions on a two-dimensional thin-layer chromatogram comparable to that in Fig. 4 and by their mass spectra. The bromine was also confirmed by the incorporation of ^{89}Br into the respective bromodiols.

C. ANALYSIS

Probably the surest and quickest method of analysis of halosulfolipids is ^{36}Cl labeling followed by thin-layer chromatography of the upper phase of a Folch extract of the tissue. This method, which will obviously not identify the nonhalogenated 1,14-docosanedilo-1,14-disulfates, is not as simple as it would appear. Since ^{36}Cl-chloride ions chromatograph up the plates, Emanuel et al. (1972) found it necessary to cospot silver nitrate on the thin-layer plate in order to prevent ^{36}Cl-chloride from moving from the origin. This modification of the above procedure permitted the rapid screening of organisms for chlorosulfatides.

The use of ^{35}S-sulfate, which was the original method of identification (Haines and Block, 1962), is also available. Since unknown sulfolipids may cochromatograph with the halosulfatides in a given solvent system, it would be advisable to solvolyze the sulfatides and chromatrograph the resulting diols for a positive identification. L,12-Octadecanediol is commercially available as a standard for thin-layer chromatography of 1,14-docosanediol. It should be noted that the halogenated diols run ahead of the unsubstituted diols on thin-layer chromatograms. The solvolysis procedure (Mayers and Haines, 1967) is especially desirable for the identification of sulfate esters because of its high specificity for this functional group.

The methods available for the analysis of sulfatides have recently been reviewed (Haines, 1971). They include oxidative, reductive, colorimetric, turbidimetric, flame photometric, infrared spectrophotometry, activation analysis, and radiometric methods. Of special note is the colorimetric method used by Haines (1965) for these compounds. This method has been improved by Kean (1968) and used by him as a very selective nondestructive method for the identification and quantitation of sulfatides in a complex mixture of crude lipids.

D. BIOSYNTHESIS AND METABOLISM

The biosynthesis of the halosulfatides is not yet understood. The structures of the various compounds provide some clues to their biosynthetic route, and perhaps these should be discussed first. There are only two series of aliphatic disulfates in the mixture—1,14-docosanediol-1,14-disulfate and 1,15-tetracosanediol-1,15-disulfate. The difference between these series is that the chain of the tetracosane series is longer by one methylene group *both proximal and distal* to that of the docosane series.

Mooney *et al.* (1972) have shown that ^{14}C-acetate and ^{14}C-octanoate are efficiently incorporated into the sulfatides (Fig. 7). This suggests that the chain is biosynthesized by the usual fatty-acid-synthesizing enzymes. It is presumed that the sulfate is derived from PAPS, as this intermediate has been reported to be the intermediate in the biosynthesis of all the sulfate esters where the biosynthesis has been investigated (Roy and Trudinger, 1970). As PAPS sulfates hydroxyl groups, it remains to determine how the hydroxyl is incorporated into the chain.

Three routes are possible for the synthesis of hydroxy fatty acids. The first is that of synthesis during the chain-lengthening process and would presumably be anaerobic. A second conceivable route is direct hydroxylation by a hydroxylase using molecular oxygen such as the microsomal P-450 hydroxylase of the mammalian liver. A third possibility is the hy-

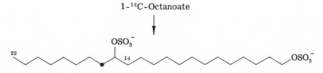

Fig. 7. Incorporation of acetate and octanoate into 1,14-docosanediol-1,14-disulfate.

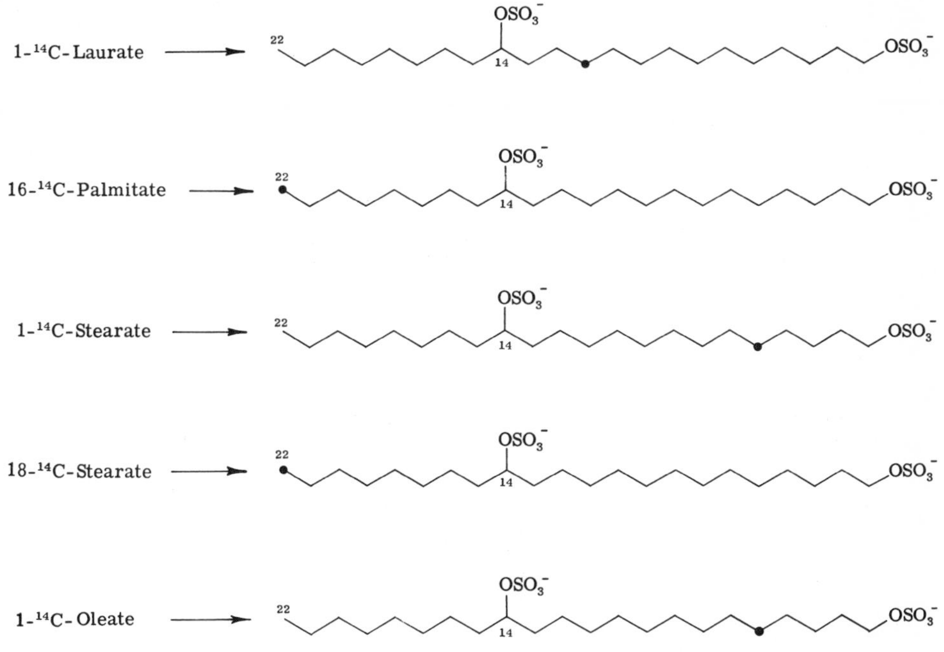

Fig. 8. Incorporation of fatty acids into 1,14-docosanediol-1,14-dusulfate.

dration of a double bond that could itself be introduced either during the chain-lengthening process (anaerobically) or into the saturated chain. An example of the latter is the biosynthesis of 10-hydroxystearic acid from oleic acid by a pseudomonad, NRRL-B-2994, (Niehaus et al., 1970). Mooney et al. (1972) have reported the rapid incorporation of laurate, palmitate, stearate, and oleate into the chain of the halosulfatides. These data (Fig. 8) show that the hydroxyl function is introduced onto the unsaturated chain and not during the chain-lengthening process.

Since the hydroxyl is on the 14 position in the docosane series and on the 15 position in the tetracosane series, the double bond of oleic acid could be hydrated after chain lengthening to C_{22} or C_{24}. In the first case the hydroxyl would be placed on C-10 of oleic acid and in the second case it would be situated on C-9. This biosynthetic route could explain the one-carbon difference in position. Mooney et al. then fed ^{14}C-oleic acid to the organism. The oleic acid was incorporated into the sulfatides as indicated in Fig. 8. It therefore appears that a double-bond intermediate is likely for the introduction of the hydroxyl.

Two important aspects of the metabolism of these substances in cells have been established. The first was noted in the very first paper by Haines and Block (1962). It was found that *Ochromonas* is unable to cleave the sulfate groups from the chain. Thus ^{35}S-sulfate is used efficiently for the biosynthesis of cystine and methionine in proteins, whereas cultures that were incubated with the ^{35}S-labeled sulfatides did not label the sulfur amino acids. In these experiments the ^{35}S-sulfatides were incorporated into the cells to the same extent as sulfatide was when the label originated as sulfate. The compounds are thus remarkably inert metabolically.

A second aspect of their metabolism was noted first by Elovson and Vagelos (1969), who observed a dramatic increase in the amount of hexachlorosulfatide in the presence of a media rich in chloride. This was confirmed by Pousada et al. (1972c), who also found that the organism survived in a chloride-free culture medium but that its morphology was changed somewhat and that it produced the nonchlorinated diol exclusively.

The sulfatases in *Pseudomonas* $C_{12}B$ described by Payne and Painter (1971) were shown to cleave 1,12-octadecanediol-1,12-disulfate and, judging from the similarity of structure, are very likely to hydrolyze the sulfate esters in these compounds.

An autoradiogram of a two-dimensional chromatogram of the diols obtained from ^{14}C-laurate-labeled sulfatides is shown in Fig. 4. The pattern is identical to that obtained from the charred diols in extracts of the cells. It is also identical to autoradiograms obtained from ^{14}C-palmitate- and ^{14}C-stearate-labeled diols. Of special importance is the fact that the C_{24} and the C_{22} halogenated diols are labeled to the same extent. This suggests that they are derived from the same intermediate and that the chloro groups are introduced *onto the saturated chain*. This implies that the ^{14}C-fatty acids are *first* incorporated into the nonhalogenated diol disulfate and that this molecule is then the substrate for the chlorinating enzyme(s). This is consistent with the location of the chloro groups around the sulfate groups on the molecule. It is also consistent with the shift of one carbon for each of the chloro groups around the 15-sulfate in the tetracosane series.

A large number of halogenated compounds are reported in the literature. Most of the chlorinated compounds have been reported in fungi or actinomycetes. Several of these are shown in Fig. 9 along with the bromo derivatives that have been obtained in each case by excluding chloride from the medium and adding bromide. It was by this approach that Pousada et al. (1972a) were able to obtain bromosulfolipids. It might be noted that to date no natural chloro compounds have been characterized or reported in marine organisms, although at least ten bromo compounds have been described. The only halogenated compound whose biosynthesis has been investigated is that of Caldariomycin in the fungus *Caldariomyces*

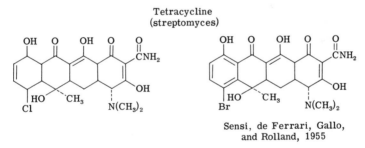

Fig. 9. Some natural chloro compounds that have also been isolated as bromo compounds. In each case the organism was grown in the absence of chloride and in the presence of bromide.

fumago by Hager *et al.*, (1970). Hager's group has isolated and characterized the chloroperoxidase that inserts the chloro group on the antibacterial product (Brown and Hager, 1967; Morris and Hager, 1966). The purified enzyme does not have a high specificity for the organic substrate and is capable of bromination as well as chlorination, although the rate of bromination is much slower than that of chlorination. This is consistent with the lower amounts of bromolipids produced and the considerably slower growth rates observed in bromide media for *Ockromonas danicd*.

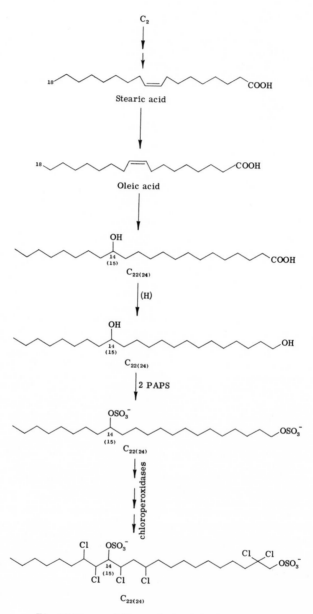

Fig. 10. Proposed biosynthesis of the halosulfatides.

An important difference between these halosulfatides and the natural and synthetic substrates for the chloroperoxidase of Hager, et al., (1970) should be noted. The Hager enzyme could use Cl^+ as a chlorinating intermediate, as it chlorinates phenyl rings or benzyllic carbons, whereas this enzyme (these enzymes) places chloro groups on a saturated hydrocarbon chain. This suggests a more energetic free radical intermediate, possibly a hydroperoxide.

These proposals are summarized in Fig. 10, which suggests a complete biosynthetic scheme for the compounds. It is assumed that there are two sulfating enzymes.

E. IN MEMBRANE VESICLES

The occurrence of these compounds in a membrane is of significance because of their unique structure as polar lipids. Recent evidence obtained from freeze-etch electron micrographs (Pinto da Silva and Branton, 1970) and from X-ray electron density profiles (Wilkins, 1972) indicates that there is a "cleavage plane" or region of very low electron density down the center of a variety of natural membranes. These data are consistent with the earlier proposals of a bilipid leaflet model for membrane structure (Gorter and Grendel, 1925; Danielli and Davson, 1935), which was subsequently demonstrated to be the structure of myelin by Schmitt and Bear (1939). The application of the freeze-etch and X-ray techniques to artificial bilayers of fatty acids or phospholipid micelles lends further support to this model of membrane structure. The picture does not appear to be as simple as the model would suggest, however, for there is much evidence that proteins are deep within the hydrophobic region of many membranes and furthermore that much of the hydrophobic region of membranes is occupied by proteins (Green and Perdue, 1966; Weier and Benson, 1967). Additionally, judging from their inability to cleave with the freeze fracture technique, some membranes such as mitochondrial membranes do not appear to contain such a cleavage plan. This is especially significant in view of the fact that the mitochondrial membrane has a trilamellar structure as viewed in thin-section electron microscopy, as do other membranes.

The bilipid leaflet as a model for membrane structure is based upon the stability of a monolayer of aliphatic hydrocarbon chains, which terminate as a methyl group at one end and a polar hydrophylic group at the other end. This is the structure of all polar lipids known to date. The halosulfo-lipids present an interesting exception to this pattern, as they all contain a sulfate group at one end of the chain and a second sulfate near the other end of the chain. These groups are negative at all aqueous pHs. Thus the lipid is not capable of forming a monolayer and likewise, presumably, a bilayer. It is thus of some value to determine whether or not these compounds are present in a membrane of the organism. Several attempts have been made to establish that this is the case, with limited success.

It was observed early that these compounds represented well over 50% of the sulfur in the cell (Haines, 1965). Elovson and Vagelos (1969) reported that they constitute over 3% of the dry weight of the cell. Aaronson and Baker (1961) and also Haines (1965) had found that lipids represent about 10% of the dry weight of the cells. This indicates that these compounds represent about one third of the lipid. In our experience the value varies from about 10% to about 40%, depending on the culture conditions. The low value would be the case for a chloride-free culture medium and the high value for a high bromide medium. It should be noted that the halogen itself may contribute substantially to the weight of the sulfatide mixture.

Early experiments by K. Kahn and T. H. Haines, which were summarized by Haines (1966), demonstrated that the sulfatide appeared in a major peak in a Ficoll density gradient that was designed for the examination of membranes by Kamat and Wallach (1965). The peak of ^{35}S-activity corresponded to a band in the density gradient that had the physical appearance of a membrane band. Numerous attempts to repeat these experiments were not successful, although Poncz and Haines (1972) have found a sharp band on a density gradient that has a very high ^{35}S-sulfatide-protein ratio.

It has been found (Haines and Block, 1962) that the sulfatides are excreted into the medium in some quantity. Subsequently, Gellerman and Schlenk (1964) reported that a pellet obtained by high-speed centrifugation of the medium (after removal of the cells) contained a large amount of a substance identified with the sulfatides (Haines, 1965). It was therefore of considerable interest to find that the organism grown under the same culture conditions excretes membrane vesicles (Orner *et al.*, 1972). These vesicles were examined as thin-section electron micrographs and have the trilamellar appearance of other biological membranes. Studies are currently under way to determine the precise lipid composition of these vesicles that are rich in halosulfatides.

V. Miscellaneous Sulfolipids of Eukaryotic Microorganisms

Sulfolipids have been reported in a variety of living systems, which include nearly the entire biosphere. Mammalian sulfatides include cerebroside sulfate, sulfo-lac ceramide, and ganglioside sulfate (Haines, 1971).

A sulfonolipid similar to that of the plant sulfolipid has been described in the sea urchin (Isono and Nagai, 1966).

There are several reports of sulfolipids in bacteria. Kates *et al.* (1967, 1968) have described a sulfate ester of 1-*O*-(glucosylmannosylgaloctosyl-2,3-di-*O*-phytanyl-L-glycerol in the extremely halophylic bacterium *Halobacterium cutirubrum.* Hancock and Kates (1972) have recently reported the 2,3-di-phytanyl-L-glycerol-1-sulfate in the same organism. Marshall and Brown (1968) have also found the glycoside in the extreme halophile *Halobacterium halobium.* Apparently it is not present in moderately halophilic or nonhalophilic bacteria but occurs in all extreme halophiles thus far examined (Kates *et al.*, 1967). Another prokaryote, *Mycobacterium tuberculosis,* contains a sulfatide mixture (Goren, 1971) that contains the 2,3,6,6'-tetraester of 2'-trehalose sulfate. The esters are of unique branched fatty acids, which is not unusual for *Tubercle bacillus.* The sulfatide's concentration in the cell of a variety of strains is apparently proportional to the virulence of the strain (Gangadharam *et al.*, 1963). In their classic study of the metabolism of *Escherichia coli* with radioisotopes, Roberts *et al.* (1957) reported a sulfolipid excreted by the bacterium. The substance constituted 20% of the sulfur in the organism. It was not present in the cells; the substance was identified only as a spot on a paper chromatogram.

A sulfolipid has been found in diatoms that does not correspond to the plant sulfonolipid (Kates and Tornabene, 1972). The substance appears to be a sulfate ester.

The conidia of the fungus *Glomerella cingulata* contain a sulfolipid (Jack, 1964). The substance was identified by thin-layer and paper chromatography of lipid extracts after incubating the organism in the presence of ^{35}S-sulfate. The compound was slightly more polar on thin-layer chromatograms than the most polar phosphatides and was not positive to ninhydrin.

Another investigation of fungal lipids was conducted by Collier and Kennedy (1963). They reported a sulfolipid in the fungus *Coprinus atramentarius* Fr. that did not cochromatograph with the sulfolipid they found in the photosynthetic microorganisms [presumably the sulfoquinovose compound of Benson *et al.* (1959a)]. A sulfolipid that cochromatographed with this new material was also reported in the fungi *Psalliota campestris* Quél and in the fruit bodies of *Clitocybe aurantiaca* Fr. It should be pointed out that the identification of these substances as sulfolipids is based solely upon their staining properties as very acidic lipids. The weakness here has already been discussed (Section III, E).

A variety of sulfolipids has been reported in insects, chicken eggs, etc., and has been reviewed by Haines (1971).

These scattered reports indicate that in our quest for sulfolipids only the surface has been scratched. Although there is a greater multiplicity of phospholipids and although they constitute a larger portion of the lipids in membranes, it appears that sulfatides are ubiquitous and that they are generally present in selected membranes.

VI. Sulfolipids and Membranes

The eukaryotic microorganisms are characterized by the presence of more than a single membrane. It is precisely this quality that makes a study of sulfolipids in eukaryotic microorganisms especially interesting. Where they have been studied, there has been but one sulfolipid in each membrane that contains them, and frequently this is the only sulfolipid in the microorganism. This substance then represents a tool for studying the membrane in question—its biogenesis, its metabolism, and its structure. For example, should the biosynthesis of the sulfolipid be inhibited selectively, then the biogenesis of that particular membrane would probably be blocked as well or its structure distorted. Studies of the structure and function of the membrane may be conducted by a replacement of a natural sulfolipid with analogs while the biosynthesis of the natural sulfolipid is blocked genetically or with an inhibitor. These approaches have been used in the study of chloroplast membranes by several investigators, as discussed earlier, but the full use of the approach has not yet been realized in this area. This has also been the case with the halosulfatides. In neither case is there a detailed understanding of the route of biosynthesis, nor have any mutants been reported to be missing appropriate enzymes. The whole area of lipid genetics is only just emerging in the prokaryotic microorganisms.

A final point should be made about a possible special role of the sulfolipid in membranes. The sulfate ester or sulfonic acid is unique in biological systems because of the extremely low pK value of the anion. It is therefore highly probable that a counterion will be present under nearly all circumstances in biological systems. It may very well be that their principal role is that of transporting the cation. Furthermore, those systems such as the stomach in which an extremely low pH is *maintained* are likely to contain a sulfate ester or sulfonic acid as the proton carrier, since few other organic groups (if any) can maintain such a low pH.

Acknowledgments

Acknowledgment is extended to the City University Research Foundation, the Office of Education, and the Petroleum Research Fund for their support of the author's work while this chapter was written. Appreciation is extended to Mrs. Clara Silver for her help in preparing the manuscript. The author appreciates the use of the space and facilities of E. Lederer at Gif-sur-Yvette, France, and E. E. Snell at Berkeley, California, during some of the work described.

References

Aaronson, S., and Baker, H. (1961). *J. Protozool.* **8,** 274–277.

Abraham, A., and Bachhawat, B. K. (1965). *Indian J. Biochem.* **1,** 192–199.

Allen, C. F., and Park, R. B. (1971). Personal communication.

Allen, C. F., Good, P., David, H. F., Chisum, P., and Fowler, S. D. (1966). *J. Amer. Oil Chem. Soc.* **43,** 223–231.

Appleman, D., Fulco, A. J., and Shugarman, P. M. (1966). *Plant Physiol.* **41,** 136–142.

Baer, E., and Stanacev, N. Z. (1964). *J. Biol. Chem.* **239,** 3209–3214.

Benson, A. A. (1963). *Proc. Nat. Acad. Sci. U.S.* **1145,** 571–574.

Benson, A. A., and Shibuya, I. (1961). *Federation Proc.* **20,** 79.

Benson, A. A., and Shibuya, I. (1962) *In* "Physiology and Biochemistry of Algae," (R. A. Lewin, ed.), p. 371–383. Academic Press, New York.

Benson, A. A., Daniel, H., and Wiser, R. (1959a). *Proc. Nat. Acad. Sci. U.S.* **45,** 1582–1587.

Benson, A. A., Wintermans, J. F. G. M., and Wiser, R. (1959b). *Plant Physiol.* **34,** 315–317.

Benson, A. A., Wiser, R., and Maruo, B. (1960). Proc. IV Intern. Congr. of Biochem., Vienna Vol. XV, 204.

Blix, G. (1933). *Hoppe-Seyler's Z. Physiol. Chem.* **219,** 82–98.

Brand, J., Krogmann, D. W., and Crane, F. L. (1971). *Plant Physiol.* **47,** 135–138.

Brown, F. S., and Hager, L. P. (1967). *J. Amer. Chem. Soc.* **89,** 719–720.

Busby, W. F. (1966). *Biochim. Biophys. Acta* **121,** 160–161.

Collier, R., and Kennedy, G. Y. (1963). *J. Mar. Biol. Ass. U.K.* **43,** 605–612.

Daniel, H., Miyano, M., Mumma, R. O., Yagi, T., Lepage, M., Shibuya, I., and Benson, A. A. (1961). *J. Amer. Chem. Soc.* **83,** 1765.

Danielli, J. F., and Davson, H. (1935). *J. Cellular Physiol.* **5,** 495–508.

Davies, W. H., Mercer, E. I., and Goodwin, T. W. (1965). *Phytochemistry* **4,** 741–749.

Davies, W. H., Mercer, E. I., and Goodwin, T. W. (1966). *Biochem. J.* **98,** 369–373.

Dreyfuss, J. (1964). *J. Biol. Chem.* **239,** 2292–2297.

Dubois, M., Gillies, K. A., Hamilton, J. K., Rebers, P. A., and Smith, F. (1956). *Anal. Chem.* **28,** 350–356.

Ellis, R. J. (1964). *Biochem. J.* **93,** 19P.

Elovson, J., and Vagelos, P. R. (1969). *Proc. Nat. Acad. Sci. U.S.* **62,** 957–963.

Elovson, J., and Vagelos, P. R. (1970). *Biochemistry* **9,** 3110–3126.

Emanuel, D., Stern, A., and Haines, T. H. (1972). *In preparation.*

Erwin, J., and Bloch, K. (1963). *Biochem. Z.* **338,** 496–511.

Erwin, J., and Bloch, K. (1964). *Science* **143,** 1006–1012.

Folch, M., Lees, M., and Sloane-Stanley, G. H. (1957). *J. Biol. Chem.* **226,** 497–509.

Gangadharam, P. R. J., Cohn, M. L., and Middlebrook, G. (1963). *Tubercle* **44,** 452–455.

Gellerman, J. L., and Schlenk, H. (1964). Personal communication.

Giovanelli, J., and Mudd, S. H. (1968). *Biochem. Biophys. Res. Commun.* **31,** 275–280.

Goldberg, I. H. (1961). *J. Lipid Res.* **2,** 103–109.

Goren, M. B. (1971). *Lipids* **6**, 40–46.

Gorter, E., and Grendel, F. (1925). *J. Exp. Med.* **41**, 439–443.

Green, D. E., and Perdue, J. F. (1966). *Proc. Nat. Acad. Sci. U.S.* **55**, 1295–1302.

Hager, L. P., Thomas, J. A., and Morris, D. R. (1970). *In* "Biochemistry of the Phago-cytic Process," (J. Schultz, ed.), pp. 67–87. North Holland Publ. Co. Amsterdam.

Hager, L. P., Thomas, J. A., and Morris, D. R. (1970).

Haines, T. H. (1965). *J. Protozool.* **12**, 655–659.

Haines, T. H. (1966). *Progr. Biochem. Pharmacol.* **3**, 184–188.

Haines, T. H. (1971). *Progr. Chem. Fats Other Lipids* **11**, 297–345.

Haines, T. H., and Block, R. J. (1962). *J. Protozool.* **9**, 33–38.

Haines, T. H., Pousada, M., Stern, B., and Mayers, G. L. (1969). *Biochem. J.* **113**, 565–566.

Helferich, B., and Ost, O. (1963). *Hoppe-Seyler's Z. Physiol. Chem.* **331**, 114–117.

Helmy, F. M., Hack, M. H., and Yaeg, R. C. (1967). *Comp. Biochem. Physiol.* **23**, 565–567.

Hodgson, R. C., Schiff, J. A., and Mather, J. P. (1971). *Plant Physiol.* **47**, 306–311.

Isono, Y., and Nagai, Y. (1966). *Jap. J. Exp. Med.* **36**, 461–476.

Ito, K. (1963). *Bull. Jap. Soc. Sci. Fish.* **29**, 771–775.

Jack, R. C. M. (1964). *Contrib. Boyce Thompson Inst.* **22**, 311–335.

Kamat, V. B., and Wallach, D. F. H. (1965). *Science* **148**, 1343–1345.

Kaplan, M. M., and Flavin, M. (1966). *J. Biol. Chem.* **241**, 5781–5789.

Kates, M. (1960). *Biochim. Biophys. Acta* **41**, 315–328.

Kates, M., and Tornabene, T. (1972). Personal communication.

Kates, M., Palameta, B., Perry, M. B., and Adams, G. A. (1967). *Biochim. Biophys. Acta* **137**, 213–216.

Kates, M., Wassef, M. K., and Kushner, D. J. (1968). *Can. J. Biochem.* **46**, 971–977.

Kean, E. L. (1968). *J. Lipid Res.* **9**, 319–327.

Kennedy, G. Y., and Collier, R. (1963). *J. Mar. Biol. Ass. U.K.* **43**, 605–612.

Kerr, D. S., and Flavin, M. (1968). *Biochem. Biophys. Res. Commun.* **31**, 124–130.

Kittredge, J. S., Simonsen, D. G., Roberts, E., and Jelinek, B. (1962). *In* "Amino Acid Pools" (J. T. Holden, ed.), p. 176–186. Elsevier, Amsterdam.

Klopfenstein, W. E., and Shigley, J. W. (1966). *J. Lipid Res.* **7**, 564–565.

Klopfenstein, W. E., and Shigley, J. W. (1967). *J. Lipid Res.* **8**, 350–353.

Kredich, N. M. (1971). *J. Biol. Chem.* **246**, 3474–3484.

Kredich, N. M., and Tompkins, G. M. (1966). *J. Biol. Chem.* **241**, 4955–4965.

Kuiper, P. J. C. (1970). *Plant Physiol.* **45**, 684–686.

Kylin, A. (1964). *Physiol. Plant.* **17**, 384–402.

Kylin, A. (1966). *Physiol. Plant.* **19**, 644–649.

Kylin, A. (1967). *Physiol. Plant.* **20**, 139–148.

Lees, M., Folch-pi, J., Sloane-Stanley, G. H., and Carr, S. (1959). *J. Neurochem.* **4**, 9–18.

Lehmann, J., and Benson, A. A. (1964a). *J. Amer. Chem. Soc.* **86**, 4469–4472.

Lehmann, J., and Benson, A. A. (1964b). *Proc. Int. Congr. Biochem., 6th, 1964* p. 66.

Lepage, M., Daniel, H., and Benson, A. A. (1961). *J. Amer. Chem. Soc.* **83**, 157.

Lichenthaler, H. K., and Park, R. B. (1963). *Nature (London)* **198**, 1070–1072.

MacMillan, J. (1954). *J. Chem. Soc., London* pp. 2585–2587.

Marshall, C. L., and Brown, A. D. (1968). *Biochem. J.* **110**, 441–448.

Mayers, G. L., and Haines, T. H. (1967). *Biochemistry* **6**, 1665–1671.

Mayers, G. L., Pousada, M., and Haines, T. H. (1969). *Biochemistry* **8**, 2981–2986.

Miyachi, S., and Miyachi, S. (1966). *Plant Physiol.* **41**, 479–486.

Miyachi, S., Miyachi, S., and Benson, A. A. (1966). *J. Protozool.* **13,** 76–78.

Miyano, M., and Benson, A. A. (1962a). *J. Amer. Chem. Soc.* **84,** 57–59.

Miyano, M., and Benson, A. A. (1962b). *J. Amer. Chem. Soc.* **84,** 59–62.

Mooney, C. L., Mahoney, E. M., Pousada, M., and Haines, T. H. (1972). *Biochemistry* In press.

Morris, D. R., and Hager, L. P. (1966). *J. Biol. Chem.* **241,** 1763.

Morris, L. J., and Wharry, D. M. (1966). *Lipids* **1,** 41–47.

Mumma, R. O. (1967). Personal communication.

Mumma, R. O., and Benson, A. A. (1961). *Biochem. Biophys. Res. Commun.* **5,** 422–423.

Mumma, R. O., and Gahagan, H. (1964). *Plant Physiol.* **39,** Suppl., XXV.

Nagai, Y., and Isono, Y. (1965). *Jap. J. Exp. Med.* **35,** 315–318.

Nichols, B. W. (1965). *Biochim. Biophys. Acta* **106,** 274–279.

Nichols, B. W., and James, A. T. (1964). *Fette, Seifen, Anstrichm.* **66,** 1003.

Niehaus, W. G., Kisic, A., Torkelson, A., Bednarczyk, D. J., and Schroepfer, G. J., Jr. (1970). *J. Biol. Chem.* **245,** 3791.

Nissen, P., and Benson, A. A. (1964). *Biochim. Biophys. Acta* **82,** 400–402.

Noriko, O., Galsworthy, P. R., and Pardee, A. B. (1971). *J. Bacteriol.* **105,** 1053–1062.

O'Brien, J. S., and Benson, A. A. (1964). *J. Lipid Res.* **5,** 432–436.

O'Brien, J. S., Filleru, D. L., and Mead, J. F. (1964). *J. Lipid Res.* **5,** 109–116.

Ohad, I., Siekevitz, P., and Palade, G. E. (1967). *J. Cell. Biol.* **35,** 521–552.

Okaya, Y. (1964). *Acta Crystallogr.* **17,** 1276–1282.

Orner, R., Haines, T. H., Aaronson, S., and Behrens, N. H. (1972). *J. Protozool.* (in press).

Oxford, A. E., Raistrick, H., and Simonart, P. (1939). *Biochem. J.* **33,** 240–248.

Pardee, A. B. (1966). *J. Biol. Chem.* **241,** 5886.

Pardee, A. B. (1967). *Science* **156,** 1627–1628.

Park, C., and Berger, L. R. (1967). *J. Bacteriol.* **93,** 221–229.

Pasternak, C. A. (1962). *Biochem. J.* **85,** 44–49.

Payne, W. J., and Painter, B. G. (1971). *Microbios* **3,** 199–206.

Pinto da Silva, P., and Branton, D. (1970). *J. Cell Biol.* **45,** 598–605.

Pohl, P., Glasl, H., and Wagner, H. (1970). *J. Chromatogr.* **49,** 488–492.

Poncz, L., and Haines, T. H. (1972). In preparation.

Pousada, M., Bruckstein, A., Das, B. P., and Haines, T. H. (1972a). In preparation.

Pousada, M., Das, B. P., and Haines, T. H. (1972b). In preparation.

Pousada, M., Roth, E., and Haines, T. H. (1972c). In preparation.

Pritchard, E. T. (1966). *J. Neurochem.* **13,** 13–21.

Pritchard, E. T. (1967a). *Arch. Oral Biol.* **12,** 1437–1444.

Pritchard, E. T. (1967b). *Arch. Oral Biol.* **12,** 1445–1456.

Radunz, A. (1968). *Hoppe-Seyler's Z. Physiol. Chem.* **349,** 303–309.

Radunz, A. (1969). *Hoppe-Seyler's Z. Physiol. Chem.* **350,** 411–417.

Roberts, R. B., Abelson, P. H., Cowie, D. B., Bolton, K. T., and Bolton, R. J. (1957). *Carnegie Inst. Wash. Publ.* **607,** 327.

Rosenberg, A. (1963). *Biochemistry* **2,** 1148–1154.

Rosenberg, A., and Gouax, J. (1967). *J. Lipid Res.* **8,** 80–83.

Rosenberg, A., and Pecker, M. (1964). *Biochemistry* **3,** 254–258.

Roughan, P. G., and Batt, R. D. (1968). *Anal. Biochem.* **22,** 74–88.

Rouser, G., Bauman, A. G., Kritchevsky, G., Heller, D., and O'Brien, J. S. (1961). *J. Amer. Oil Chem. Soc.* **38,** 544–555.

Rouser, G., Kritchevsky, G., and Yamamoto, A. (1967). *Lipid Chromatogr. Anal.* **1,** p. 99.

Roy, A. B., and Trudinger, P. A. (1970). "The Biochemistry of Inorganic Compounds of Sulfur." Cambridge Univ. Press, London and New York.

Russell, G. B. (1966). *Anal. Biochem.* **14**, 205–214.

Russell, G. B., and Bailey, R. W. (1966). *N.Z. J. Agr. Res.* **9**, 22.

Sastry, P. S., and Kates, M. (1963). *Biochim. Biophys. Acta* **70**, 214–216.

Schmitt, F. O., and Bear, R. S. (1939). *Biol. Rev. Cambridge Phil. Soc.* **14**, 27–50.

Sensi, P., DeFarrari, G. A., Gallo, G. G., and Rolland, G. (1955). *Farmaco, Ed. Sci.* **10**, 337–345.

Shibuya, I., and Benson, A. A. (1961). *Nature (London)* **192**, 1186.

Shibuya, I., and Hase, E. (1965). *Plant Cell Physiol.* **6**, 267–283.

Shibuya, I., Yagi, T., and Benson, A. A. (1963). *In* "Microalgae and Photosynthetic Bacteria" (Jap. Soc. Plant Physiol., ed.), pp. 627–636. Univ. of Tokyo Press, Tokyo.

Smith, C. G. (1958). *J. Bacteriol.* **75**, 577–583.

Stoffyn, P. (1966). *J. Amer. Oil. Chem. Soc.* **43**, 69–74.

Stoffyn, P., and Stoffyn, A. (1963). *Biochim. Biophys. Acta* **70**, 218–220.

Thierfelder, H., and Klenk, E. (1930). "Chemie der Cerebroside und Phosphatide," p. 63. Berlin and New York. Springer-verlag.

Thomas, D. R., and Stobart, A. K. (1970). *J. Exp. Bot.* **67**, 274–285.

Thudichum, J. L. W. (1874). *Rep. Med. Officer Privy Council, London* [N.S.] **3**, Append. 5, 134–247 (cited by Thierfelder and Klenk, 1930).

Wedding, R. T., and Black, M. K. (1960). *Plant Physiol.* **35**, 72–80.

Weenink, R. O. (1962). *Biochem. J.,* **82**, 523–527.

Weenink, R. O. (1963). *Nature (London)* **197**, 62–63.

Weibers, J. L., and Garner, H. R. (1967). *J. Biol. Chem.* **242**, 5644–5649.

Weier, T. E., and Benson, A. A. (1967). *Amer. J. Bot.* **54**, 389–402.

Wickberg, B. (1957). *Acta Chem. Scand.* **11**, 506–511.

Wilkins, M. H. F. (1972). *Ann. N.Y. Acad. Sci.* **195**, 291–292.

Wilson, L. G., and Bandurski, R. S. (1958). *J. Biol. Chem.* **233**, 975–981.

Wintermans, J. F. G. M. (1960). *Biochim. Biophys. Acta* **44**, 49–54.

Wood, J. B., Nichols, B. W., and James, A. T. (1965). *Biochim. Biophys. Acta* **106**, 261–273.

Yagi, T., and Benson, A. A. (1962). *Biochim. Biophys. Acta* **57**, 601–603.

Yamakawa, T., Kiso, N., Handa, S., Makita, A., and Yokoyama, S. (1962). *J. Biochem. (Tokyo)* **52**, 226–227.

Yamamoto, L. A., and Segel, I. H. (1966). *Arch. Biochem. Biophys.* **114**, 523–538.

Zill, L. P., and Cheniae, G. M. (1962). *Annu. Rev. Plant. Physiol.* **13**, 225–264.

CHAPTER 5

Chloroplast Lipids of Photosynthesizing Eukaryotic Protists

Abraham Rosenberg

I. Introduction

Protists have served as powerful tools for the study of chloroplast development and chloroplast function under controllable conditions. Many classes of photosynthesizing protists, commonly grouped together as *algae*, have been used for such studies. Most notable among them are hardy euglenids (Euglenineae) and other green algae (Chlorophyceae), which may be grown almost as effortlessly as weeds. The genealogical range of photosynthesizing protists that have been under study is extensive. It reaches from the apparently primitive Chlorophyceae of the order Volvocales (such as *Chlamydomonas*) and the primitive Chrysophyceae of the order Chrysomonadales (such as *Ochromonas*) through assorted freshwater monads, marine dinoflagellates, diatoms, and red and brown to green algae. This section treats some of the findings concerning complex lipids as related to studies of chloroplast development and function made with these eukaryotic photosynthesizing protists. The findings must be compared on a sliding scale. Chloroplast development, composition, structure, and function all have been found to change dramatically with conditions, and from one eukaryote to another—changes in degree but generally not in kind. Similarities outweigh differences. The current growing body of knowledge tends to indicate that one may, in a useful manner, combine and then extrapolate to a generalized concept the bits of information regarding the nature of the chloroplast of the eukaryote obtained through investigations made with various protistan individuals. The information covers three broad categories of investigation: (1) the nature of the major lipid components of protistan chloroplasts, (2) the routes whereby these lipid components are biosynthesized and metabolized, and (3) postulated developmental processes whereby these components go to build chloroplasts under controlled conditions.

II. Nature of the Lipid Components

Genealogical Progression

Going from the photosynthetic bacteria to the prokaryotic blue-green algae and then to the eukaryotic photosynthesizing protists, one finds an interesting change in organismic lipid components. For an excellent documentation of this progression, the reader is referred to reviews by James and Nichols (1966a) and Nichols and James (1968). Lipids of five strains of photosynthetic bacteria (Antiorhodaceae) have been extracted by Wood, Nichols, and James (1965) and identified by one- and two-dimensional thin-layer chromatography on silica gel. Of these strains, *Rhodopseudomonas gelatinosa* alone has been found to contain sulfoquinovosyl diglyceride, a compound that is, however, intimately related to eukaryotic chloroplast structure. See Fig. 1 for the structure of this and other chloroplast-typical lipids. *Rhodopseudomonas spheroides*, *capsulata*, and *palustris* contain phosphatidyl choline, but this lipid is absent in *Rhodopseudomonas gelatinosa* and *Rhodospirillum rubrum*. However, five of the bacteria contain phosphatidyl glycerol and phosphatidyl ethanolamine. Galactosyl diglycerides, which appear to constitute components sine qua non for oxygen-evolving eukaryotic chloroplasts, have not been detected in these bacteria.

In a further study, involving the more highly developed blue-green algae, Nichols and Wood (1968) extracted total lipids from six different organisms and examined them by thin-layer chromatography on silica gel and by column chromatography on diethylaminoethyl cellulose. They found that the Myxophyceae—*Spirulina platensis*, *Myxosarcina chroococcoides*, *Chlorolgloea fritschii*, *Anabena cylindrica*, *Anabaena flosaquae*, and *Mastigocladus laminosus*—had neither detectable phosphatidyl choline nor phosphatidyl ethanolamine, in contrast with the bacteria. These algal prokaryotes, however, now contained mono- and digalactosyl and sulfoquinovosyl diglycerides. In common with the photosynthesizing bacteria, they still possessed phosphatidyl glycerol. Along with chlorophylls, these three classes of lipid compounds, that is, sulfoquinovosyl diglycerides, galactosyl diglycerides, and phosphatidyl glycerols, and greater or lesser amounts of phosphatidyl choline appear to typify the structural lipid complement of the chloroplast of the eukaryote. Whether or not phosphatidyl ethanolamine and phosphatidyl inositol are chloroplast components is undecided for the moment. Consideration of other lipophilic components, for example, carotenoids, quinones, and tocols (Crane *et al.*, 1966), are beyond the purview of this chapter. As James and Nichols have pointed out (1966b), the ubiquitous lipid component for photosynthesis thus may be phosphatidyl glycerol, along with specific lipid or nonlipid photoreceptor molecules. For higher forms (eukaryotes) that can convert radiant energy into chemical potential and can also (as lower forms—bacteria and many prokaryotic algae—cannot) reduce strong oxidants with the concomitant release of molecular oxygen—the so-called chloroplast or Hill (1951) reaction—galactosyl diglycerides and sulfoquinovosyl diglycerides may be obligate structural components of the functioning photochemical transducer. Implication that these lipids are restricted to photosynthesizing organisms is not intended. Small amounts of galactosyl diglycerides occur in mammalian brain (Rouser and Yamamoto, 1969), and various glycosyl diglycerides (Lennarz, 1966) and phosphatidyl glycerol are (Kates, 1964), for example, found in many bacteria. The fatty acyl compositions, quantities, and types of membrane structure in which they are included differ vastly for such bacteria and for the eukaryotic photosynthesizing algae.

III. Major Lipid Components

Structures of the major chloroplast lipids are shown in Fig. 1. The major lipids of eukaryotic chloroplasts are seen to cover the range from polar nonionic (galactosyl diglycerides) to strongly anionic (sulfoquinovosyl diglycerides). While the gross morphology of their chloroplasts, when

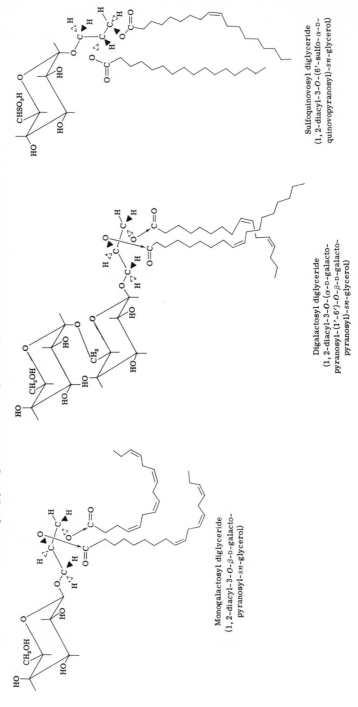

Major glycolipids of the chloroplasts of photosynthesizing eukaryotic protists.

Sulfoquinovosyl diglyceride
(1, 2-diacyl-3-*O*-(6′-sulfo-α-D-quinovopyranosyl)-*sn*-glycerol)

Digalactosyl diglyceride
(1, 2-diacyl-3-*O*-(α-D-galacto-pyranosyl-(1′-6′)-*O*-β-D-galacto-pyranosyl)-*sn*-glycerol)

Monogalactosyl diglyceride
(1, 2-diacyl-3-*O*-β-D-galacto-pyranosyl-*sn*-glycerol)

Major phospholipids of the chloroplasts of photosynthesizing eukaryotic protists.

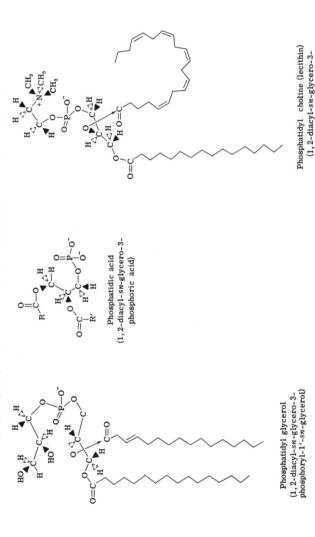

Phosphatidyl glycerol
(1, 2-diacyl-*sn*-glycero-3-
phosphoryl-1'-*sn*-glycerol)

Phosphatidic acid
(1, 2-diacyl-*sn*-glycero-3-
phosphoric acid)

Phosphatidyl choline (lecithin)
(1, 2-diacyl-*sn*-glycero-3-
phosphorylcholine)

Fig. 1. Major lipid components of the chloroplasts of eukaryotic protists. The fatty acyl groups are shown as typical for each lipid, although many other kinds of fatty acids may occur as indicated in the text. The fatty acids shown in the diagrams are as follows: For monogalactosyl diglyceride, position 1, octadecatrienoic (Δ9, 12, 15) acid; position 2, hexadecatetraenoic (Δ4, 7, 10, 13) acid. For digalactosyl diglyceride, position 1, octadecamonoenoic (Δ9) acid; position 2, hexadecadienoic (Δ9, 12) acid. For sulfoquinovosyl diglyceride, position 1, hexadecanoic acid; position 2, octadecenoic (Δ9) acid. For phosphatidyl glycerol, position 1, hexadecanoic acid; position 2, hexadecenoic (Δ3-*trans*) acid. For phosphatidyl choline, position 1, hexadecanoic acid; position 2, eicosipentaenoic (Δ5, 8, 11, 14, 17) acid. Fatty acid groups for phosphatidic acid are not specified since they may be variable, as indicated by their possible precursor role for the other lipids.

fully formed and functional, may differ greatly from protist to protist (Kirk and Tilney-Bassett, 1967), the lipid compositions are notably similar. Chloroplasts per se have rarely been isolated (Rosenberg, 1963) from protistan organisms. The organellar standard for reference perforce has been chloroplasts of higher plants and the chloroplast lamellae, substructural membranous components of the chloroplasts most closely associable with the phenomenon of photosynthesis (Allen et al., 1966a). Insofar as protists are concerned, information on chloroplast lipid composition has mostly been indirect and has usually been derived from differences between the photosynthesizing forms and the nonphotosynthesizing, essentially achloroplastic forms that may be obtained from such organisms as Euglena (Epstein and Schiff, 1961), Ochromonas (Gibbs, 1962), or Chlamydomonas (Ohad et al., 1967) by withholding light or Chlorella (Shibuya and Hase, 1965) by deprivation of available nitrogen in a glucose-rich medium.

For cellular components, it appears that photosynthetic membrane structures are among the most rich in lipid. They are generally very roughly half lipid on a mass basis and preponderantly lipid on a molecular basis (Park and Pon, 1961, 1963; Menke, 1966). It is remarkable that the total lipid of fully photosynthetic protists, for example, Chlorella, Anacystis, and Euglena (Allen et al., 1966a; Ferrari and Benson, 1961; Nichols, 1965; Rosenberg, 1963), resembles the isolated higher-plant chloroplast itself in composition insofar as the quantitatively major lipid components are concerned. The inference here is that chloroplast membranes constitute a great part of the total membranous elements of the photosynthesizing eukaryotic protist, which lends validity, in a sense, to the assumption that data derived from comparisons between nonphotosynthesizing and photosynthesizing protistan forms are directly informative of protistan chloroplast structure. This is not to say that the chloroplast membranes obligatorily increase the overall lipid burden of the cell. Chloroplast membranes may be synthesized at the expense of storage lipid (Rosenberg, 1963). In all chloroplasts (Nichols and James, 1968), galactosyl diglycerides, mostly the monogalacto compound, predominate in quantity among the lipids, and therefore, in all probability, these compounds comprise the basic bulk structural lipid substance of the lamellae of the eukaryotic organism's chloroplast (Rosenberg, 1967).

A. GALACTOSYL DIGLYCERIDES

1. Biosynthetic Pathway

Spinach chloroplasts have been shown to utilize UDP-gal in the synthesis of mono- and digalactosyl diglycerides and, possibly, higher homologs (Ongun and Mudd, 1968; Neufeld and Hall, 1964). The lipid acceptor for the biosynthetic transfer of galactose is unknown but is presumed to be diglyceride:

$$sn\text{-}1,2\text{-diglyceride} \xrightarrow{\text{(UDP-gal)}} \text{monogalactosyl diglyceride} \xrightarrow{\text{(UDP-gal)}}$$
$$\rightarrow \text{digalactosyl diglyceride, etc.}$$

Although kinetic data that indicate that label from $^{14}CO_2$ or [^{14}C]acetate is incorporated by *Chlorella* more rapidly into the mono- than the digalactosyl diglycerides of the organism (Ferrari and Benson, 1961) are consistent with the hypothesis that the monogalacto compound is a precursor for the digalacto compound, it appears that for eukaryotic photosynthesizing protists, the galactosylation of diglyceride molecules to form monogalactosyl diglycerides and the further galactosylation of the latter to form digalactosyl diglyceride is, for the moment, only with difficulty reconcilable with other experimental observations. The fatty acyl compositions of the two lipids are significantly different in all organisms studied. This finding would seem either to preclude a precursor role for the monogalactosyl compound or to require an as yet undemonstrated, relatively specific de-acylation and reacylation of the digalactosyl compound (Nichols and James, 1968). Renkonen and Bloch (1969) have produced ^{14}C-labeled monogalactosyl diglycerides utilizing a crude cell-free extract from *Euglena gracilis* and 1-^{14}C-oleate or stearate in thioester linkage to acyl carrier protein of *Escherichia coli* (Goldfine et al., 1967) as fatty acyl donor. Although in some experiments high levels of UDP-gal were added in addition to the labeled fatty acid, only monogalactosyl diglycerides appeared to be labeled. It is difficult to interpret these findings, since phosphatidic acid seemed to be the precursor for the monogalactosyl diglyceride, the implication being that the following pathway is operative for the biosynthesis of monogalactosyl diglyceride in *Euglena*:

$$2 \text{ (Fatty acyl-ACP)} + 3\text{-phosphoglycerol} \xrightarrow{\text{transferase}}$$
$$\text{phosphatidic acid} \xrightarrow{\text{hydrolase}} \text{diglyceride}$$
$$\text{Diglyceride} + \text{UDP-gal} \xrightarrow{\text{transferase}} \text{monogalactosyl diglyceride}$$

It is of interest that chloroplasts obtained from *Euglena gracilis* (Matson et al., 1970) have been found to incorporate label more rapidly from UDP-[^{14}C]gal into the digalactosyl diglycerides within 1 hour than into the monogalactosyl diglycerides. Also, during the early stage of light-induced chloroplast formation in *Euglena gracilis*, it has been shown by Rosenberg and Gouaux (1967) that digalactosyl diglycerides are synthesized more rapidly than monogalactosyl diglycerides (with a different fatty acid composition), although monogalactosyl diglyceride synthesis eventually predominates. Nevertheless, in the experiment by Matson *et al.* the chloroplast material presumably must have contained, to begin with, a great deal of structural galactosyl diglyceride, so that rapid labeling of the digalacto compound could be, possibly, the result of the availability of much monogalactosyl diglyceride but little free diglyceride. In the latter experiment by Rosenberg and Gouaux, rapid galactosylation of newly formed monogalactosyl diglyceride could explain the findings. Although the current information is still fragmentary, a common thread would appear, namely that structural placement, that is, specific lipid-protein association regulated by the nature of the fatty acyl components of the lipid (Weier and Benson, 1966), predetermines the availability of the lipid precursor for mono-, or for di- or multiple, galactosyl transfer from UDP-gal, possibly by regulating the physicochemical substrate nature of the precursor lipid.

2. *Quantitative Considerations*

Shortly after their description in green *Euglena* (Carter *et al.*, 1961), it was adduced (Rosenberg and Pecker, 1964) from their appearance along with chlorophyll that the galactosyl diglycerides are related to chlorophyll as partners in structuration of the functioning chloroplast. The finding that a relatively invariant ratio of chlorophyll to galactosyl diglyceride is maintained in *Euglena* (Rosenberg and Gouaux, 1967) during light-induced accretion of chloroplasts supports the contention that galactosyl diglycerides, in essence, function as a stabilizing matrix for chlorophyll molecules in the photosynthetic lamellae and that the constant ratio of chlorophyll to galactosyl diglyceride reflects a dilution of the former by the latter to afford the requisite degree of separation of chlorophyll molecules for efficient photoreception (Rosenberg, 1967). Temporally stable combinations of the lipophilic diglyceride and phytol residues of the galactosyl diglyceride and chlorophyll molecules, respectively, could serve, as well, as unit substrates for the specific transfer and hydrolytic removal of the hydrophilic galactose or chlorophyllide end groups (Patton, 1968). Trosper and Sauer (1968) have reported on the interaction between chlorophyll and mono- and digalactosyl diglycerides and sulfoquinovosyl diglyceride in three-dimensional solution and as monolayers. The results of their circumscribed study with the isolated lipophilic chloroplast components, based upon spectral properties and compression behavior, show that galactosyl diglycerides and chlorophyll form an ideal solution and that the interaction of chlorophyll and galactosyl diglyceride is powerful enough to dissociate chlorophyll dimers and to effectively do so in an environment that contains water. Observations that point to the maintenance of a relatively constant molar ratio of galactosyl diglycerides to chlorophyll by the photosynthesizing organism under a variety of environmental and developmental conditions continue to accrue. Light deprivation of photobiotic euglenas in a mineral medium for a week occasioned only a small loss of chlorophyll and a diminution of galactosyl diglyceride until a unimolecular ratio of these lipids to chlorophyll obtained. Thereafter, light starvation occasioned chloroplast degeneration and a simultaneous, equivalent, loss of both chlorophyll and galactosyl diglyceride, so that despite the loss, a unimolecular ratio remained (Rosenberg, 1967). Interference with light-induced chloroplast formation in euglenas by growth under manganese deficiency (Constantopoulos, 1970) occasioned a similar reduction in chlorophyll and in galactosyl diglyceride levels, so that a constant ratio was obtained regardless of the degree of inhibition of chloroplast formation by the reduced availability of manganese. Bishop and Smillie (1970) have investigated the effect of inhibition of protein synthesis during chloroplast development in *Euglena* by chloramphenicol and cycloheximide. Etiolated cells were first starved to deplete the wax esters (Rosenberg, 1963, 1967; Guehler *et al.*, 1964). Over 48 hours of constant illumination to produce chloroplasts, increasing the inhibition of chlorophyll production (which was taken as an index of the inhibition of chloroplast biosynthesis), by increasing the amount of chloramphenicol, was synchronous with corresponding decreases in galactosyl diglycerides. The total phospholipid content of the cell remained relatively unchanged. With cycloheximide, there was reversible inhibition of phospholipid biosynthesis as well as of cell division. The initial lag phase noted in the production of chlorophyll when etiolated euglenas are illuminated (Rosenberg and Gouaux, 1967) was increasingly augmented with increasing but low (2.5–5.0 μg/5 $\times$ 10^6 cells) concentrations of cycloheximide. A synchronous and proportionate lag in the production of galactosyl diglycerides was noted; a simultaneously accelerated rate of synthesis of both components was observed to occur after the lag. Total phospholipid production was severely inhibited, but no correspondence with chlorophyll was noted. Melandri *et al.* (1970), in an investigation of the effect of adaptation of *Euglena* to change in population density, measured the quantitative relationship between chlorophyll concentration and the level of other chloroplast components. This group had previously reported (Böger and San Pietro, 1967) that the increase in cellular chlorophyll with population expansion is not accompanied by a corresponding increase either in ferredoxin or in cytochrome-552, which are key proteins for the photosynthetic electron transport machinery. They now have additionally observed that particle-bound cytochrome-559 increases along with chlorophyll, as does plastoquinone A, but that ferredoxin-NADP reductase remains unchanged. Cellular chlorophyll content increases linearly with cell number, ranging from 7 to 25 $\mu g/10^6$ cells, during the logarithmic phase of growth. In the expanding population of organisms, synthesis of monogalactosyl diglyceride paralleled cellular chlorophyll, so that a 2:1 ratio of the former to the latter was always maintained. Under the conditions employed, digalactosyl diglycerides at first remained essentially static in amount and then began to increase as cell population grew, but their quantity was always sufficiently low to exert only a minor influence on the ratio of total galactosyl diglyceride to chlorophyll (far less than an order of magnitude). Carotenoids increased moderately. Along with cytochrome-559, Hill activity (Hill, 1951), as measured by the photoreduction of ferricyanide, increased in parallel with cellular chlorophyll accretion. As measured by the photoreduction of 2,6-dichlorophenolindophenol, it increased even more greatly. As noted above, in the genealogical progression from photosynthesizing bacteria to blue-green algae to green algae, the first appearance of galactosyl diglycerides along with chlorophyll, in the blue-green algae, coincides with the appearance of the ability to carry out the Hill reaction. The aforementioned study by Melandri et al. also gives indication of the possibility of an independent structuration of photosystems I and II (Racker, 1970) in the chloroplast; the reactions attributed to the latter appeared to increase more markedly. The possible relationship of complex formation between specific chlorophylls and galactosyl diglycerides to individual photosystems becomes, by inference, an intriguing open question. Although these studies have been mostly restricted to *Euglena*, one may note in the data of Lichtenthaler and Park (Lichtenthaler and Park, 1963) for higher-plant (spinach) chloroplasts a similar molecular ratio of 2:1 for galactosyl diglycerides and chlorophyll. Bloch et al. (1967) have observed that the synchrony of chlorophyll and galactosyl diglyceride production also occurs in illuminated *Chlorella*.

Great variations in the relative proportions of mono- and digalactosyl diglycerides are usually found in the very same organism with changing conditions of growth and illumination. In most of the studies already described, where quantitative measurement of mono- and digalactosyl diglyceride has been made, variable ratios of these two major chloroplast glycolipids have been found. These variable ratios reflect clear differences between the two lipids in the ways in which changing cellular requirements may affect the structuration of the chloroplast lamellae where they comprise the major lipid components. As mentioned previously, the usually differing fatty acyl compositions of the mono- and digalacto compounds would seem to preclude their ready interchangeability by the concerted action of galactohydrolases and galactosyl transferases unless a complex metabolic pattern involving fatty acyl exchange, concurrently with glycosyl transfer, also occurs. When dark-grown *Euglena* was illuminated in a mineral medium under air to induce chloroplast formation (Rosenberg and Gouaux, 1967), lipid-bound galactose and chlorophyll accrued at almost the same rate over 160 hours of illumination. The amount of galactose in the digalactosyl diglyceride fraction, before chloroplast synthesis occurred, was roughly half that in the monogalactosyl diglyceride; that is, there was roughly a 4:1 ratio of the two compounds. As greening progressed over the first 80 hours, both compounds accumulated, but the digalacto compound increased more rapidly until it almost equaled the monogalacto compound in amount; then the latter increased more rapidly until, after 180 hours of illumination, the ratio of mono- to digalactosyl diglyceride was 2:1, and thus the number of molecules of lipid-bound galactose was distributed equally between the two compounds. When the greening of *Euglena* was carried out in a mineral medium, under air containing 5% carbon dioxide, there was almost an equal distribution of galactose between the two lipids as they increased in amount over some 25 hours of illumination, again indicating a continuous 2:1 molecular ratio of mono- to digalactosyl diglyceride. But monogalactosyl diglyceride then began, and continued, to increase more rapidly for 45 hours thereafter (Bloch *et al.*, 1967). In the same investigation, *Chlorella* grown in a glucose-supplemented medium and illuminated under air containing 5% CO_2 started out with roughly a 2:1 molecular ratio of mono- to digalactosyl diglyceride, but the monogalacto compound began to accumulate far more rapidly until, after 60 hours of illumination, the ratio of the mono- to the digalacto compound was close to 5:1. In the effect of population density of *Euglena* on cellular chlorophyll accretion, by Melandri *et al.* (1970), the mono- and digalacto lipids began with a 2:1 ratio, but after roughly a 7.5-fold increase in cell population, which was accompanied by a doubling of the amount of chlorophyll per organism, the ratio of mono- to digalactosyl diglyceride rose one order of magnitude to reach approximately 3:1. In a different kind of study, illumination of dark-grown, etiolated, *Euglena* for 12 hours in the presence of 2.5 $\mu g/ml$ of the inhibitor of organismic protein synthesis, cychloheximide, halted population increase, but chlorophyll and galactosyl diglyceride accumulated nevertheless. Under these conditions, the ratio of mono- to digalactosyl diglyceride rose from 2.5 for controls to 3.7 for cycloheximide-treated organisms (Bishop and Smillie, 1970). In *Chlorella pyrenoidosa* illuminated for 6 days in an inorganic medium, O'Brien and Benson (1964) found somewhat higher levels of di- than monogalactosyl diglyceride. They also found the same for August Alfalfa leaves, but for spinach lamellae, Allen *et al.* (1966a) found a value for mono- to digalactosyl diglyceride of close to 2:1. Whether the mono- or digalacto lipids impart different hydrophilic surface characteristics to the lamellar membrane by virtue of the larger number of polar groups in the latter compounds and whether the mono- or digalacto compounds associate with different proteins or other cellular components by virtue of their differing fatty acyl residues are completely open questions for the time being. It is not unreasonable, however, to presume that the degree of change observed in the ratio of the mono- and digalacto compounds under various conditions reflects changes in the nature of the lamellae, be they subtle or gross. There are not yet any real clues to the biological significance of the changes that are observed.

3. *Fatty Acid Composition*

A very thorough and detailed accounting of the total polyunsaturated fatty acid composition of microorganisms, including eukaryotic protists, has been prepared by Shaw (1966), and there also are available several excellent compilations of the fatty acyl compositions of specific lipids, including the monogalactosyl and digalactosyl diglycerides, for example, those of Nichols and James (1968) and Allen *et al.* (1966b). From the compiled information and from work published subsequently, some of which will be discussed below, a few salient features of the fatty acyl compositions of the mono- and digalactosyl diglycerides of eukaryotic photosynthesizing protists emerge. The fatty acids are mostly restricted to the 16- and 18-carbon variety, and in functioning chloroplasts, most of the fatty acyl groups of the galactosyl diglycerides contain one, two, three, or four double bonds, and more of the multiple than the unitary degree of unsaturation. The digalactosyl diglycerides tend to have more saturated fatty acid, mostly 16-carbon (palmitic) but little 18-carbon (stearic) fatty acid (Constantopoulos and Bloch, 1967; Nichols, 1965; Allen *et al.*, 1964; Rosenberg *et al.*, 1966), than the monogalactosyl diglycerides. For contrast, the prokaryotic algae (blue-green), for example *Anacystis nidulans*, have mostly 16-carbon saturated fatty acid and only monoenoic 16-carbon fatty acid (a double bond at position 9 from the carboxyl group) and monoenoic 18-carbon fatty acid (double bond at either position 9 or 11) in either the mono- or digalactosyl diglycerides (Allen *et al.*, 1966b). Rosenberg *et al.* (1966) have found a similar relatively saturated fatty acid composition in the small and sometimes overlooked (Hulanicka *et al.*, 1964) quantity of digalactosyl diglycerides that appear always to occur in achloroplastic (either etiolated dark-grown, or bleached mutant) *Euglena*

(Rosenberg, 1963; Rosenberg and Pecker, 1964; Rosenberg and Gouaux, 1967; Helmy et al., 1967; Bishop and Smillie, 1970). It has not been determined whether the galactosyl diglycerides of the achloroplastic organisms are components of the spherical structural primordia for chloroplasts that have been observed in Euglena (Ben-Shaul et al., 1964). Upon illumination under photobiotic conditions, dark-grown euglenas accumulate galactosyl diglycerides with di-, tri-, and tetraenoic unsaturated 16- and 18-carbon fatty acids and develop functioning chloroplasts after a short lag period (Rosenberg and Gouaux, 1967). 16-Carbon, dienoic fatty acid becomes quantitatively the major fatty acyl component of both the mono- and digalactosyl diglycerides. In monogalactosyl diglycerides, 16-carbon tetraenoic and trienoic fatty acids and 18-carbon trienoic and dienoic fatty acids all run a close second to 16-carbon dienoic fatty acid as the major component. In the digalactosyl diglycerides, 16-carbon dienoic and 18-carbon di- and trienoic fatty acids do likewise. Although there are distinct variations in relative quantity from organism to organism, and intraorganismic changes occur with changed conditions, the aforementioned fatty acids appear to comprise the major unsaturated fatty acyl components in the galactosyl diglycerides of the photosynthetic organelles of eukaryotic organisms. Although exceptions may be found, the double bonds usually are uniformly of the cis configuration and are methylene-interrupted. Their placement, counting from the carboxyl, is generally as follows: for 16-carbon acids—7,10 for the dienes, 7,10,13 for the trienes, and 4,7,10,13 for the tetraenes (mostly but not exclusively located in the monogalacto compound); for 18-carbon acids—all-cis, methylene-interrupted, with double-bond placement for dienes of 9,12, for trienes of 9,12,15, and for tetraenes of 6,9,12,15. Smaller amounts of other fatty acids in addition to 16- and 18-carbon saturated and 9- and 11-monoenoic fatty acids also are found, notably 14-carbon saturated and 17-carbon monoenoic fatty acid whose double-bond placement is mostly at position 9 (Rosenberg et al., 1965) despite the odd number of carbon atoms, and 20-carbon 5,8,11,14-unsaturated (arachidonic) acid and higher 20-carbon unsaturated homologs. The review by Nichols and James (1968) gives a useful listing of many of the fatty acyl components of photosynthetic tissues, including those of the photosynthesizing protistan eukaryotes. The relative proportions of fatty acyl components of the galactosyl diglycerides are observed to change not only with organismic population growth, organellar development, and nutritional conditions, described in various studies as indicated previously, but also with environmental stress, for example, destruction of chloroplast components (Constantopoulos and Bloch, 1967) at excessive light intensity, growth in the presence of inhibitors of protein synthesis (Bishop and Smillie, 1970), and elevated temperature (Kleinschmidt and McMahon, 1970).

4. *Functional Considerations*

Of the four major chloroplast lipids (the monogalactosyl diglycerides, digalactosyl diglycerides, sulfoquinovosyl diglycerides, and glycerophosphoryl diglycerides), the galacto lipids comprise not only the bulk of the organellar lipid but also their fatty acid components tend to be the most unsaturated, so that they contribute the large mass of methylene-interrupted, *cis* unsaturated hydrocarbon chains to the membranes, presumably the lamellae, of the chloroplasts. Aside from their potential physicochemical role in providing an ideal mixing phase as stabilizing matrix for the lipophilic photoreceptive pigments (Rosenberg, 1967), arguments and counter-arguments have been advanced for the role of certain of their fatty acyl components, notably 9,12,15-octadecatrienoic (α-linolenic) acid and 4,7,11,15-hexadecatetraenoic acid, in the functioning of the system responsible for noncyclic photophosphorylation (Arnon, 1959) or, the electron transport system of photosystem II (Witt *et al.*, 1965) as it relates to the Hill reaction and the photosynthetic evolution of oxygen in eukaryotes. It is of more than passing interest that photosynthetic saltwater as well as freshwater monads have been discovered by Beach *et al.* (1970) to contain in their monogalactosyl diglycerides a very high percentage of 18-carbon tetraenoic fatty acid, rather than 16-carbon tetraenoic acid as is the case for *Euglena*, that diatoms are reported to have only 20-carbon polyenes (Kates and Volcani, 1966), and that a marine dinoflagellate, *Gonyaulax*, has polyenoic fatty acids of the 20-carbon variety as well as 18-carbon tetraenoic fatty acids (Patton *et al.*, 1966). *Anacystis nidulans*, a prokaryote, has none of these fatty acids and only monoenoic fatty acids for the unsaturated fatty acid complement of its galactosyl diglycerides, yet it is capable of Hill activity, and this has been taken as the exception that destroys the rule (Holton *et al.*, 1964; Constantopoulos and Bloch, 1967). Furthermore, many photosynthesizing eukaryotic organisms reportedly may lack one or the other or even all of the polyenoic fatty acids and still fully carry on the Hill reaction. Nevertheless, there appears to be a certain validity in the basic concept, that is, that a hydrocarbon matrix that is rich generally in *cis* double bonds possibly, although not necessarily, in methylene-interrupted sequence within each hydrocarbon chain, as provided by the galactosyl diglycerides of the lamallae, acts as a lamellar factor that is both variable and controllable metabolically and can in some way influence photosynthetic electron transport, the photolysis of water, and oxygen evolution in the eukaryotic organism.

5. *In Situ Molecular Changes*

There appears to be a compositionally related difference between the kind of positional placement of the fatty acyl residues in the galactosyl diglycerides in green algae (e.g., *Chlorella*) compared with higher plants (Safford and Nichols, 1970). In the former, chain length seems to be the determining factor, with 18-carbon fatty acid preferentially accumulating at the 1 position of the *sn*-glycerol backbone of the compound and 16-carbon fatty acid at position 2. In higher plants, there is often little 16-carbon fatty acid in the galactosyl diglycerides, so that, generally speaking, positional placement now relates more closely to degree of unsaturation. In a previous study and in the study cited (Nichols and Moorehouse, 1969), intriguing evidence has been presented that the degree of unsaturation of the fatty acyl residues of the monogalactosyl diglycerides may be increased subsequent to the total synthesis of the molecule and that the fatty acids at both positions 1 and 2 may be further desaturated. Further evidence along these lines has been produced by Gurr (1970). Incorporation studies with ^{14}C-labeled acetate indicate that newly synthesized monogalactosyl diglyceride molecules have fatty acyl groups that for the most part are saturated, but that subsequent to synthesis of monogalactosyl diglycerides their fatty acyl groups are desaturated. These findings open an intriguing set of possibilities: (1) that the mono- and digalacto compounds are found to differ in their fatty acyl composition because of specific desaturations subsequent to molecular synthesis and (2) that the biosynthetic machinery as regards precursor specificity for these compounds is relatively independent of chloroplast structuration but that, in the latter process, fatty acyl desaturations of the galactosyl diglycerides occur that can tailor the physicochemical nature of the hydrocarbon matrices of the chloroplast lamellae to correspond to proper photoreceptor dispersal in accordance with inner and outer environmental dictates. As mentioned previously, numerous studies have shown that before chloroplast development, at least in *Euglena*, there is a recognizable cellular galactosyl diglyceride component and that its fatty acyl residues tend to be mostly saturated and monoenoic, but that with chloroplast development and accretion of galactosyl diglycerides, these compounds have a high unsaturated fatty acyl complement.

B. SULFOQUINOVOSYL DIGLYCERIDES (SULFOLIPIDS)

1. *Biosynthesis*

Sulfur compounds as disparate as sulfate (Abraham and Bachhawat, 1963; Shibuya and Hase, 1965) and cysteic acid (Davies *et al.*, 1966) have been observed to be incorporated readily into protistan sulfoquinovosyl diglycerides. A major pathway for biosynthesis *in vivo* of these molecules has not been worked out conclusively. Sulfolactaldehyde, sulfolactic acid, a sulfophosphoketohexose, and disulfoquinovose (Shibuya *et al.*, 1963) have all been indicated to occur in *Chlorella*. These findings, coupled with the likely possibility of deamination or transamination of cysteic acid to sulfopyruvate, have led to the following postulated biosynthetic route:

Phosphoenolpyruvate $\xrightarrow{\text{phosphoadenosyl phosphosulfate}}$

2-phospho-3-sulfolactate $\xrightarrow{-(\text{H}_3\text{PO}_4)}$ 3-sulfolactate $\xrightarrow{\text{ATP}}$ 1-phospho-3-sulfolactate $\rightarrow$

3-sulfolactaldehyde $\xrightarrow{\text{dihydroxyacetone phosphate}}$ 6-sulfo 6-deoxyfructose-1-phosphate $\rightarrow$

6-sulfoquinovose

and then (Benson, 1963)

Nucleosidediphospho-6-sulfoquinovose $\xrightarrow{sn-1,2-\text{diglyceride}}$ 6-sulfo-α-D-quinovopyranosyl-$(1 \rightarrow 3')$-sn-1,2-diglyceride

For this scheme, cysteic acid presumably may be converted to 3-sulfopyruvate by transamination and the latter compound indirectly to 2-phospho-3-sulfolactate. Another partial pathway has been proposed (Zill and Cheniae, 1962) that involves the dehydration of a nucleosidediphosphoglucose to the corresponding 5,6-glucosene, which then may be sulfonated by phosphadenosyl phosphosulfate, through the action of a kinase, to give the requisite nucleosidediphospho-6-sulfo-6-deoxyglucose, which then may be utilized as outlined in the preceding scheme.

2. *Quantitative Considerations*

Sulfoquinovosyl diglycerides appear to be present in substantial quantity in achloroplastic photosynthesizing eukaryotes, as first shown in dark-grown *Euglena* (Rosenberg and Pecker, 1964). In this organism, upon illumination, sulfolipids begin initially to increase in quantity before chlorophyll accretion becomes detectable. Then a parallel increase in both chlorophyll and sulfoquinovosyl diglyceride characterizes the light-induced development of chloroplasts. Similar, corroborative, findings have been made with tissue from a higher plant (*Kalanchoe crenata*) developed under controlled conditions in tissue culture (Thomas and Stobart, 1970). Although the evidence is still sparse, it may be that these lipids are involved in a preliminary structural response to the photoinduction of chloroplast formation. Their discovery (Benson *et al.*, 1959) has opened the possibility of informative basic studies of organellar development along these lines. These lipids appear to be in greater quantity than galactosyl diglycerides, as they are found in small amounts in the achloroplastic forms of photo-synthesizing eukaryotes (Rosenberg, 1963; Miyachi et al., 1966). With chloroplast development, the galactosyl diglycerides usually greatly super-cede in quantity, although under some conditions (e.g., Euglena in enriched organic medium and brown algae) (Rosenberg, 1963; Radunz, 1969), the quantity of the sulfolipids may actually approach that of the galacto lipids.

The fatty acid compositions of the sulfoquinovosyl diglycerides in higher plants, and in protists as well, differs significantly from the highly un-saturated fatty acid complement of the galactosyl diglycerides. A very substantial portion (usually at least 25% and sometimes over 50%) of the fatty acid complement of the sulfoquinovosyl diglycerides is saturated fatty acid, usually of the 16-carbon variety (O'Brien and Benson, 1964; James and Nichols, 1966a; Kates, 1970). Desaturation of the fatty acyl groups of sulfoquinovosyl diglyceride molecules subsequent to total molecu-lar biosynthesis, as for monogalactosyl diglycerides, has not been observed, and furthermore, the sulfoquinovosyl diglycerides in the organism com-pared to the galactosyl diglycerides appear to display relative temporal stability (Nichols et al., 1967). As shown for Chlorella, illumination of the organism induces only a moderate change in the content in unsaturated fatty acids, which are mostly of the 18-carbon variety, in contrast with the behavior displayed by galactosyl diglycerides. As a clue to their physico-chemical attributes in the chloroplast, pure sulfoquinovosyl diglycerides, as well as monogalactosyl diglycerides, display the characteristics of an ideal solution when admixed with chlorophyll a (Trosper and Sauer, 1968).

C. PHOSPHATIDYL GLYCEROL

1. Biosynthesis

The discovery and structural elucidation of phosphatidyl glycerol by Benson and Maruo (1958) has led to a number of studies that have indi-cated that this compound is at once found to be common to all photosyn-thesizing organisms in which its presence has been investigated and that it undergoes dynamic and perhaps unique metabolism during photosynthesis in eukaryotic organisms. The pathway for the biosynthesis of phosphatidyl glycerol in such organisms actually has not yet been elucidated. The path-way proposed for its biosynthesis in animal and bacterial organisms (Ken-nedy, 1961; Kanfer and Kennedy, 1964; Haverkate and van Deenen, 1965),

$$\text{Cytosine diphosphate diglyceride} \xrightarrow{sn-3-\text{phosphoglycerol}}$$

$$sn\text{-3-phosphatidyl-1'-}sn\text{-glycero-3'-phosphate} \xrightarrow[\text{hydrolase}]{-H_3PO_4}$$

$$sn\text{-3-phosphatidyl-1'-}sn\text{-glycerol (phosphatidyl glycerol)}$$

has been implicated to operate in photosynthesizing organisms (Douce and DuPont, 1969). It has been demonstrated that the chemical structure of phosphatidyl glycerol from such organisms is the same as that of phosphatidyl glycerol in animals and bacteria and is the structure stated in the biosynthetic pathway outlined above (van Deenen and Haverkate, 1966). This pathway will produce the proper stereochemical configuration. Alternatively, phosphatidyl glycerol may be synthesized by interchange of glycerol for choline on phosphatidyl choline by the action of phospholipase D (Yang *et al.*, 1967), although, for the stereospecific synthesis, one must postulate that enzyme-substrate binding provides a selectivity for the proper structure from the racemic mixture that one would expect to be produced in the interchange reaction. Allowance must also be made for the disparity between the fatty acyl compositions of the phosphatidyl choline and phosphatidyl glycerol in the chloroplast.

2. *Quantitative Considerations*

Detailed studies of the relationship of the increase of the phosphatidyl glycerol component in photosynthetic organisms as related to chlorophyll accumulation and chloroplast development apparently are yet to be recorded. A recent study by Calvayrac and Douce (1970) has shown that only traces of phosphatidyl glycerol are found in dark-grown, presumably achloroplastic euglenas, while light-grown, photosynthesizing, euglenas have a full complement of this lipid. It is clear that there is rapid turnover during photosynthesis—in most cases more rapid than that of the other chloroplast-typical lipids. Phosphatidyl glycerol comprises a major proportion of the phospholipid component in practically all photosynthesizing protists, ranging from a reported 8% in *Euglena* to over 50% in *Chlorella* (Haverkate *et al.*, 1965; Ferrari and Benson, 1961; Sastry and Kates, 1965). In the lamellae of the chloroplast, at least those from higher plants, this lipid appears to be the major phosphatide. For example, in spinach chloroplast lamellae it comprises almost two thirds of the total phospholipid (Allen *et al.*, 1966a). Similar to the sulfoquinovosyl diglycerides, phosphatidyl glycerol tends to be rich in 16-carbon saturated palmitic acid—usually about one third of the total fatty acids of this class of complex lipids in photosynthesizing organisms and more in nonphotosynthesizing organisms. It occurs along with 18-carbon di- and trienoic acids and the unusual fatty acid, *trans*-3-hexadecenoic acid. Bartels' study (1969) has added greatly to an understanding of the metabolism of phosphatidyl glycerol with relationship to this fatty acid (Debuch, 1961), which appears to be almost specific to the phosphatidyl glycerol molecules in photosynthesizing eukaryotes in general (Weenink and Shorland, 1964; Haverkate *et al.*, 1964; James and Nichols, 1966a; Haverkate and van Deenen, 1965), although small amounts are reported to occur in phosphatidyl choline in a higher plant (Rotsch and Debuch, 1965), and it may be stored in the triglycerides in the seeds of such higher organisms (Wolf, 1966). It is not found in photosynthesizing prokaryotes and dwindles to trace amounts in nonphotosynthesizing eukaryotic protists as well as higher organisms (Nichols and James, 1968) and in *Chlorella* grown in light in an organic medium (Nichols, 1965). Location of the *trans*-3-hexadecenoic acid appears to be exclusively at position 2 of glycerol in the phosphatidyl glycerol molecule. All the available evidence, based upon the fate of labelled palmitate and 3-*trans*-hexadecenoate, indicates that 3-*trans*-hexadecenoate is biosynthesized from saturated hexadecanoic (palmitic) acid *in situ* on the phosphoglyceride molecule and that phosphatidyl glycerol (1-fatty acyl-2-hexadecanoyl-*sn*-glycero-3-phosphoryl-1'-*sn*-glycerol) may be the required substrate (Bartels, 1969; Bartels *et al.*, 1967) for the formation of this acid. Practically all of the precursor palmitic acid has been shown to be esterified to phosphatidyl glycerol, mainly at position 2, in nonphotosynthesizing *Chlorella*. Also, upon illumination of the organisms, a rapid conversion of the linked palmitate to 3-*trans*-hexadecenoate, with quantitative diminution of the former acid, is found to occur (Bartels, 1969).

3. *Functional Considerations*

In the aforementioned process, there is evidence for an intimate participation of the hydrocarbon residues of the phosphatidyl glycerol molecules in the photosynthetic machinery. The positional placement of the 3-*trans* double bond relative to the hydrophilic glycerophosphoglycerol residue of the molecule, as compared with a similar placement of the double-bonded first branch of the hydrocarbon phytol residue of chlorophyll relative to the ester-linked hydrophilic chlorophyllide residue of the molecule, may provide a clue to the role of 3-*trans*-hexadecenoic acid in phosphatidyl glycerol and the production of this acid during active photosynthesis in eukaryotic organisms. It should be pointed out that, as a biochemical mechanism, transformations of the hydrocarbon chains of intact lipids in nonphotosynthesizing bacterial and animal cells have also been reported (Chung and Law, 1964; Malins, 1968; Thompson, 1968) to occur.

D. PHOSPHATIDYL CHOLINE (LECITHIN)

1. *Biosynthesis*

As proposed previously for phosphatidyl glycerol, biosynthesis of phosphatidyl choline may occur by a transphosphatidylation through the action of phospholipase D (Sastry, and Kates, 1965), which in this case would exchange choline for glycerol on the phosphatidic acid residue. The same considerations regarding racemization of the molecule and disparate fatty acid compositions apply here. The animal pathway for biosynthesis of phosphatidyl choline via diglyceride and CDP-choline (Kennedy, 1961) has not been demonstrated to occur in photosynthesizing organisms so far. The bacterial pathway (Kates, 1966),

$$\text{Cytidine diphosphate diglyceride} \xrightarrow{\text{serine}} \text{phosphatidyl serine} \xrightarrow{-CO_2} \text{phosphatidyl}$$

$$\text{ethanolamine} \xrightarrow{\text{S-adenosyl methionine}} \text{phosphatidyl-}N\text{-methyl-}$$

$$\text{ethanolamine} \xrightarrow{\text{S-adenosyl methionine}} \text{phosphatidyl-}N,N\text{-dimethyl-}$$

$$\text{ethanolamine} \xrightarrow{\text{S-adenosyl methionine}} \text{phosphatidyl choline}$$

appears to be operative in photosynthesizing protists as well as in non-photosynthesizing bacteria. N-methylethanolamine (Kates and Volcani, 1966) has been found in diatoms. In higher-plant tissue and in a photosynthetic prokaryote (Willemot and Boll, 1967; Gorchein et al., 1968), evidence has been presented that the above pathway is used in the synthesis of phosphatidyl choline, and in Euglena, S-adenosyl methionine and ATP induce a rapid synthesis of this lipid (Tipton and Swords, 1966). However, the finding that, by the kinetics of labeling, the fatty acyl groups of phosphatidyl ethanolamine do not appear to be the precursors for those in phosphatidyl choline (Nichols and James, 1968) in Chlorella remains to be reconciled with the above findings, and more knowledge concerning the biosynthesis of phosphatidyl choline in eukaryotic protists is still required.

2. Quantitative Considerations

The amount of phosphatidyl choline in the chloroplast appears to be variable, and no relationship appears to hold, insofar as present evidence is concerned, with regard to organismic phosphatidyl choline content and the formation or function of chloroplasts. This compound can be the major chloroplastic phospholipid, although phosphatidyl glycerol often approaches or exceeds it in quantity. Like the phosphatidyl glycerols and sulfoquinovosyl diglycerides, phosphatidyl choline is rich in saturated, hexadecanoic (palmitic) acid, which occurs with considerable but variable quantities of mono-, di-, and trienoic 18-carbon and 20-carbon polyenoic fatty acids. Compared with animal phosphatidyl cholines, which are reported to contain roughly half saturated (at position 1) and half unsaturated (at position 2) fatty acids, the chloroplastic phosphatidyl cholines have a higher complement of unsaturated fatty acids, often 75% or more

(Nichols and James, 1968; Kates, 1970). A specific function for phosphatidyl choline, which is an ubiquitous membrane component, has not yet been profered. It is of great interest that further desaturation of monoenoic 18-carbon fatty acid to dienoic fatty acid in the intact lipid may occur in the phosphatidyl cholines of photosynthesizing protists, for example, *Chlorella* (Gurr et al., 1968), similarly to the desaturations noted for monogalactosyl diglycerides.

IV. Overall Functional Considerations

In spite of the influx of information from various quarters regarding the close relationship between the specific kinds of lipids, and their typical fatty acid components, and the structuration and function of the protistan chloroplast as outlined in previous sections, it is too soon to dissect the various observations and relate them to defined photosynthetic phenomena. It appears almost certain that phosphatidyl glycerol is the common denominator between the less well-defined chromatophore of the prokaryote and the well-delineated semiautonomous (Gibor and Granick, 1964; Hoober et al., 1969; Carell, 1969; Goodenough and Levine, 1970; Hoober, 1970) photosynthetic organelle of the eukaryotic protist. It may be common to the biosynthesis of spatially organized cellular systems, that is, biological membranes, that productions of the individual lipid, protein, and related molecules are interdependent and that their assembly into viable temporally stable structures (or their primordia) provides in the long run essential controlling biosynthetic factors. But within this dynamic framework, so many degrees of variation are possible that to relate carefully obtained detailed analytical data regarding the lipid components and their variable componential substructures to chloroplastic function in an abstract sense is indeed a difficult problem. Yet it is from such exercises that the basic knowledge of the phenomenon of photosynthetic membrane function eventually must emerge. It will be useful to categorize observations according to two interdependent but separate phenomena related to the lipid components of the protistan chloroplast: (1) the role of lipids in the assembly of the functional organellar structure and (2) the responses of the lipid components to operation of the photosynthetic machinery.

For category 1, it appears that the assembly of fully developed chloroplasts possessed of both photosystems I and II and capable of photosynthetic oxygen evolution requires, in addition to phosphatidyl glycerol, which is common to all photosynthetic organisms, two classes of glycolipid: the (mono- and di-) galactosyl diglycerides and sulfoquinovosyl diglycerides, the former being a polar but nonionic lipid and the latter being strongly anionic as well, by virtue of its sulfonyl group. Both these lipids are capable of dissolving and therefore dispersing, chlorophyll with a zero free energy change on mixing (Gaines *et al.*, 1964). When examined in detail, the mono- and digalactosyl and sulfoquinovosyl diglycerides are seen, as described previously, to accumulate at different rates along with chlorophyll in the developing chloroplast structure, and it may be that they take part in differing substructures. The kinetic studies of Ohad *et al.* (1967) with a mutant of *Chlamydomonas* suggest that chloroplast membrane elements are assembled in concert, but the development of membranous details can result from a multistep process of assembly (Goldberg and Ohad, 1970a) by the assimilation of preformed molecular complexes into a somewhat morphologically differentiated groundwork of preexisting membranous structures, which is not to say that the quantitative ratio of total photoreceptor pigments to one or another complex lipid may not become invariant when a constant rate of chloroplast membrane assembly is in progress (Goldberg and Ohad, 1970b). The galactosyl diglycerides diminish in light-starved photobiotic *Euglena* (Rosenberg, 1967), with concomitant chloroplast shrinkage and, eventually, chloroplast destruction. In *Chlorella*, it has been observed (Shibuya and Hase, 1965) that glucose-bleached chloroplast destruction results in degradation of sulfolipid and the casting off of sulfoquinovosyl glycerol into the medium.

For category 2, the pioneering work of Ferrari and Benson (1961) has shown that during active steady-state photosynthesis there is rapid turnover of the polar moieties of the chloroplast-typical lipids due to the assimilation of carbon dioxide but that the hydrocarbon residues of the lipids are less active in this regard. In *Chlorella* (Sastry and Kates, 1965) a far more rapid turnover of orthophosphate and α-glycerophosphate is observed in phosphatidyl glycerol than in other phospholipids during photosynthesis. In photosynthetic diatoms (Kates and Volcani, 1966), studies on the incorporation of labeled carbon dioxide, orthophosphate, and sulfate again showed the active turnover of the polar residues of the chloroplast-typical lipids during photosynthesis. With regard to fatty acyl components of *Chlorella* (Nichols *et al.*, 1967), a rapid synthesis of fatty acids was observed during photosynthesis with more active turnover in monogalactosyl diglyceride and phosphatidyl glycerol than in the other major chloroplast-typical lipids, but this was not the case for prokaryotic bluegreen algae (Nichols, 1968). It is too early to derive a generalized concept regarding the rapid turnover of the chloroplast lipids during photosynthesis, but they appear to undergo intriguingly high metabolic activity. Equally intriguing are the *in situ* desaturations of the fatty acyl components of monogalactosyl diglycerides to more highly unsaturated fatty acyl components during photosynthesis and the conversion of palmitic acid in phosphatidyl glycerol to 3-*trans*-hexadecenoic acid during photosynthetic activity. Whether these conversions, which are bound to influence the physicochemical nature of the lamellar matrices in the chloroplast, precede efficient photoreception and the transduction of radiant energy to chemical potential by establishing the proper environment for this process or whether they are the ongoing products of certain as yet undetermined steps in the overall photosynthetic process remains for a great deal of future investigation to determine.

Acknowledgment

Work reported from the author's laboratory has been supported by U.S. Public Health Service Research Grants GM09041 and NS08258.

References

Abraham, A., and Bachhawat, B. K. (1963). *Biochim. Biophys. Acta* **70**, 104.

Allen, C. F., Good, P., Davis, H. F., and Fowler, S. D. (1964). *Biochem. Biophys. Res. Commun.* **15**, 424.

Allen, C. F., Good, P., Davis, H. F., Chisum, P., and Fowler, S. D. (1966a). *J. Amer. Oil Chem. Soc.* **43**, 233.

Allen, C. F., Hirayama, O., and Good, P. (1966b). *In* "The Biochemistry of Chloroplasts" (T. W. Goodwin, ed.), Vol. 1, p. 195. Academic Press, New York.

Arnon, D. I. (1959). *Nature* (*London*) **184**, 10.

Bartels, C. T. (1969). Thesis, University of Utrecht, Rotterdam.

Bartels, C. T., James, A. T., and Nichols, B. W. (1967). *Euro. J. Biochem.* **3**, 7.

Beach, D. J., Harrington, G. W., and Holz, Jr. G. G. (1970). *J. Protozool.* **17** (3), 501.

Ben-Shaul, Y., Schiff, J. A., and Epstein, H. T. (1964). *Plant Physiol.* **39**, 231.

Benson, A. A. (1963). Advan. *Lipid Res.* **1**, 387.

Benson, A. A., and Maruo, M. M. (1958). *Biochim. Biophys. Acta* **27**, 89.

Benson, A. A., Daniel, H., and Wiser, R. (1959). *Proc. Nat. Acad. Sci. U.S.* **45**, 1582.

Bishop, D. G., and Smillie, R. M. (1970). *Arch. Biochem. Biophys.* **137**, 179.

Bloch, K., Constantopoulos, G., Kenyon, C., and Nagai, J. (1967). *In* "The Biochemistry of Chloroplasts" (T. W. Goodwin, ed.), Vol. 2, p. 197. Academic Press, New York.

Böger, P., and San Pietro, A. (1967). *Z. Pflanzenphysiol.* **58**, 70.

Calvayrac, R., and Douce, R. (1970). *FEBS Lett.* **7**, 259.

Carell, E. F. (1969). *J. Cell Biol.* **41**, 431.

Carter, H. E., Ohno, K., Nojima, S., Tipton, C. L., and Stanacev, N. Z. (1961). *J. Lipid Res.* **2**, 215.

Chung, A. E., and Law, J. H. (1964). *Biochemistry* **3**, 967.

Constantopoulos, G. (1970). *Plant Physiol.* **45**, 76.

Constantopoulos, G., and Bloch, J. (1967). *J. Biol. Chem.* **242**, 3538.

Crane, F. L., Henninger, M. O., Wood, P. M., and Barr, R. (1966). *In* "The Biochemistry of Chloroplasts" (T. W. Goodwin, ed.), Vol. 1, p. 133. Academic Press, New York.

Davies, W. H., Mercer, E. I., and Goodwin, T. W. (1966). *Biochem. J.* **98,** 369.

Debuch, H. (1961). *Z. Naturforsch. B* **16,** 56.

Douce, R., and DuPont, J. (1969). *C. R. Acad. Sci.* **268,** 1657.

Epstein, H. T., and Schiff, J. A. (1961). *J. Protozool.* **8,** 427.

Ferrari, R. A., and Benson, A. A. (1961). *Arch. Biochem. Biophys.* **93,** 185.

Gaines, G. L., Jr., Bellamy, W. D., and Tweet, A. G. (1964). *J. Chem. Phys.* **41,** 538.

Gibbs, S. P. (1962). *J. Cell Biol.* **15,** 343.

Gibor, A., and Granick, S. (1964). *Science* **145,** 890.

Goldberg, I., and Ohad, I. (1970a). *J. Cell Biol.* **44,** 563.

Goldberg, I., and Ohad, I. (1970b). *J. Cell Biol.* **44,** 572.

Goldfine, H., Ailhaud, G. P., and Vagelos, P. R. (1967). *J. Biol. Chem.* **242,** 4466.

Goodenough, V. W., and Levine, R. P. (1970). *J. Cell Biol.* **44,** 547.

Gorchein, A., Neuberger, A., and Tait, G. H. (1968). *Proc. Roy. Soc., Ser. B* **170,** 279.

Guehler, P. F., Peterson, L., Tsuchiya, H. M., and Dodson, R. M. (1964). *Biochim. Biophys.* **106,** 294.

Gurr, M. I. (1970). *Lipids* **6,** 266.

Gurr, M. I., Robinson, M. P., Sword, R. W., and James, A. T. (1968). *Biochem. J.* **110,** 49p.

Haverkate, F., and van Deenen, L. L. M. (1965). *Biochim. Biophys. Acta* **106,** 78.

Haverkate, F., De Gier, J., and van Deenen, L. L. M. (1964). *Experientia* **20,** 511.

Haverkate, F., Teulings, F. A. G., and van Deenen, L. L. M. (1965). *Proc., Kon. Ned. Akad. Wetensch.* **68,** 154.

Helmy, F. M., Hack, M. H., and Yaeger, R. G. (1967). *Comp. Biochem. Physiol.* **23,** 565.

Hill, R. (1951). *Advan. Enzymol.* **12,** 1.

Holton, R. W., Blecker, H. H., and Onore, M. (1964). *Phytochemistry* **3,** 595.

Hoober, J. K. (1970). *J. Biol. Chem.* **245,** 4327.

Hoober, J. K., Siekevitz, P., and Palade, G. E. (1969). *J. Biol. Chem.* **244,** 2621.

Hulanicka, D., Erwin, J., and Bloch, K. (1964). *J. Biol. Chem.* **239,** 277.

James, A. T., and Nichols, B. W. (1966a). *Nature (London)* **210,** 366.

James, A. T., and Nichols, B. W. (1966b). *Nature (London)* **210,** 372.

Kanfer, D., and Kennedy, E. P. (1964). *J. Biol. Chem.* **239,** 1720.

Kates, M. (1964). *Advan. Lipid Res.* **2,** 17.

Kates, M. (1966). *Annu. Rev. Microbiol.* **20,** 13.

Kates, M. (1970). *Advan. Lipid Res.* **8,** 225.

Kates, M., and Volcani, B. E. (1966). *Biochim. Biophys. Acta* **116,** 000.

Kennedy, E. P. (1961). *Fed. Proc., Fed. Amer. Soc. Exp. Biol.* **20,** 934.

Kirk, J. T. O., and Tilney-Bassett, R. A. E. (1967). *In* "The Plastids," Part I, pp. 1–62. Freeman, San Francisco, California.

Kleinschmidt, M. G., and McMahon, V. (1970). *Plant Physiol.* **46,** 290.

Lennarz, W. (1966). *Advan. Lipid Res.* **4,** 175.

Lichtenthaler, H. K., and Park, R. B. (1963). **198,** 1070.

Malins, D. C. (1968). *J. Lipid Res.* **9,** 687.

Matson, R. S., Fei, M., and Chang, S. B. (1970). *Plant Physiol.* **45,** 531.

Melandri, B. Z., Baccarini-Melandri, A., and San Pietro, A. (1970). *Biochim. Biophys. Acta* **138,** 598–605.

Menke, W. (1966). *In* "The Biochemistry of Chloroplasts" (T. W. Goodwin, ed.), Vol. 1, p. 3 *et seq.* Academic Press, New York.

Miyachi, S., Miyachi, S., and Benson, A. A. (1966). *J. Protozool.* **13,** 76.

Neufeld, E. H., and Hall, E. W. (1964). *Biochem. Biophys. Res. Commun.* **14,** 503.

Nichols, B. W. (1965). *Biochim. Biophys. Acta* **106,** 174.

Nichols, B. W. (1968). *Lipids*, **3**, 354.

Nichols, B. W., and James, A. T. (1968). *Progr. Phytochem.* **1**, 1.

Nichols, B. W., James, A. T., and Breuer, J. (1967). *Biochem. J.* **104**, 486.

Nichols, B. W., and Moorhouse, R. (1969). *Lipids*, **4**, 311.

Nichols, B. W., and Wood, B. J. B. (1968). *Lipids* **3**, 46.

O'Brien, J. S., and Benson, A. A. (1964). *J. Lipid Res.* **5**, 432.

Ohad, I., Siekevitz, P., and Palade, G. E. (1967). *J. Cell Biol.* **35**, 553.

Ongun, A., and Mudd, J. B. (1968). *J. Biol. Chem.* **243**, 1558.

Park, R. B., and Pon, N. G. (1961). *J. Mol. Biol.* **3**, 1.

Park, R. B., and Pon, N. G. (1963). *J. Mol. Biol.* **6**, 105.

Patton, S. (1968). *Science* **159**, 219.

Patton, S., Fuller, G., Loeblich, A. R., III, and Benson, A. A. (1966). *Biochim. Biophys. Acta* **116**, 577.

Racker, E. (1970). *In* "Membranes of Mitochondria and Chloroplasts" (E. Racker, ed.), p. 145. Van Nostrand-Reinhold, Princeton, New Jersey.

Radunz, A. (1969). *Hoppe-Seyler's Z. Physiol. Chem.* **350**, 411.

Renkonen, O., and Bloch, K. (1969). *J. Biol. Chem.* **244**, 4899.

Rosenberg, A. (1963). *Biochemistry* **2**, 1148.

Rosenberg, A. (1967). *Science* **157**, 1191.

Rosenberg, A., and Gouaux, J. (1967). *J. Lipid Res.* **8**, 80.

Rosenberg, A., and Pecker, M. (1964). *Biochemistry* **3**, 254.

Rosenberg, A., Pecker, M., and Moschides, E. (1965). *Biochemistry* **4**, 680.

Rosenberg, A., Gouaux, J., and Milch, P. (1966). *J. Lipid Res.* **1**, 733.

Rotsch, E., and Debuch, H. (1965). *Hoppe-Seyler's Z. Physiol. Chem.* **843**, 135.

Rouser, G., and Yamamoto, A. (1969). *In* "Handbook of Neurochemistry" (A. Lajtha, ed.), Vol. 1, p. 121. Plenum, New York.

Safford, R., and Nichols, B. W. (1970). *Biochim. Biophys. Acta* **210**, 57.

Sastry, P. S., and Kates, M. (1965). *Can. J. Biochem.* **43**, 1445.

Shaw, R. (1966). *Advan. Lipid Res.* **4**, 107.

Shibuya, I., and Hase, E. (1965). *Plant Cell Physiol.* **6**, 267.

Shibuya, I., Yagi, T., and Benson, A. A. (1963). *In* "Studies on Microalgae and Photosynthetic Bacteria," p. 627, Japan Soc. Plant Phys., ed., Univ. of Tokyo Press, Tokyo.

Thomas, D. R., and Stobart, A. K. (1970). *J. Exp. Bot.* **21**, 274.

Thompson, G. A. (1968). *Biochim. Biophys. Acta* **152**, 409.

Tipton, C. L., and Swords, M. D. (1966). *J. Protozool.* **13**, 469.

Trosper, T., and Sauer, K. (1968). *Biochim. Biophys. Acta* **162**, 97.

van Deenen, L. L. M., and Haverkate, F. (1966). *In* "The Biochemistry of Chloroplasts" (T. W. Goodwin, ed.), Vol. 1, p. 117. Academic Press, New York.

Weenink, R. O., and Shorland, F. B. (1964). *Biochim. Biophys. Acta* **84**, 613.

Weier, T., and Benson, A. A. (1966). *In* "The Biochemistry of Chloroplasts" (T. W. Goodwin, ed.), Vol. 1, p. 91. Academic Press, New York.

Willemot, C., and Boll, W. G. (1967). *Can. J. Bot.* **45**, 1863.

Witt, H. T., Rumberg, B., Schmidt-Mende, P., Siggel, U., Skerra, B., Vater, J., and Weikard, J. (1965). *Angew. Chem.* **4**, 799.

Wolf, I. A. (1966). *Science* **154**, 1140.

Wood, J. B., Nichols, B. W., and James, A. T. (1965). *Biochim. Biophys. Acta* **106**, 261.

Yang, S. F., Freer, S., and Benson, A. A. (1967). *J. Biol. Chem.* **242**, 477.

Zill, L. P., and Cheniae, G. (1962). *Annu. Rev. Plant Physiol.* **13**, 225.

CHAPTER 6

Membranes of Yeast and Neurospora: Lipid Mutants and Physical Studies

A. D. Keith, B. J. Wisnieski, S. Henry, and J. C. Williams

I. Introduction

The nature of membranes has intrigued man since 1899 when Overton, on the basis of permeability considerations, postulated that membranes were impregnated by a fatty oil substance. In 1925 Gorter and Grendel performed their first simple calculations to ascertain the amount of lipid in a red blood cell ghost and postulated enough for a bilayer. Since then, the number of models put forth to suggest the organizational basis of membranes has mounted. There are scientists in every field looking for clues to the structure, function, and biosynthetic processes of membranes. Those interested primarily in the lipid moiety have benefited from the isolation of lipid-requiring strains of various microbes. These have provided the first source of biological membranes with relatively homogeneous fatty acid compositions.

This chapter deals with lipid mutants, physical properties of lipids, the general relationship of lipids to proteins, and spin label studies on membranes.

II. Membrane Lipids

Yeasts, as other organisms, have membranes composed primarily of lipid and protein. Wild-type yeasts are capable of growth on medium containing inorganic salts, a few vitamins, and glucose. Consequently, they are capable of synthesizing all the structural molecules necessary for cell replication and metabolism. Most yeasts are adaptable to either aerobic or anaerobic growth conditions. For our purposes, anaerobic conditions preclude the presence of oxygen, which can act as an electron acceptor for the electron transport system of mitochondria. Under these conditions, mitochondria appear as membranous shells with little, if any, internal membrane structure. Aerobic versus anaerobic growth is an important consideration in the study of yeast membranes since the primary membrane type can be regulated by controlling growth conditions. *Neurospora* and many other eukaryotic microbes can grow and function aerobically and consequently have mitochondrial-dependent growth and metabolism. Since mitochondria have extensive infolding of their inner membranes, the presence or absence of inner mitochondrial membrane can have a major influence on the primary membrane type of a given cell system.

The fatty acid composition of yeast has been reported several times. Some representative values are shown in Table I (Suomalainen and Keränen, 1968; Keith *et al.*, 1969). These values serve to illustrate that growth conditions and supplementation can extensively alter the fatty acid composition of an organism. Supplementation of an enriched medium with fatty acids does not alter the composition as much as fatty acid supplementation of minimal medium. Two cases of extreme fatty acid enrichment are presented: Supplementation with palmitoleate (16:1) resulted in its becoming 91% of the total fatty acid composition, and supplementation with oleate (18:1) resulted in its becoming 90% of the total (Table I). This ability allows for the enrichment of lipids and membranes with a single fatty acid component, thus making the lipid moieties more homogeneous. Supplementation with Tween 80 (sorbitan monooleate; however, the fatty acid composition is similar to olive oil) did not lead to any sizable enrichment.

Fatty acid composition also depends on such variables as growth temperature, amount of aeration of the culture, duration of the growth period, size of the culture, age of the culture, other medium supplementations besides fatty acids, the carbon source, and so on. Since most membrane properties are probably affected by lipid composition, as many variables as possible should be controlled.

It is generally envisioned that the majority of the membrane lipids are phospholipids; however, in fungi, neutral lipids may be present in substantial quantities (Horecker, 1967; Letters, 1967). Ergosterol is the predominant sterol in yeast and *Neurospora*. Both of these organisms possess triglycerides and also carotenoids. The latter occur in greater quantities in *Neurospora*.

The phospholipid composition of the yeast *Saccharomyces cerevisiae* is shown in Table II (Letters, 1967; Deierkauf and Booij, 1968; Suomalainen and Nurminen, 1970). Fractionating yeast usually results in a similar phospholipid composition for the different cellular fractions, with the mitochondria having perhaps some more cardiolipin. Whole yeast cells are about 5% phospholipid by dry weight (Letters, 1967). Yeast mitochondria are reported to have a somewhat higher phospholipid content, with about 12 to 14% being phospholipid by dry weight (Klein, 1957; Letters, 1967).

TABLE I

FATTY ACID COMPOSITION OF *Saccharomyces cerevisiae*

	<16:0	16:0	16:1	18:0	18:1	Other
Wild type, enriched medium[a]	3.2	14.3	47.6	2.5	32.5	—
Same + 18:1[a]	3.9	12.1	23.6	1.4	52.9	3.9
Same + 18:0[a]	5.3	7.3	51.2	10.9	22.6	—
Same + 16:1[a]	0.9	7.9	84.9	3.9	2.5	—
S288C + Tween 80[b]	0.4	15.0	11.9	8.9	64.4	
S288C on minimal	—	16	24	9	50	1
S288C + 16:1, 1 g/liter	—	3	91	3	2	1
S288C + 18:1, 1 g/liter	—	4	4	1	90	1
S288C + Tween 40 (16:0)	—	16	48	6	30	—
S288C + Tween 20 (12:0)	1	18	43	6	32	—

[a] Suomalainen and Keränen (1968).

[b] Distribution of ^{14}C from acetate-1-^{14}C with growth on YEPD supplemented with Tween 80. Other data are unpublished from the authors' laboratory.

TABLE II

PHOSPHOLIPID COMPOSITION OF YEAST

	Other[a]	PC	PE	PI	PS	CL
Whole cells[b]		48.4	23.2	16.3	6.9	3.2
Mitochondria[b]		47.8	22.4	16.4	7.2	8.2
$10^5 \times g$ pellet		48.2	22.1	18.0	6.4	5.3
Cell wall[b]		48.6	26.6	15.2	6.8	<1
Whole cells[c]	15	55	20	15		
Heavy sediment[c] fraction $10^3 \times g/5$ min	20	45	10	35		
Light sediment[c] fraction $1.5 \times 10^4 \, g/15$ min	20	20	20	35		
Whole yeast[d]	9	45.2	16.7	19.8	9.3	

[a] The phosphatidyl serine and phosphatidyl glycerol fractions. PC, phosphatidyl choline; PE, phosphatidyl ethanolamine; PI, phosphatidyl inositol; PS, phosphatidyl serine; and CL, cardiolipin.

[b] Suomalainen and Nurminen (1970).

[c] Letters (1967).

[d] Deierkauf and Booij (1968).

Glycerophospholipids in fungi and most other organisms are very flexible, as demonstrated by the following glycerol skeleton:

where FA_1 is either a saturated or an unsaturated fatty acid, FA_2 is usually an unsaturate, and R is a nitrogen base, a hydrogen, or a sugar. Since palmitate, palmitoleate, and oleate account for most of the fatty acid composition of yeast and since palmitate, oleate, and linoleate do so for *Neurospora*, permutation results in nine different diglycerides if fatty acid acylation is random. Since there are three major polar groups to occupy the R position, the number of random structural species is brought to 27. However, from the work of Selinger and Holman (1965), Lands (1965), Van Deenen (1966), and others, we know that the distribution is not random; still, the organism is left with a great deal of structural flexibility in its phospholipid composition.

The distribution of acetate and dietary oleate into different lipid classes is shown in Table III (Keith *et al.*, 1969). Yeast strains S288C and S288C:XL687-101B are both wild type; the KD strains are unsaturated fatty acid requirers. In wild-type yeast, acetate-1-^{14}C is incorporated into the fatty acids of phospholipids and nonphospholipids about equally; labeled oleate (18:1) is incorporated into neutral lipids in slightly higher proportion. These yeasts were grown for 48 hours at 30°C on yeast extract peptone dextrose (YEPD) medium. When growth is on minimal medium, the neutral lipid content is reduced; when growth is on very rich medium (2 × YEPD), the proportion of neutral lipid is increased. The several lipid mutants shown in Table III will be discussed shortly.

Ergosterol is a minor fraction of the total lipid in yeast yet is probably

TABLE III

PERCENT DISTRIBUTION OF FATTY ACID INTO LIPID CLASSES FROM
18:1-9,10-^{3}H AND ACETATE-1-^{14}C[a]

Yeast Strain	Phospholipid		Free Fatty Acid		Neutral Lipid	
	^{3}H	^{14}C	^{3}H	^{14}C	^{3}H	^{14}C
S288C	36	49	15	13	49	38
S288C:X1687-101B[b]	62	60	6	6	32	34
KD20	6	10	1	4	93	86
KD115	26	24	16	20	58	56
KD18	10	20	28	26	62	54
KD144	5	14	35	35	60	51
KD46	9	21	21	16	70	63
KD91	11	21	27	25	62	54
KD46:KD46[b]	55	45	25	11	20	44

[a] Acetate-1-^{14}C and 18:1-9,10-^{3}H were employed simultaneously. The lipid classes were purified, and each class was saponified. Fatty acids from these classes were analyzed for tracer content employing the discriminator ratio method for simultaneous ^{3}H and ^{14}C analysis in a liquid scintillation spectrometer (Selinger and Holman, 1965). All analyses yielded a total of > 12,000 counts/min in the three lipid classes analyzed for each tracer.

[b] Diploid: S288C:X1687-101B data are presented for comparison with that of S288C, since both are *ol*+ and grow on minimal medium, and with that of the homozygous mutant diploid KD46:KD46.

necessary for growth. It was first demonstrated by Andreasen and Stier (1953) that ergosterol is a growth requirement for yeast grown under anaerobic conditions. The antibiotic nystatin is reported to interact with ergosterol (Lampen, 1966). Yeast mutants that are resistant to nystatin have altered sterol compositions (Bard, 1971). The role of ergosterol in yeast membranes is undoubtedly important, and studies dealing with this sterol and interacting antibodies may offer a new profitable approach to the study of membrane phenomena.

III. Lipid-Requiring Organisms

A. Lipid Requirers

Over the years, there has been a gradual recognition that certain fatty acids are dietary requirements for the growth of a wide variety of organisms. For example, a component of casein hydrolysate that seemed to be a requirement for growth in *Sarcina* sp. (a *Staphlococcus*) later proved to be a requirement for long-chain unsaturated fatty acids (Demain *et al.*, 1959). Fatty acids are extremely difficult to completely extract from casein and were present in casein hydrolysates. In 1947, cultures of lactobacilli were shown to require oleic acid (Hutchings and Boggiana, 1947; W. L. Williams *et al.*, 1947). These findings were extended to several other species of lactobacilli by Kitay and Snell in 1950. Several years later, Hofmann *et al.* (1959) demonstrated that several other fatty acids could replace the oleic acid requirement in *Lactobacillus*: lactobacillic acid, *cis*-vaccenic acid, $14:1\Delta^7$ *cis*, and also $12:1\Delta^5$ *cis* if it were added together with a small amount of biotin. This was the earliest experiment to suggest that fatty acid requirements may not be rigid and that some degree of flexibility is permitted.

A requirement for unsaturated fatty acids was generally accepted for many years in the medical literature. The essential fatty acids of mammalian systems are thought to be linoleic ($18:2\Delta^{9,12}$ *cis, cis*) and arachidonic ($20:4\Delta^{5,8,11,14}$ all *cis*) acids (Holman, 1969). Rats raised on $18:2$-deficient diets show abnormalities probably because $18:2$ is the precursor of $20:4$. The omission of $18:3$ from the diet of weanling rats results in skin lesions. The nature of the requirements for $18:2$ and $18:3$ has not been established.

Mycoplasma laidlawii is a lipid requirer. It has been the subject of many membrane experiments for two apparent reasons: First, it has no detectable (easily detectable) cell wall, and, second, it has only trace synthesis of long-chain fatty acids, particularly of unsaturated fatty acids. Consequently, its lipid composition can be extensively controlled. Observations of structural changes in response to 18:0 versus 18:1 enrichment have been made at both physiological and electron microscope levels. Enrichment with 18:1 results in cells that tend to form spheres, resist osmotic lysis, and display a high frequency of 80-Å particles embedded in the membranes when viewed by freeze-fracture electron microscopy (Tourtellotte *et al.*, 1970). Enrichment with 18:0 results in filamentous cells, a decreased tolerance to osmotic change, and a lower frequency of 80-Å particles. Mycoplasma has been the subject of membrane experiments employing differential scanning calorimetry (Stein *et al.*, 1969), X-ray diffraction (Engelman, 1972), freeze-fracture electron microscopy (Tourtellotte *et al.*, 1970), electron-spin resonance (ESR) spectroscopy (Tourtellotte *et al.*, 1970), and several physiological approaches.

B. Conditional Lipid Requirers

In 1953, Andreasen and Stier reported that anaerobically grown yeasts require ergosterol and oleic acid. Some years later Bloomfield and Bloch (1960) explained the requirements of anaerobic yeasts on the basis of their need for molecular oxygen to carry out the enzymic desaturation of palmitic and stearic acids. Certain other fatty acids, besides oleic acid, were found to satisfy the growth requirement (Meyer and Bloch, 1963; Meyer *et al.*, 1963) of anaerobic yeast.

Yeast grown aerobically on biotin-deficient medium mimic chain elongation mutants. Biotin is a required cofactor for the production of malonic acid via acetyl-CoA carboxylase. Since acylated malonate is the 2-carbon donor for chain elongation, a fatty acid supplement must be supplied in the absence of biotin (Suomalainen and Keränen, 1963). Biotin, however, is not a cofactor for desaturation, and a saturated fatty acid of appropriate chain length is as effective for growth as an unsaturated one, provided that aspartic acid is also added to the growth medium. The precise role of the aspartic acid is unclear.

An oleate auxotroph of *Neurospora crassa* was reported by Lein and Lein in 1949 on the basis of its requirement for oleic acid. Another *Neurospora* auxotroph that grew in the presence of Tween 80 was described by Perkins *et al.* in 1962. Recently, a rigorous characterization of the latter mutant (Henry and Keith, 1972) has revealed that it is a chain elongation mutant. Growth of this mutant has been attained on medium supplemented with highly purified saturated fatty acids, as well as with certain combinations of *cis* and *trans* unsaturated fatty acids.

The isolation of an *E. coli* auxotroph defective in the synthesis of unsaturated fatty acids was reported by Silbert and Vagelos (1967). *Escherichia coli* normally produces long-chain unsaturates by the synthesis and subsequent elongation of $10:1\Delta^3$ *cis*. The *E. coli* auxotroph was unable to synthesize this precursor. Silbert *et al.* (1968) later demonstrated that several unsaturated and cyclopropane-containing fatty acids could replace the naturally occurring fatty acids as growth requirements. Some of these replacements restricted growth rate, and in a few instances, cell morphology was altered.

In maize (Poneleit and Alexander, 1965), an inheritance pattern was reported for high and low levels of the unsaturated fatty acid $18:2\Delta^{9,12}$ *cis, cis*. The following year, Resnick and Mortimer (1966) isolated eight yeast mutants that required an unsaturated fatty acid for growth. All but two of these mutants were petite (respiratory-deficient). Of seven mutants subjected to genetic analysis, six proved to be recessive and one proved to be cytoplasmically inherited. The respiratory-competent mutants, KD20 and KD115, were allelic at the *ol* 1 locus. In the yeast mutants, both the petite and the desaturase phenotypes segregate together, and in two of the mutants that have been examined, they revert together; consequently, both probably arose by a single mutational event. The specificity of the fatty acid requirements of three of these desaturase mutants has been tested (Table V) (Wisnieski and Kiyomoto, 1972; Wisnieski *et al.*, 1970).

A search for yeast fatty acid chain elongation mutants was successful (Henry, 1971). Some analyses of these mutants with respect to growth-supporting fatty acids have been accomplished.

All organisms having biological membranes composed of protein and lipid appear to have at least one thing in common: They all have a heterogeneous lipid composition in the native or wild state. This lipid heterogeneity is reflected in the presence of several species of phospholipids, a sterol component(s) in all eukaryotes, sometimes other neutral lipids, and always a heterogeneous fatty acid composition. The heterogeneity of fatty acid composition is usually expressed as a mixture of straight-chained saturates and unsaturates. Sometimes fatty acids with a methyl branch, cyclopropane, or cyclopropene group may largely replace the unsaturates. A double bond, methyl branch, or cyclopropane serves to lower the melting point of a fatty acid. These groups all act as interrupters of ordered structure that resist crystal formation and therefore help to maintain the lipids in a relatively fluid state.

Several strains of organisms and mutants have now been described that require various fatty acids to achieve growth, some under restricted growth conditions. *Escherichia coli* and *S. cerevisiae* have been examined most closely with respect to fatty acid structural requirements. The *E. coli* and yeast mutants are both capable of synthesizing long-chain saturated fatty acids but cannot grow on them alone. Thus it is important to know what other fatty acids are able to serve as adequate replacements for the "native" unsaturated fatty acids of these organisms. Since yeast differs from *E. coli* in the method of synthesis of unsaturated fatty acids and thus in the kinds of unsaturates normally present, we would not necessarily expect the fatty acid "replacements" for both organisms to be identical. Yeast directly desaturates long-chain saturated fatty acids to Δ^9 *cis* analogs and then metabolizes these to no longer chain length. Thus the proportion of unsaturates gradually increases in aging cultures. Unsaturated fatty acid requirers in yeast are deficient for an enzyme that inserts a Δ^9 *cis* double bond into palmitic and stearic acids. Bloomfield and Bloch (1960) have investigated the nature of this enzyme and characterized it as an oxygenase. *Escherichia coli*, as mentioned earlier, produces long-chain unsaturates via an extensive elongation of $10:1\Delta^3$ *cis*. The longest of these unsaturates, *cis*-vaccenic acid ($18:1\Delta^{11}$ *cis*), is then converted into a cyclopropane derivative in stationary-phase cultures. The *E. coli* auxotroph has lost enzyme activity for β-hydroxydecanoyl thioester dehydrase, which is responsible for the formation of *cis*-Δ^3-decanoyl-acyl carrier protein (Silbert and Vagelos, 1967). The fatty acid synthetic pathways of yeast and *E. coli* are compared in Fig. 1.

Another difference between yeasts and *E. coli* is that yeasts, unlike *E. coli*, do not appear to have any oxidative pathway for metabolizing fatty acids as an energy source. Thus the yeast mutants are able to incorporate 16:1, 18:1, 18:2, and 18:3 intact, and when these compounds are radioactive, most of the counts are recoverable in the same component as

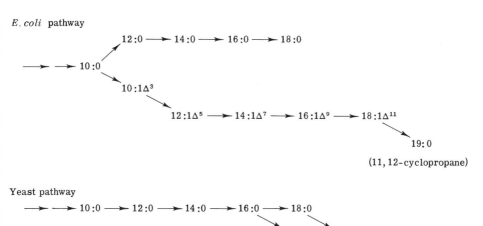

Fig. 1. Top, *Escherichia coli* pathway; bottom, yeast pathway.

TABLE IV

FATE OF EXOGENOUSLY SUPPLIED TRACER-CONTAINING FATTY ACIDS

Fatty Acid Recovered	Supplemented Fatty Acid							
	16:1-³H-R		18:1-9,10-³H		18:2-¹⁴C-1		18:3-¹⁴C-1	
	cpm Net	%	cpm Net	%	cpm Net	%	cpm Net	%
<16:0	32[a]	1.7	34	0.1	26	0.5	0	0
16:0	0	0	8	0	3	0.1	0	0
16:1	1646	87.6	307	0.5	0	0	0	0
18:0	69	3.7	555	0.9	12	0.2	39	0.3
18:1	91	4.8	60,078	96.9	141	2.9	56	0.4
18:2	41	2.2	819	1.3	4704	95.2	64	0.4
18:3	0	0	194	0.3	54	1.1	14,463	98.9

[a] Tracer analysis. The 18-carbon radioactive fatty acids used were as follows: 9,10-³H-18:1Δ^9 cis, specific activity 46.5 mc/mmole, New England Nuclear Corp., Boston, Mass.; 1-¹⁴C-18:2$\Delta^{9,12}$ cis, cis, specific activity 52.9 mc/mmole, Nuclear-Chicago Corp., Des Plains, Ill.; and 1-¹⁴C-18:3$\Delta^{9,12,15}$ all cis, specific activity 41.5 mc/mmole, Nuclear-Chicago Corp. All were verified to be at least 99% pure by preparative gas-liquid chromatography followed by scintillation spectroscopy. The fatty acid "random"-³H-16:1Δ^9 cis (³H on even-numbered carbons) was prepared by growing wild-type yeast on 25 mc of ³H-acetate and recovering the 16:1 by preparative gas-liquid chromatography. The column used for this purpose was discarded because it retained a high background. The yeast cells used for tracer analysis were grown for 20 hours on 2×10^{-4} M tracer-containing minimal medium with the following specific activities: "random"-³H-16:1Δ^9 cis, 3 μc; 9,10-³H-18:1Δ^9 cis, 20 μc; 1-¹⁴C-18:2$\Delta^{9,12}$ cis, cis, 0.5 μc; and 1-¹⁴C-18:3$\Delta^{9,12,15}$ all cis, 1 μc. Erlenmeyer flasks (250 ml) fitted with Klett side arms were employed. Centrifugation yielded pellets that were washed twice. The pellet was directly saponified in 2N KOH and methylated in HCl-methanol as before (Selinger and Holman, 1965). The background for each radioactive component collected by preparative gas-liquid chromatography was corrected for by subtracting the counts obtained from each component of a carrier fatty acid mixture collected in the same manner. No appreciable buildup of background counts was noted. Liquid scintillation was carried out on a Beckman liquid scintillation spectrometer equipped with an external quenching standard.

supplied (Table IV) (Wisnieski et al., 1970). This convenience when controlling the fatty acid composition is important.

C. YEAST FATTY ACID DESATURASE MUTANTS

We shall now present an analysis of two types of lipid mutants used in our own membrane studies: the fatty acid desaturase mutants of yeast and the chain elongation mutants of Neurospora and yeast.

With respect to the first class of mutants, the desaturase, we would like first to correct some published inaccuracies (Wisnieski *et al.*, 1970). We initially tested several unsaturated and branched-chain fatty acids on one of the yeast unsaturated fatty acid requirers, KD115 (respiratory-competent), in an attempt to determine the specificity of the requirement (Wisnieski *et al.*, 1970). At that time growth occurred only with those fatty acids that contained a Δ^9 *cis* double bond, as shown in Table V. Several months later, we repeated these experiments, supplementing minimal medium with fatty acids obtained from the Hormel Institute, and obtained considerably different results. Several other double-bond positions were adequate for growth. Since this contradiction had arisen, we then tried two other mutants: KD20, also respiratory-competent, and KD46, a petite (respiratory-deficient). These mutants generally confirmed the results obtained from the "new" KD115 experiments. We believe that our first set of experiments on KD115 was carried out properly, and we have not been able to find any experimental error or technique that was deficient. We offer two possible reasons for the discrepancies in growth requirements between the "old" and "new" experiments: We feel that this mutant, KD115, may have initially had some permeability barrier or some other metabolic restriction that it later lost or that some of the fatty acids employed in the early experiments contained toxic impurities that are not seen by gas chromatographic analysis.

Table VI presents the results of tracer experiments on several yeast unsaturated fatty acid requirers, two revertants, and two wild types (Keith *et al.*, 1969). All the mutants are relatively "clean" in that little, if any, unsaturated fatty acids are synthesized. Table V is a summary of the growth response of KD115 (early and recent), KD20, and the petite KD46 as a function of fatty acid supplement. The results are in reasonably good agreement with those for anaerobically grown yeast and the *E. coli* mutant. *Escherichia coli* demonstrates a greater flexibility than yeast. At this point in our analysis the features of a growth-supporting fatty acid for yeast are (1) a double bond in the *cis* configuration at either the Δ^5, Δ^6, Δ^9, or Δ^{11} position, or (2) a triple bond at the Δ^9 position, or (3) the Δ^9 *trans* double bond of $16:1\Delta^9$ *trans*, $18:2\Delta^{9,12}$ *trans, trans*, and, to some extent, $18:1\Delta^9$ *trans*, 12-hydroxy ($18:1\Delta^9$ *trans* did not support growth). Yeast grown on *trans* components show no detectable isomerization of the double bonds, as demonstrated by thin-layer chromatography on AgNO₃-impregnated plates (Suomalainen and Keränen, 1963; Wisnieski and Kiyomoto, 1972). The *cis* configuration is not a strict requirement [Fig. 2(a) and (b)]; however, the 18-carbon *trans* fatty acid, $18:1\Delta^9$ *trans*, is not adequate for yeast growth. We think that this difference in ability to grow

TABLE V

FATTY ACID GROWTH RESPONSE[a]

Supplemented Fatty Acid	Anaerobic Yeast[b]	E. coli[c]	KD115, Early[d]	KD115, Recent[e]	KD20[e]	KD46[e]	KD115, Lactic Acid[e]	KD20, Lactic Acid[e]	Neurospora cel	Yeast Mutant 1[a]
12:0				0		0			+*	+*
14:0			0	0		0			+	+
15:0			0	0					+	+
16:0			0	0		0			+	+
17:0									+	+
18:0			0	0		0			0	0
19:0									0	
20:0									0	
14:1Δ5 cis		+	0	+	+	+			0	
14:1Δ9 cis		+	+	+	+	+			0	0
16:1Δ9 cis		+	+	+	+	+			0	0
16:1Δ9 trans		+	0	+		+			0	
18:1Δ6 cis		+	+	+		+	+	+	0	
18:1Δ9 cis	+	+	0	+	+	+	+	+	0	0
18:1Δ9 trans	0	+	0	0		0	+	+	+*	
18:1Δ11 cis	+	+	0	+	+	+	+	+	0	
18:1Δ11 trans		+	0	0		0	+	+		
18:2Δ9,12 cis, cis		+	+	+	+	+	+	+	0	0
18:2Δ9,12 trans, trans		+	0	+		+	+	+	0	0

18:3Δ⁶,⁹,¹² *cis, cis, cis*				+	+	+	
18:3Δ⁹,¹²,¹⁵ *cis, cis, cis*		+	+	+	+	+	O
20:1Δ⁵ *cis*	O	O					
20:1Δ¹¹ *cis*	+*	O	+*	O	O	O	
22:1Δ¹³ *cis*	O	O	O	O	O		
24:1Δ¹⁵ *cis*	O	O	O	O			
20:2Δ¹¹,¹⁴ *cis, cis*	+*	+	+*	O	O		
20:3Δ¹¹,¹⁴,¹⁷	+	+	+*	O			
20:4Δ⁵,⁸,¹¹,¹⁴ *cis, cis, cis, cis*		O	+	+	+	+	
18:1Δ⁹ *cis-ol*			O	O			
18:1Δ⁹ *cis-ol-PO₄*			O				
18:1⁹ᴱ		+*	+*	+	+	O	
18:1Δ⁹ *cis*, 12OH	+	+*	O				
18:1Δ⁹ *trans*, 12OH		O	O	O			
18:1Δ⁹ *cis*, 12-acetoxy	+						

[a] Symbols: +, essentially wild-type growth; +*, less than wild type growth; O, no growth.

[a] Data from Bloch et al. (1961), Light et al. (1962), Meyer and Bloch (1963), and Meyer et al. (1963).

[b] Data from Silbert et al. (1968).

[c] Data from Wismieski et al. (1970).

[d] Data from Wismieski and Kiyomoto (1972).

[e] Data from Henry and Keith (1971).

[f] Data from Henry (1971).

on $18:1\Delta^9$ *trans* between yeast and *E. coli* suggests the existence of a correlation between the optimal growth temperature of an organism (yeast, about 30°C; *E. coli*, about 37°C) and the melting points of the fatty acids that will support the growth of that organism. Thus the higher growth temperature for *E. coli* brings the hydrocarbon portion of $18:1\Delta^9$ *trans* into a more fluid state. Esfahani *et al.* (1969), employing the *E. coli* auxotroph, reported that at 37°C $18:1\Delta^9$ *trans* supports growth and is incorporated into phospholipids at levels higher than those attained with $18:1\Delta^9$ *cis*. With $18:1\Delta^9$ *trans* a shift to 27°C causes loss of viability and cell lysis.

All the mentioned growth studies on KD115, KD20, and KD46 were carried out with 2% glucose as a carbon source. Additional tests of KD115 and KD20 were run on 1% DL-sodium lactate, a nonfermentable substrate. With this carbon source, the relative proportion of the yeast membranes that are mitochondrial in nature is approximately 80% (calculations based upon visual inspection of electron micrographs). The fatty acid requirements under these conditions did not change. In wild type, growth on sodium lactate was enhanced about 6- to 12-fold when the minimal medium was supplemented with a fatty acid (Fig. 3).

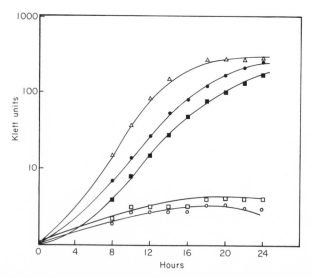

Fig. 2(a). Growth curves were carried out on yeast nitrogen base (2% glucose) for respiratory-competent desaturase mutants of yeast. △, wild type (S288C) on $16:1\Delta^9$ *trans*; ●, KD115 on $16:1\Delta^9$ *trans*; ■, KD20 on $16:1\Delta^9$ *trans*; ○, KD115 on minimal medium; and □, KD20 on minimal medium.

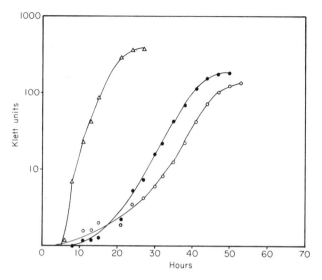

Fig. 2(b). Growth curves on YEPD (2% glucose) for a respiratory-deficient desaturase mutant of yeast. △, wild type (S288C); ●, petite, KD46 on 16:Δ⁹ *trans*; and ○, petite, KD46 on 16:1Δ⁹ *cis*. Growth curves were obtained by inoculating Klett tubes containing 8 ml of 2×10^{-4} M fatty acid in either minimal medium [Fig. 2(a); yeast nitrogen base plus 2% glucose and 1% Tergitol] or yeast extract peptone dextrose medium [Fig. 2(b); yeast extract peptone plus 2% glucose and 1% Tergitol] with about 10^5 cells/ml. The tubes were aerated on a rotor at constant speed in an incubator at 30°C. Fatty acids are from the Hormel Institute.

A sometimes ignored factor that may be of considerable importance is the concentration of the supplemented fatty acid. The outcome of one experiment designed to test the effect of fatty acid molarity on the doubling time of the yeast KD115 is shown for five growth-supporting fatty acids in Table VII. Thus a fatty acid may be inhibitory at high concentrations (approximately 10^{-3} M) and yet growth-supporting or neutral at lower molarities. Others may be able to support growth only at the higher concentrations. Effects of this type appear to be related to chain length and/or degree of unsaturation.

D. FATTY ACID CHAIN ELONGATION MUTANTS

If yeast and other organisms require a heterogeneous fatty acid composition, then it is clear that there may be other types of fatty acid mutants

TABLE VI

DISTRIBUTION OF [14]C- FROM ACETATE-1-^{14}C INTO FATTY ACIDS[a]

Yeast Strain	Genotype	Fatty Acid (% of total cpm)					
		<16:0	16:0	16:1	18:0	18:1	16:1/18:1
KD20	ol 1	2.3	79.0	0.7	17.8	0.2	
KD115	ol 1	4.1	78.4	1.0	16.2	0.3	
KD118	ol 3; ol 4(?)	10.0	74.1	0.7	13.4	1.8	
KD144	Cytoplasmic (?)	4.1	80.7		15.2		
KD46	ol 2-1	5.6	75.5	2.1	13.5	3.3	
KD91	ol 2-2	1.6	83.4		15.0		
KD180	ol 2-3	2.1	83.1	1.4	12.5	0.8	
KD20r₁[b]	?	2.0	13.7	8.7	6.2	69.4	0.12
KD115r₁[c]	?	1.7	9.4	48.6	2.4	37.9	1.3
KD46:KD144	ol 2-1/cytoplasmic(?)	1.1	9.4	39.3	2.7	47.5	0.7
S288C		0.4	15.0	11.9	8.3	64.4	0.22

[a] GLC conditions and tracer analysis are described in Keith *et al.* (1969). All numbers represent an average of two independent determinations, except as noted; 18:2 and 18:3 are not considered in these analyses, since they are minor components in strains derived from the wild type S288C. All determinations were represented by a total of > 5000 cpm in the five fractions shown; 16:1/18:1 ratios are valid only where Δ^9-desaturase activity is clearly present. The following strains are respiratory-deficient and also require a nonlipid supplement: KD18, KD144, KD46, KD91, and KD180. The diploid KD46: KD144 does not require an unsaturate or a nonlipid supplement for growth, nor is it respiratory-deficient.

[b] Average of three determinations.

[c] Average of four determinations.

TABLE VII

EFFECT OF CONCENTRATION OF FATTY ACIDS ON DOUBLING TIME
OF KD115

Concentration (moles/l)	Fatty Acid				
	14:1Δ^9 *cis*	16:1Δ^9 *cis*	18:1Δ^9 *cis*	18:2$\Delta^{9,12}$ *cis, cis*	18:3$\Delta^{9,12,15}$ all *cis*
10^{-3}	> 24.0[a]	1.4	1.8	> 24.0	1.7
10^{-4}	1.6	1.4	3.1	1.4	1.6
10^{-5}	1.8	2.4	24.0	3.5	4.4

[a] Doubling time in hours.

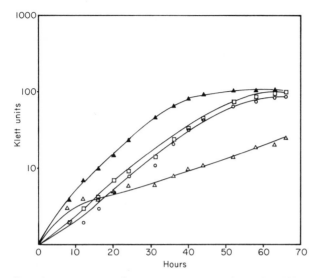

Fig. 3. Growth curves were taken on yeast nitrogen base plus 1% DL-sodium lactate for respiratory-competent desaturase mutants of yeast. △, wild type (S288C) on minimal medium; ▲, wild type (S288C) on 14:1Δ⁹ *cis*; □, KD20 on 14:1Δ⁹ *cis*; and O, KD115 on 14:1Δ⁹ *cis*. Growth curves were obtained by inoculating Klett tubes containing 8 ml of $2 \times 10^{-4} M$ fatty acid in minimal medium (yeast nitrogen base plus 1% DL-sodium lactate and 1% Tergitol) about 10^5 cells/ml. The tubes were aerated on a rotor at constant speed in an incubator at 30°C. The fatty acid is from Applied Sciences Laboratory, Inc.

in addition to the unsaturated fatty acid requirers. In the fall of 1968, a search was begun in this laboratory for a palmitate requirer in *Neurospora* or yeast. Such a mutant was found in *Neurospora crassa*, although it had been previously characterized as a Tween 80 requirer (Henry and Keith, 1971). Upon close examination of this *Neurospora* mutant with oleate and other highly purified fatty acids, it was found that its desaturase activity was normal but that chain elongation was impaired to the extent that the growth rate was about 1% that of wild type.

Work with this mutant clearly indicated the need for a new fatty acid solubilizing agent, as detergents commonly used to solubilize fatty acids either contained fatty acids themselves or were toxic to *Neurospora*. After considerable effort, we found Tergitol NP-40 (Union Carbide) to be adequate. Although wild-type growth was detectably reduced, the growth curves had normal shapes, and the mutant was able to grow on Tergitol and palmitate. Tergitol effectively solubilized all fatty acids tested, did

TABLE VIII

FATTY ACID COMPOSITION OF *cel* AND WILD-TYPE *Neurospora*[a]

Strain and Supplementation	(Moles %)									
	<16:0	16:0	16:1	17:0	17:1	18:0	18:1	18:2	18:3	% Unsat./% Sat.
Wild type unsupplemented	4	21	3	—	—	8	27	32	5	(2.0)
cel unsupplemented	5	35	9	—	—	16	17	18	—	(0.8)
Wild type 18:1Δ⁹ *cis* + 18:1Δ⁹ *trans*	6	13	4	—	—	3	61*	11	2	(3.5)
cel 18:1Δ⁹ *cis* + 18:1Δ⁹ *trans*	5	10	2	—	—	4	61*	18	—	(4.3)
Wild type 17:0	2	15	3	38	8	8	13	13	—	(0.6)
cel 17:0	11	3	13	29	6	17	15	6	—	(0.7)
cel 16:0	Traces	64	2	—	—	15	11	8	—	(0.3)

[a] Fatty acid compositions, obtained by analytical GLC, of *Neurospora* mutant *cel* and wild type in response to fatty acid supplementation. Fatty acids were added to minimal liquid *Neurospora* medium at a concentration of 10^{-3} M with 3% Tergitol. Growth was at 34°C for 48 hours, except for *cel* unsupplemented, which was harvested after 24 days.

not serve as a carbon source or fatty acid supplement, and did not appear to be metabolized in any way.

The first striking observation with this mutant was that no *cis* unsaturated fatty acid tested proved adequate for growth by itself (Table V). On the other hand, saturates with chain lengths from 12 to 18 carbons permitted growth. Some examples of resulting fatty acid compositions are presented in Table VIII. Wild type grown under the same conditions and on the same supplementations, in most cases, resulted in similar fatty acid compositions.

At an extremely reduced rate, the chain elongation mutant was capable of incorporating acetate into long-chain fatty acids (Table IX). Under conditions that allowed the mutant to mimic wild-type growth, acetate was, however, incorporated into other nonlipid cellular components (Table X); consequently, impermeability to acetate was not the source of its exclusion from the fatty acid components. An acylase deficiency did not seem to be involved, as the chain elongation mutant could incorporate stearate, oleate, and 12-nitroxide stearate into lipid classes in about the same distribution as wild type (Table XI). Under some supplementation conditions, the mutant could grow without a *cis* unsaturate being present. Table XII summarizes these findings. Both $18:1\Delta^9$ *trans* and $18:1\Delta^{11}$ *trans*

TABLE IX

DISTRIBUTION OF ^{3}H FROM ^{3}H-ACETATE INTO FATTY ACIDS OF *Neurospora*[a]

Fatty Acid Fraction	Counts/mg Dry Weight								
	12	12:0	14:0	16:0	16:1	18:0	18:1	18:2	18:3
Wild type									
Exp. I	11	4	21	1613	47	406	2165	480	37
Exp. II	6	2	12	1257	33	313	1758	392	26
cel									
Exp. I	3	4	5	14	1	6	25	12	5
Exp. II	1	3	1	8	0	3	17	9	2

[a] Distribution of ^{3}H from acetate into fatty acids in *Neurospora* wild type and mutant *cel*. Both *cel* and wild type were grown for 24 hours on 20 ml of minimal medium + 16:0 (10^{-3} M) + 3% Tergitol and then transferred for an additional 12 hours to 20 ml of minimal medium with no supplement; 100 μc of ^{3}H-acetate was then added for one-half hour, during which the cultures were agitated. Following the labeling period the flasks were plunged into a water bath at 85°C and the fatty acids were subsequently extracted and analyzed.

TABLE X

Counts from ^{14}C-Acetate in
Solubilized Whole *Neurospora*
Tissues[a]

	Counts/mg Dry Weight
Wild type	
Exp. I	81,496
Exp. II	61,924
cel	
Exp. I	67,343
Exp. II	74,891

[a] Counts from ^{14}C-acetate in solubilized whole tissues. The growth conditions were the same as described in Table IX except that 4 μc of ^{14}C-acetate was added 1 hour prior to harvesting. The tissues were subsequently solubilized with Soluene TM 100 obtained from Packard.

gave restricted growth when present as the only supplementation. However, when an additional Δ^9 *cis* or other appropriate unsaturate was added, growth was considerably enhanced. Both 19:0 and 20:0 proved inadequate for growth by themselves, but when an unsaturate was also added, some growth occurred. Although this mutant is somewhat "leaky," the data strongly indicate that a saturated fatty acid is a growth requirement. We postulate that this requirement resides in the membrane structure, or that a membrane containing only *cis* unsaturated fatty acids does not possess the necessary properties to allow physiological function.

Several chain elongation mutants of yeast have now been isolated in this laboratory (Henry, 1971). These, in general, are very similar to the *Neurospora* chain elongation mutant with respect to their growth requirements.

Haploid yeast strains X2180 1Aa and X2180 1B were mutagenized by exposure to ethyl methane sulfonate. Mutants were identified by their failure to grow when replica-plated to YEPD medium containing no Tween 40. Thirty-nine mutants that failed to grow on unsupplemented YEPD were obtained. Complementation tests of these strains demonstrated the existence of several complementation groups. All the mutants that complement mutant 1, for example, fail to complement mutant 13, and those that complement 13 fail to complement 1. Within the groups of mutants that complement either mutant 1 or 13, there are complex patterns of complementation. These results indicate that several genetic loci may be involved.

In separate experiments, ^{14}C-acetate and ^{14}C-12:0 were added to the

TABLE XI[a]

Lipid Fraction	Counts/mg Dry Weight		
	Phospholipids	Neutral Lipids	Free Fatty Acids

A. Distribution of ^{14}C from 18:0-1-^{14}C into Lipid Classes of *Neurospora*

Wild type			
Exp. I	989 (77%)	211 (17%)	78 (6%)
Exp. II	607 (75%)	174 (21%)	32 (4%)
cel			
Exp. I	625 (47%)	387 (29%)	330 (24%)
Exp. II	382 (51%)	119 (16%)	243 (33%)

B. Distribution of ^{14}C from 18:1Δ^9 *cis*-1-^{14}C into Lipid Classes

Wild type			
Exp. I	19,330 (39%)	23,390 (47%)	6,796 (14%)
Exp. II	15,778 (35%)	24,095 (53%)	5,781 (12%)
cel			
Exp. I	17,104 (34%)	21,638 (43%)	11,244 (23%)
Exp. II	12,222 (30%)	16,870 (42%)	11,027 (28%)

C. Distribution of ^{3}H from 12NS-9,10-^{3}H into Lipid Classes

Wild type			
Exp. I	7,842 (26%)	19,123 (64%)	2,838 (10%)
Exp. II	6,400 (19%)	24,084 (73%)	2,418 (8%)
cel			
Exp. I	7,715 (25%)	18,192 (59%)	5,023 (16%)
Exp. II	6,052 (23%)	15,092 (58%)	4,768 (19%)

[a] Distribution of tracer into lipid classes from 18:0, 18:1Δ^9 *cis*, and 12NS in *Neurospora* mutant *cel* and wild type. A, Distribution of ^{14}C from 18:0-1-^{14}C into lipid classes. Growth took place under the conditions described in Table IX; 2 μc of 18:0-1-^{14}C was added one-half hour before harvesting. B, Distribution of ^{14}C from 18:1Δ^9 *cis*-1-^{14}C into lipid classes. Growth took place under the conditions described in Table IX except that 2 μc of 18:1Δ^9 *cis*-1-^{14}C was added 1 hour before harvest. C, Distribution of ^{3}H from 12NS-9,10-^{3}H into lipid classes. Growth took place under the conditions described in Table IX except that 5 μc of 12NS-9,10-^{3}H was added 1 hour prior to harvest.

TABLE XII

INDUCED GROWTH OF *Neurospora cel* BY COMBINATIONS OF SUPPLEMENTED FATTY ACIDS[a]

Fatty Acid	Additional Fatty Acid	Growth (mg Dry Weight)				
		3 Days	6 Days	12 Days	20 Days	24 Days
$18:1\Delta^9$ *trans*	None	0	2.4	8.0	—	—
	$18:1\Delta^9$ *cis*	15.1	52.0	—	—	—
	$16:1\Delta^9$ *cis*	0	10.5	39.5	—	—
	$18:1\Delta^6$ *cis*	0	0	1.4	—	—
	$18:1\Delta^{11}$ *cis*	0	0	0	0	0
	$18:2\Delta^{9,12}$ *cis, cis*	0	1.5	—	—	—
	$18:3\Delta^{9,12,15}$ *cis, cis, cis*	0	0	0	0	0
	19:0	0	0.3	22.5	—	—
	$18:1\Delta^{11}$ *trans*	0	6.5	29.5	—	—
$18:1\Delta^{11}$ *trans*	None	0	0	3.0	—	—
	$18:1\Delta^9$ *cis*	0	5.0	44.4	—	—
	$16:1\Delta^9$ *cis*	0	0.7	29.2	—	—
19:0	None	0	0	0	—	4.5
	$18:1\Delta^9$ *cis*	0	4.8	63.7	—	—
	$16:1\Delta^9$ *cis*	0	0.5	4.6	—	—
	$18:1\Delta^{11}$ *cis*	0	0	0	0	0
	$18:1\Delta^{11}$ *trans*	0	4.7	46.8	—	—
	$18:2\Delta^{9,12}$ *cis, cis*	0	0	0	0	0
	$18:3\Delta^{9,12,15}$ *cis, cis, cis*	0	0	0	0	0
20:0	None	0	0	0	0	0
	$18:1\Delta^9$ *cis*	0	—	18.0	60.3	0
	$16:1\Delta^9$ *cis*	0	0	—	32.7	34.9
No fatty acid, no tergitol	None	0	0	0	1.3	3.0
No fatty acid + +3% Tergitol	None	0	0	0	0	0

[a] Induced growth of *cel*, *Neurospora* mutant, by combinations of supplemented fatty acids. Growth took place without agitation at 34°C for the stated periods of time.

TABLE XIII

DISTRIBUTION OF ^{14}C FROM ACETATE-1-^{14}C IN YEAST AFTER ONE DOUBLING[a]

Lipid Source	Fatty Acid								
	12	12:0	14:0	16:0	16:1	18:0	18:1	18:2	18:3
Nonlabeled carrier	42	35	43	34	45	33	31	29	32
Wild type 2180	32	33	52	667	676	146	557	48	39
Mutant 1	31	32	31	66	41	53	88	28	35

[a] Distribution of ^{14}C from 12:0 into fatty acid fractions in mutant 1 and X2180 1A; 2 μc of ^{14}C-12:0 was added to 8 ml of YEPD + 1% Tween 40 during logarithmic growth for exactly one doubling time, corresponding to the period between 50 and 100 Klett units. Unlabeled 12:0 was added to bring the total concentration to 10^{-4} M.

TABLE XIV

DISTRIBUTION OF ^{14}C FROM 12:0 – 1-^{14}C IN YEAST[a]

Lipid Source	Fatty Acid								
	12	12:0	14:0	16:0	16:1	18:0	18:1	18:2	18:3
Nonlabeled carrier	42	35	43	34	45	33	31	29	32
Wild type 2180	34	157	98	235	255	58	192	62	36
Mutant 1	30	359	683	885	520	97	300	90	51

[a] Distribution of ^{14}C from 12:0 into fatty acid fractions in mutant 1 and X2180 1A; 2 μc of ^{14}C-12:0 was added to 8 ml of YEPD + 1% Tween 40 during logarithmic growth for exactly one doubling time, corresponding to the period between 50 and 100 Klett units. Unlabeled 12:0 was added to bring the total concentration to 10^{-4} M.

liquid growth medium (YEPD + 1% Tween 40) of both mutant 1 and X2180 1Aa (the strain from which mutant 1 was derived). The tracer was added for exactly one doubling time, corresponding to the period between 50 and 100 Klett units (Table XIII).

Mutant 1 demonstrates a considerably reduced ability to incorporate acetate into fatty acids (Table XIII). However, it is able to elongate 12:0 and to desaturate the elongation products (Table XIV). This result is somewhat inexplicable since 12:0-supplemented minimal medium will not support the growth of mutant 1 under conditions in which 12:0 is supplied to the mutant as the only fatty acid, indicating that it is toxic in some way. 12:0, at these concentrations, is not highly inhibitory to wild type but wild type is also producing relatively large quantities of other fatty acids that may compete with 12:0. These results indicate that mutant 1 can desaturate long-chain fatty acids and that it has reduced ability to incorporate acetate into its fatty acids.

Mutants 1, 3, 4 and 13 were selected for further examination of the growth requirement. No differences in the growth requirement were apparent among these four mutants even though each of the four had a different complementation pattern with respect to the other mutants. (Mutants 3 and 4 complement mutant 1 and do not complement 13; however, 3 and 4 complement each other and also certain other mutants within the group of mutants that fail to complement 13.) Saturated fatty acids from 14 to 17 carbons in length supported growth at 10^{-3} M at 30°C [Fig. 4(a)–(d)]. These fatty acids were solubilized with 3% Tergitol NP-40 in liquid YEPD. The ability of 15:0 and 17:0 to support growth demon-

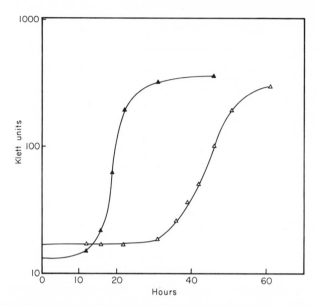

Fig. 4(a). Growth curves for saturated fatty acid requirers show yeast mutant 1 and wild type X21801A. ▲, X21801A in yeast extract peptone dextrose plus 1% Tween 40 (16:0); △, mutant 1 in yeast extract peptone dextrose and 1% Tween 40.

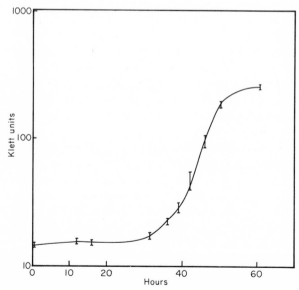

Fig. 4(b). Mutant 1 in yeast extract peptone dextrose plus 16:0 at 10^{-3} M plus 3% Tergitol NP40.

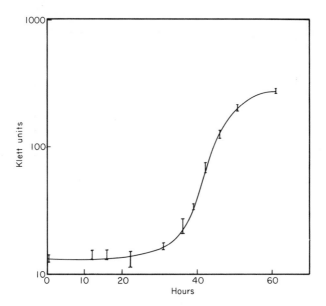

Fig. 4(c). Mutant 1 in yeast extract peptone dextrose plus 14:0 at 10^{-3} plus 3% Tergitol NP40.

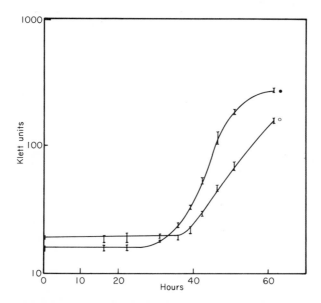

Fig. 4(d). ●, Mutant 1 in YEPD plus at 15:0 at 10^{-3} M plus 3% Tergitol NP40; ○, mutant 1 in yeast extract peptone dextrose plus 17:0 at 10^{-3} M plus 3% Tergitol NP40. Growth curves were obtained by inoculating Klett tubes containing 8 ml of medium with 10^4 cells. The tubes were aerated on a rotor at constant speed and 30°C.

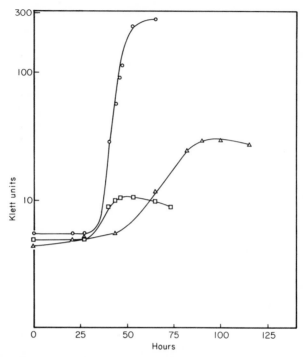

Fig. 5. Growth curves for saturated fatty acid-requiring yeast mutant 13. ○, Mutant 13 on yeast extract peptone dextrose plus 1% Tween 40; △, yeast extract peptone dextrose plus 10^{-4} M $18:1\Delta^9$ *cis* and 1% Tergitol NP40; and □, yeast extract peptone dextrose only (no supplemented fatty acid). Same conditions hold as for Fig. 4.

strated that odd-chain fatty acids satisfy the fatty acid requirement even though such fatty acids are not normally found in yeast. In this regard, the yeast mutants are similar to the *Neurospora* mutant *cel*. At 30°C, 18:0 did not support growth, while at 35°C, the mutants grew in the presence of this fatty acid. This may mean that at 30°C a membrane constructed primarily from 18-carbon fatty acids was not sufficiently fluid to be functional or desaturase activity may also have been lowered. Schweizer and Bolling (1970) isolated a mutant of yeast defective in saturated fatty acid biosynthesis, but it was able to grow on oleate as a fatty acid supplement. This mutant, no doubt, was adequately "leaky" to supply its own saturated fatty acid requirement. $18:1\Delta^9$ *trans* (10^{-3} M) was tried but did not satisfy the growth requirement at 30°C. However, this fatty acid has not been tried in combination with *cis* unsaturated fatty acids. Combinations of *cis* and *trans* unsaturated fatty acids were growth-supporting in the *Neurospora* mutant *cel*.

No growth occurred in the presence of $16:1\Delta^9$ *cis* at concentrations of 10^{-3} M, 5×10^{-4} M, or 10^{-4} M at 30°C; 10^{-3} M $18:1\Delta^9$ *cis* also failed to support growth. However, at 5×10^{-4} M and 10^{-4} M several doublings took place beyond the growth obtained in unsupplemented YEPD (Fig. 5). Since $18:1\Delta^9$ *cis* at lower concentrations allows several divisions to occur, it may indicate that any saturated fatty acid that is present in these mutants at the start of an experiment can be diluted to some extent but that at some point a membrane containing a high percentage of unsaturated fatty acids becomes inadequate for growth. No growth occurs with $16:1\Delta^9$ *cis*, perhaps because its shorter chain length and lower melting point create an even more fluid membrane than $18:1\Delta^9$ *cis*. We postulate that a membrane with only one fatty acid is too homogeneous in lipid content to be physiologically functional or flexible. Figure 6 shows data reported by

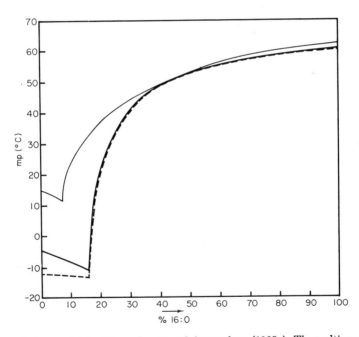

Fig. 6. Data are obtained from Lyons and Asmundson (1965a). The melting point of mixtures of oleate and palmitate is shown by the light line, mixtures of linoleate and palmitate are shown by the dark line, and mixtures of linolenate and palmitate are shown by the dotted line. Percentages are expressed as mole percent.

Lyons some years ago (Lyons and Asmundson, 1965b). Oleate has about the same melting point when highly purified or when containing small quantities of palmitate. As the percentage of palmitate goes above a critical minimum, the melting point becomes very sensitive to small changes in the percentages of palmitate. Linoleate and linolenate give a similar relationship except that they both hold their melting point until each contains about 16% palmitate. Then the rise in melting point is very steep, so that when the ratio of palmitate to either unsaturate changes between about 16 and 20% palmitate, the change in melting point is drastic. We feel that these data illustrate an important heterogeneity effect. Supplementation with fatty acids can result in greater than 90% of one fatty acid or about 95% in unsaturated fatty acid content. We feel that 95% approaches the critical limit of unsaturation. Supplementation of chain elongation mutants with unsaturated fatty acids may exceed the tolerable unsaturated fatty acid limit.

A long-chain saturated fatty acid is required for optimal growth in these yeast mutants, as it is in the *Neurospora* chain elongation mutant. The existence of a requirement for saturated fatty acids in these two organisms strongly suggests a generalized requirement for both saturated and unsaturated fatty acids in the formation of functional membranes. Furthermore, saturated fatty acids are not required simply as precursors of unsaturated fatty acids. Expressed in more general terms, we say that the fatty acid composition of a functional membrane appears to require some heterogeneity.

IV. Physical Properties Conferred to Fatty Acids by Double Bonds

Since double bonds are vital to most organisms, we present the following discussion with respect to the alterations and physical properties of fatty acids imposed by double-bond position as relevant to understanding the role of double bonds in the fatty acids of membrane elements. Figure 7 shows data obtained from Gunstone (1967) concerning the melting point of 18-carbon unsaturated fatty acids as a function of double-bond positions. Of course, *cis* double bonds have lower melting points than their respective *trans* analogs and that seems reasonable, since a *cis* double bond is more of an interruption to stacking (crystallization) than a *trans*. That the melting point is lowest when the double bond is near the center of the chain is reasonable on similar grounds relating to the ease of crystal formation. Fatty acids having double bonds that span carbons odd–even have considerably lower melting points than those that span even–odd. As shown in Fig. 7, two examples are $18:1\Delta^9$ *cis* and $18:1\Delta^{10}$ *cis*, where

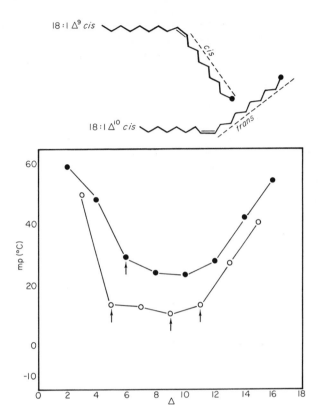

Fig. 7. The abscissa shows the double-bond position in the delta notation on 18-carbon fatty acids.

the melting points differ by some 13° or 14°C. This relates to the *cis-trans* nature of the double-bond carbons to the end groups (methyl and carboxyl terminals) and the energy level at which crystal formation occurs. The terminal CH=CH *cis* has the lower melting point and is by far the most common occurring in nature. The Δ⁹ *cis* and Δ¹¹ *cis* are widespread in occurrence. The membrane fatty acids of vertebrates are polyunsaturated; however, the depot fats are largely monoenes of the Δ⁹ *cis* variety. In the yeast desaturase mutants, we have determined that certain double-bond-containing fatty acids are growth-supporting (those with arrows in Fig. 7) and that certain ones are not. For example, 18:1Δ⁹ *cis* supports growth (melting point = 10°C) but 18:1Δ⁹ *trans* does not (melting point = 45°C). We think that this difference mainly relates to the melting point difference. Most fatty acids we have tested are consistent with this frame of thinking.

Stearolic acid ($18:1^{9Z}$) also has a melting point at about 45°C yet supports growth well. We do not have an explanation for this inconsistency unless these phospholipid ester-bearing regions of biological membranes have an intrinsically lower melting point than those of $18:1\Delta^9$ *trans*. The observation was made several years ago that anaerobic yeast could use $18:1^{9Z}$ as a growth requirement. Therefore our same general argument still holds— that it is a requirement in membrane lipid elements to have interruptions or impurity zones such that there is some deviation from complete homogeneity in fatty acid composition. This heterogeneity imposes a more fluid condition upon the biological membrane and allows for localized variation in physical properties.

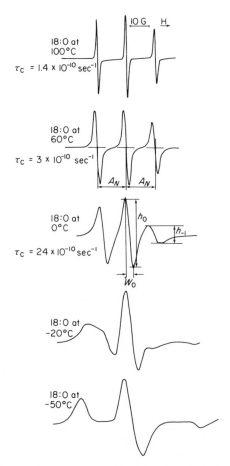

Fig. 8. The spin label 12NS (I) was used at 10^{-4} M in approximately 98% stearic acid.

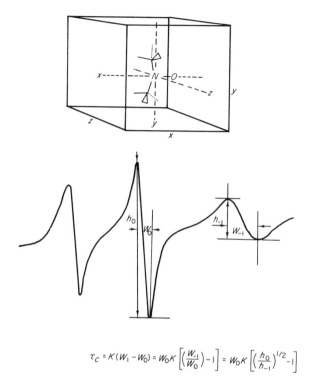

$$\tau_C = K(W_1 - W_0) = W_0 K \left[\left(\frac{W_{-1}}{W_0}\right) - 1\right] = W_0 K \left[\left(\frac{h_0}{h_{-1}}\right)^{1/2} - 1\right]$$

Fig. 9. A schematic, partially immobilized ESR spectrum is shown together with spectral measurements relevant for τ_c calculations. The box above shows a schematic nitroxide with the orientation of the three principal axes of the $2p\pi$ orbital bond relating to the unpaired electron. The x axis is parallel to the base of the box, the y axis is parallel to the upright edges of the box, and the z axis projects into the plane of the paper.

V. Spin-Labeled Lipids

Stable chemical radicals have been known for many years, and a great deal of chemistry has been carried out on these molecules [for reviews of stable radical chemistry, see Forrester *et al.* (1968) and Rozantsev (1970)]. Many of these chemicals are nitroxides and have a three-electron bond of N to O, which we denote as N → O. This three-electron bond has a net unpaired electron and is paramagnetic. Several reviews are now available that deal with chemical, biochemical, and biophysical applications of nitroxides (Hamilton and McConnell, 1968; Griffith and Waggoner, 1969; Metcalfe, 1970; Smith, 1970; McConnell and McFarland, 1970; Snipes and Keith, 1970).

Figure 8 shows a series of spectra taken of 98% stearic acid with 12-nitroxide stearate (12NS, I) at temperatures ranging from 100° to −50°C.

We begin our treatment of molecular motion with a discussion of the information contained in Fig. 8 and shown somewhat more clearly as to spectral measurement in Fig. 9. Figure 8 shows the range of spectra that results from very free motion at 100°C to highly restricted motion at −50°C. The nitroxide molecule acts as an impurity and therefore may be expected to yield an impurity effect. When 12NS is added to highly purified stearic acid at about 10^{-3} M, the spectra are about the same as those shown in Fig. 8 at temperatures above the melting point of stearic acid. At temperatures below the melting point, concentration effects occur that indicate that the 12NS (I) molecules are moving together to form impurity pools of several 12NS molecules each. At lower concentrations of 12NS (about 10^{-5} M or lower) the spectra remain approximately like those in Fig. 8 over the entire temperature range.

For the case shown in Fig. 8, where 12NS is 10^{-4} M in 98% stearic acid, we have the approximate concentration relationships shown in Table XV. The generalization arrived at from this indicates that the concentration effects, reflecting 12NS molecules in close proximity to each other, do not occur when there is a large number of impurity molecules per each 12NS molecule. It is well known that impurities resist crystal formation. If 12NS

TABLE XV

IMPURITY EFFECTS IN STEARIC ACID[a]

Compound	Molar Concentration
18:0 (98%)	3.4
Impurity fatty acids	0.068
12NS (10^{-4} M)	0.0001
Impurity fatty acids/12NS	680
18:0 (99.9%)	3.4
Impurity fatty acids	0.0034
12NS (10^{-3} M)	0.001
12NS (10^{-5} M)	0.00001
Impurity fatty acids/12NS (10^{-3} M)	3.4
Impurity fatty acids/12NS (10^{-5} M)	340

[a] Relative molar concentrations are shown for 98% stearic acid with respect to impurity fatty acids and 12NS. The same relationships are shown for stearic acid at 99.9%

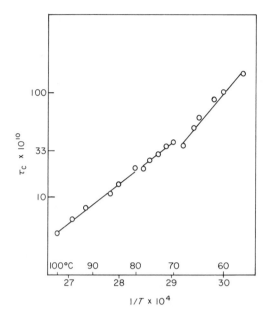

Fig. 10. Arrhenius plot for 10^{-5} M 12NS (I) in approximately 99.9% stearic acid.

behaved structurally as a stearic acid molecule, then all spectra below the melting point would reflect a state of rigid viscosity. Since there is a gradual change in motion in the spectra presented in Fig. 8, as the motion at 0°C is 24×10^{-10} second for a "90-degree arc of motion," the 12NS molecules are quite free. We imagine the 12NS molecule as existing in liquid impurity pools. The restraint on motion placed on the boundaries of the impurity pool is linear on a log τ_c versus °K^{-1} plot for a single bulk phase; consequently, the slope of the line is linear both above the melting point and below the melting point, but the slopes are unequal, reflecting different motion barriers. Figure 10 shows the plot of 12NS (10^{-5} M) in 99.9% stearic acid.

We stated earlier that the fatty acid composition of biomembranes is always heterogeneous; consequently, we never have a situation where 12NS (I) or any other nitroxide is the only heteromolecule. Phospholipids and the lipids of membranes exist in physical states that are different from those of fatty acids or hydrocarbons. Phospholipids in a bulk aqueous medium are usually represented by bilayers where the hydrocarbon elements are laminar and the polar groups are stacked side by side; these may be multilayers or coiled myelinlike vesicles, but the unit structure is still lamillar. Phospholipids at lower states of hydration go through at least three mesomorphic liquid crystal states as a function of temperature. We imagine that a nitroxide solubilized in a phospholipid or membrane is treated somewhat like 12NS solubilized in stearic acid in that the local domain of the N → O moiety is treated as an impurity. In the hydrocarbon interior of phospholipids there is probably a melting point at some temperature depending on the fatty acid composition. The physical state below the melting point must have relatively ordered hydrocarbon zones, and at the melting point the hydrocarbon zones acquire enough energy to become more "fluid." Impurity nitroxides reflect this change, and the same general argument holds for nitroxides in phospholipids and membranes as for 12NS in stearic acid. Of course, sometimes other considerations must be taken into account such as the size of the impurity pool caused by the nitroxide and specific interactions between certain matrices and certain nitroxides.

Rotational correlation time (τ_c) based upon the Kivelson (1960) treatment has been used several times in the past few years. The equation shown below (and in Fig. 9) is valid for isotropic motion in the fast tumbling range. The range of motion we employ varies from about 10×10^{-10} to about 50×10^{-10} second. Even though the absolute times may be in error, particularly those stated at the slower values, the relative values are valid. In single-phase media, a plot of log τ_c versus $°K^{-1}$ varies smoothly; therefore, errors in τ_c are probably errors in absolute τ_c values, which would be reflected by translation of the line or by τ_c errors that cause changes in slope. Neither of these error sources would affect relative values where the same nitroxide is used in the same type of matrix. The equation for τ_c shown here can be derived from Kivelson (1960):

$$\tau_c = K[W_{-1} - W_0] = KW_0 \left[\frac{W_{-1}}{W_0} - 1 \right] = KW_0 \left[\left(\frac{h_0}{h_{-1}} \right)^{1/2} - 1 \right]$$

K is a constant that depends on the anisotropic hyperfine coupling values (A_x, A_y, A_z) and the anisotropic g values (g_x, g_y, g_z). The three characteristic ^{14}N hyperfine lines are denoted by $1, 0$, and -1; therefore the first derivative line widths are denoted by W_1, W_0, and W_{-1}, and the numerical values are expressed in gauss. By assuming Lorentzian line shapes, the ratio W_{-1}/W_0 can be replaced by $(h_0/h_{-1})^{1/2}$; usually line heights can be measured with great accuracy.

The nitroxides used and referred to in this chapter are shown in Fig. 11.

One of the earliest studies employing spin labels to study biological membranes was carried out by Chapman et al. (1969). These authors employed the much-used maleimide spin label (X), which alkylates free amines and sulfhydryl groups. They showed that the spin label alkylated

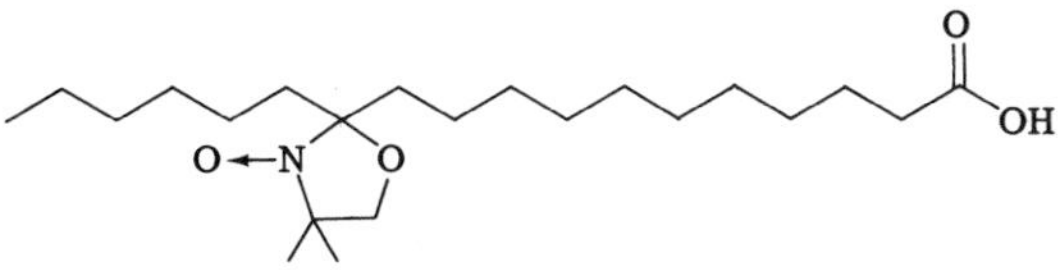

(I)

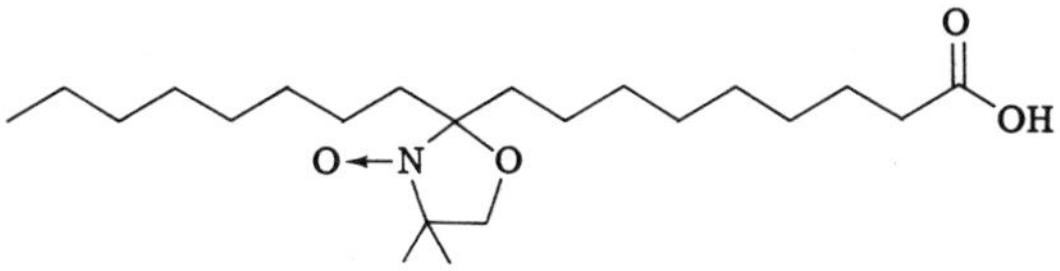

(II)

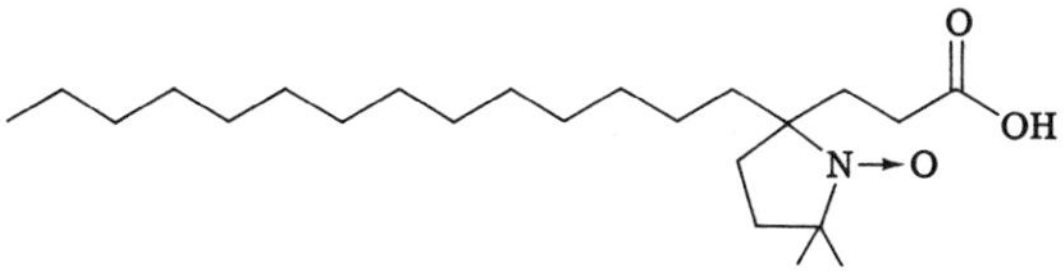

(III)

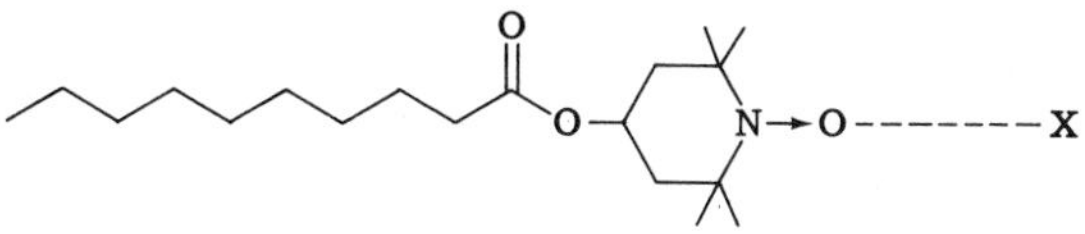

(IV)

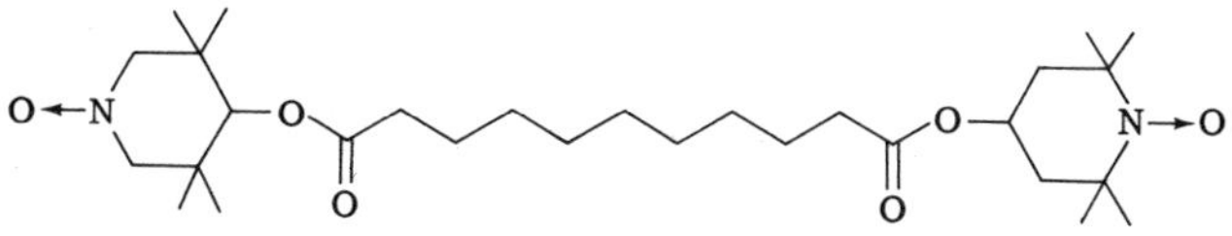

(V)

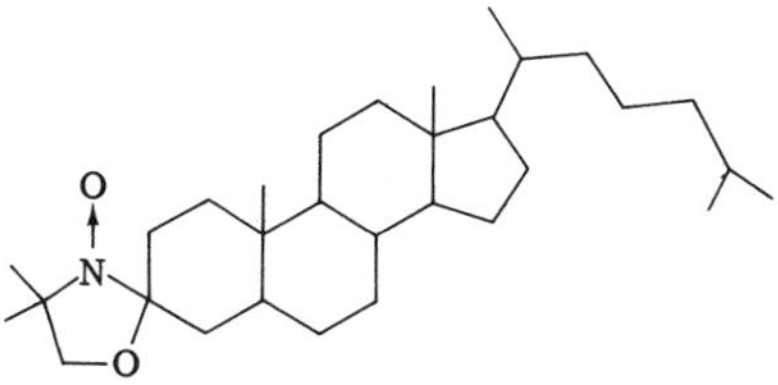

(VI)

Fig. *11* (a). See page 295 for legend.

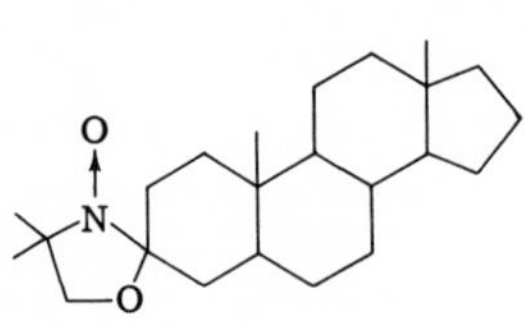

(VII)

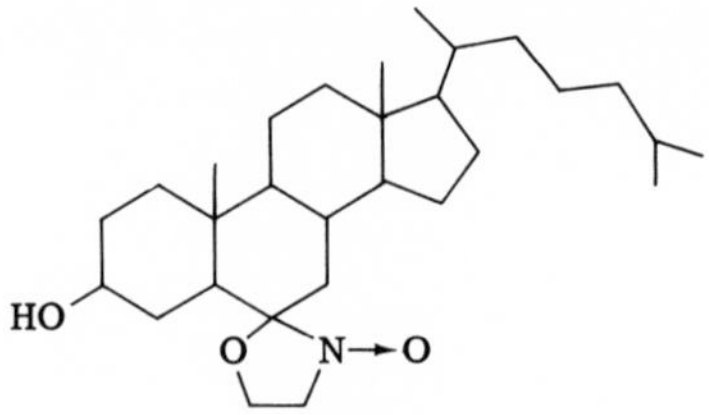

(VIII)

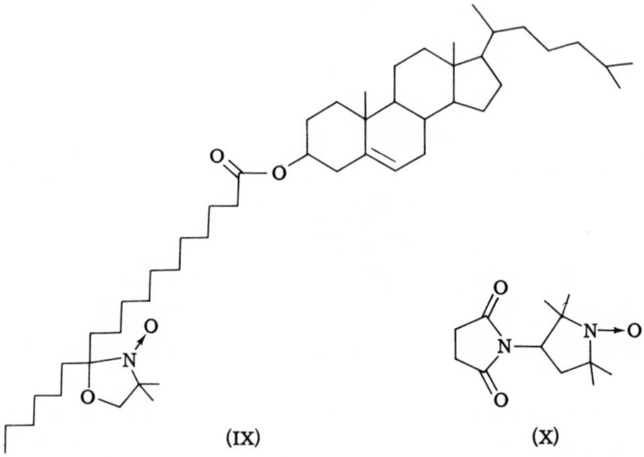

(IX)

(X)

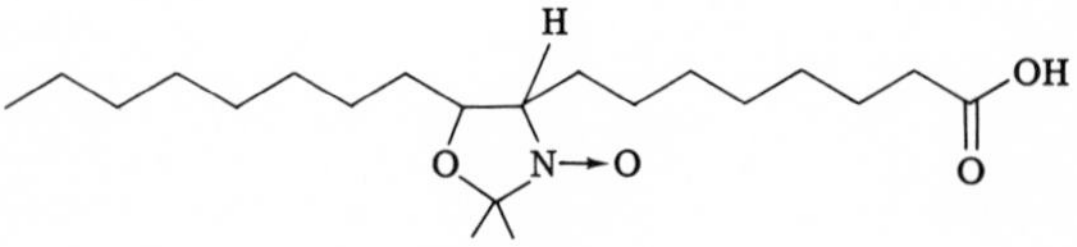

(XI)

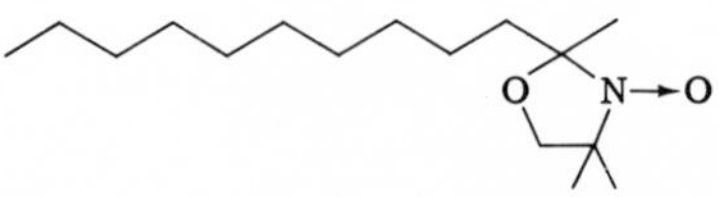

(XII)

Fig. *11*(b). See facing page for legend.

on red blood cells resulted in a two-component signal: One was free in solution, and the other was strongly immobilized. Pretreatment of the red blood cells with maleimide, followed by treatment with spin label maleimide, prevented the formation of any of the strongly immobilized component.

The novelty of the Keana synthesis (Keana *et al.*, 1967) for synthesizing nitroxides on ketone sites proved to be a major step in allowing experimenters to produce chemicals with desired structural shapes and other features essential for particular problems. The synthesis of a long-chain fatty acid containing a nitroxide at carbon 12 proved important in developing the use of spin label molecules for biological membrane systems. This was carried out by Keith *et al.* (1968), who demonstrated that growing *Neurospora* hyphae incorporated the spin label fatty acid (12NS, I) into mitochondrial lipids. Resulting spectra from the isolated mitochondria demonstrated that the nitroxide-containing fatty acid was in a rather fluid lipid environment. Later, this same spin label was shown to be useful in the study of phospholipid vesicles and other lipid systems (Waggoner *et al.*, 1969). Since then, this nitroxide-containing fatty acid has been used in several studies.

Keith *et al.* (1970) carried out further spin-labeling experiments on *Neurospora* mitochondria and showed that the signals originating from isolated phospholipids and intact mitochondria were very similar, with the membrane preparations always showing slightly more restriction of the spin label motion. This work also demonstrated that when a spin label fatty acid is bound to a protein moiety such as bovine serum albumin, the spin label is strongly immobilized. No such immobilization was found at comparable temperatures originating from biological membranes or phospholipid vesicles. An ESR signal of 12NS (I) originating from mitochondria is shown in Fig. 12, together with a signal showing the state of motion when 12NS is bound to bovine serum albumin. Both signals were taken at 30°C.

Fig. 11. Structures of spin labels. I. 12NS, the di-methyl oxazolidine derivative of 12 keto-stearic acid. II. 10NS, the di-methyl oxazolidine derivative of 10 keto-stearic acid. III. 4NS, the di-methyl oxazolidine derivative of 4 keto-stearic acid. IV. 10:OT, the decanoic acid ester of 2,2,6,6-tetra methyl-4-hydroxyl-1-N-oxyl piperidine. V. 11:OT, the bi-radical of nonane dicarboxylic acid ester of 2,2,6,6-tetra methyl-4-N-oxyl piperidine. VI. 3NC, the di-methyl oxazolidine derivative of 3 keto-cholestane. VII. 3NA, the di-methyl oxazolidine derivative of 3 keto-androstane. VIII. 6NC, the di-methyl oxazolidine derivative of 6 keto-cholestane. IX. 12NS-C, the 12NS ester of cholesterol. X. maleimide spin label, the di-methyl pyrolidine derivative of maleimide. XI. 9,10NS, the di-methyl oxazolidine derivative of oleic acid (the oxazolidine ring is co-planer with that of the alkyl pleat). XII. the di-methyl oxazolidine derivative of 2 keto-tetradecane.

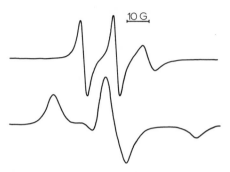

Fig. 12. ESR signal of 12NS originating from mitochondria is shown (upper) together with a signal showing the state of motion when 12NS is bound to bovine serum albumin (lower). Both signals were taken at 30°C.

Tourtellotte *et al.* (1970) used 12NS as a spin label to probe the membranes of *Mycoplasma laidlawii.* The motion of 12NS in the whole *Mycoplasma* and in extracted polar lipids showed a temperature dependency. These data plotted on Arrhenius graphs showed an activation energy of around 4 kcal/mole and demonstrated that fatty acid composition had a measurable effect on the rotational correlation time (τ_c). The spin label in *Mycoplasma* and in the extracted phospholipids gave similar τ_c values and varied in the same way with temperature changes for a given fatty acid composition. When the fatty acid composition was largely unsaturated, the motion was more free than when it was largely saturated (Fig. 13). In every case the motion was more free in the extracted phospholipids than in the whole cell preparation, indicating that the protein elements of the membranes served in some way to restrict the motion of the substituent lipid elements. It was concluded that 12NS goes largely into the hydrophobic lipid zones of *Mycoplasma,* most of which are probably in the membranes. Since freeze-etch electron microscopy was also carried out on the same preparations, it was possible to state that the inclusion of the spin label fatty acid into hydrophobic zones of *Mycoplasma* membranes did not cause gross abnormalities.

These authors also showed that gramicidin-D in dispersions immobilized 12NS and that when *Mycoplasma* was first treated with 12NS and then with gramicidin-D immobilization of the 12NS signal also occurred (Fig. 14). This observation was not dealt with in detail but probably illustrates an important action by certain antibiotics. This mechanism of growth or function inhibition would be by binding to membrane hydrophobic zones and rendering them nonfunctional by placing them in a state of rigid viscosity.

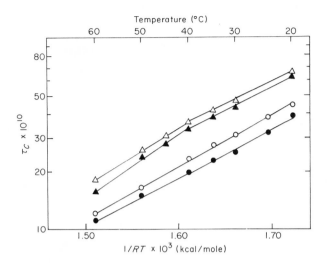

Fig. 13. △, Extracted polar lipids of *Mycoplasma* grown on 18:1Δ⁹ *cis*; ▲, *Mycoplasma* grown on 18:1Δ⁹ *cis*; ○, extracted polar lipids of *Mycoplasma* grown on 18:0; and ●, *Mycoplasma* grown on 18:0.

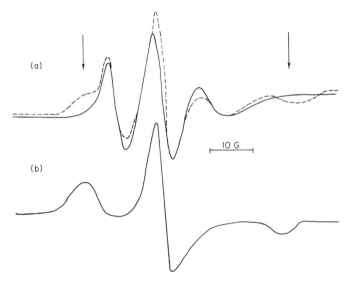

Fig. 14. (a) Typical spectrum of 12NS in *Mycoplasma* with no gramicidin. (b) The solid line shows 12NS in a 10^{-4} *M* suspension of gramicidin-D. The dotted line shows 12NS in *Mycoplasma* membranes after gramicidin-D has been added.

Libertini *et al.* (1969) showed that 3-nitroxide cholestane (3NC, VI) and 12-nitroxide stearic acid (I) would orient in an oriented egg lecithin multilayer. Under the conditions employed, the two spin labels were sufficiently immobilized that the three principal axes did not time-average to give an isotropic signal. Because of the way 3NC and 12NS intercalate into a phospholipid multilayer, their two z axes of symmetry are perpendicular to each other. Therefore, when 3NC gave its maximum hyperfine coupling, 12NS resulted in its minimum coupling. This experiment was very convincing and showed how the spin label sterol and spin label fatty acid intercalated into a natural product phospholipid. This study was made more convincing and more sophisticated by including computer simulations of the anisotropic spectra. Shortly thereafter, Hsia *et al.* (1970) demonstrated that egg lecithin multilayers oriented a spin-labeled sterol (3NC) when partially hydrated but not when dehydrated. These authors also showed that egg lecithin-cholesterol multilayers oriented 3NC whether hydrated or dehydrated. Hubbell and McConnell (1969b) showed orientation of spin-labeled fatty acids having the oxazolidine ring close to the carboxyl group in red blood cell ghosts and in the walking-leg nerve fibers of *Homarus americanus*. Nitroxides of this type all have the z principal axis of symmetry parallel to the long axis of the fatty acid. They also demonstrated orientation of a spin-labeled sterol (the y principal axis of symmetry is parallel to the long axis of the sterol molecule) in red blood cells using hydrodynamic shear as the mechanical method of orientation (Hubbell and McConnell, 1969a).

The last few paragraphs have shown that partially hydrated multilayers of egg lecithin, dehydrated egg lecithin-cholesterol, and red blood cell ghosts and nerve fibers will orient certain spin labels with respect to the applied magnetic field when the structure containing the nitroxide is also oriented with respect to the magnetic field. Such orientation is of a static nature and has the requirement that the molecular motion of the spin label be sufficiently slow such as not to time-average away the anisotropic character of the ESR signal. Consequently, rigid positional anisotropy is not equivalent to anisotropic motion. As a further example of positional anisotropy, a mixture of cytochrome c-egg lecithin-water (1:1:1 by weight) mixed with 4NS (III) is shown in Fig. 15 (Gulik-Krzywicki *et al.*, 1971). This clearly demonstrates that a protein (cytochrome c) and a phospholipid (egg lecithin) partially hydrated will orient under appropriate conditions.

The partitioning effect of a small nitroxide soluble in organic solvents and water was used by Hubbell and McConnell (1968). This small nitroxide gave different hyperfine couplings in organic and aqueous media. In a

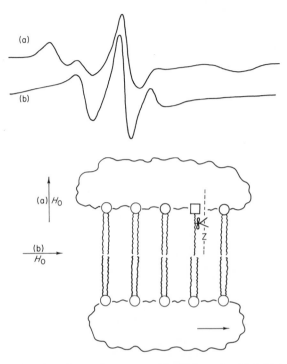

Fig. 15. A preparation of cytochrome c–egg lecithin–water (1:1:1) is shown that is doped with the spin label fatty acid 4NS. (a) Orientation of the planes of the multilayer perpendicular to the applied magnetic field; consequently, the z principal axis of the $2p\pi$ orbital of the nitroxide on 4NS is parallel to the applied magnetic field. (b) An orientation 90 degrees from (a) with the plane of the multilayer parallel to the applied magnetic field; consequently, the z principal axis of the nitroxide is perpendicular to the applied magnetic field.

heterogeneous medium composed of an aqueous phase and some biological membranes, it was possible to see two high-field lines, one characteristic of the organic environment of the biological membrane and the other characteristic of the more polar aqueous environment.

Other paramagnetic species such as many heavy metal ions result in spin-spin interactions with nitroxides that broaden the nitroxide signal in a distance-dependent manner. If the other paramagnetic species is spaced close to the nitroxide, then the nitroxide signal is broadened proportionate to the distance. Since these species interact with nitroxides in a distance-dependent manner, they will only broaden a signal in the same environment. For example, some nitroxides will partition in an oil-water emulsion, and if a paramagnetic ion such as copper is added in appropriate concentration, then only the signals originating from nitroxides in the aqueous environment will be broadened.

Lyons and Raison (1969, 1970) showed that some homeothermic animals and chilling-sensitive plants have mitochondria that give nonlinear Arrhenius plots. These curves are all the same type, having a straight-line relationship above and below some "critical temperature." The temperature represented by the perturbation was thought to be the temperature where some membrane phase transition altered membrane-bound enzyme activity. These findings were made more interesting by the observations that a poikilothermic animal (trout) and a chilling-resistant plant (potato) had mitochondria that gave a straight-line relationship for oxygen uptake on an Arrhenius plot over the same temperature range. Subsequent work was carried out on these same systems using the spin label 12NS (I) added *in vitro* to mitochondrial suspensions (Raison *et al.*, 1972). The homeothermic animal (rat) and chilling-sensitive plant (sweet potato) showed physical phase transitions at the same temperatures as previously recorded for oxygen uptake (Fig. 16). The poikilothermic animal (trout) and the

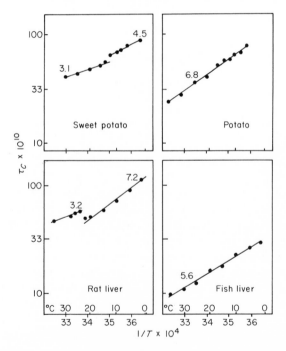

Fig. 16. Arrhenius plots for *in vitro* 12NS-labeled mitochondria. The numerical values represent E_a in kilocalories per mole.

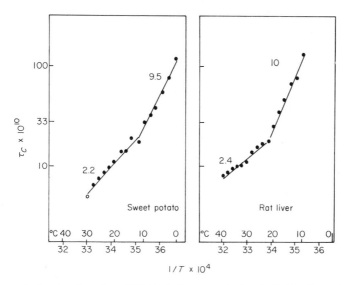

Fig. 17. Arrhenius plots for 12NS-labeled extracted phospholipids. Numerical values represent E_a values.

chilling-resistant plant (potato), as before with oxygen uptake, showed no nonlinearity on an Arrhenius plot. These data indicate that the respiratory enzyme activity of mitochondria is altered by the physical state of membrane lipids and that the ability of organisms to tolerate temperature extremes may be largely controlled by the physical state of membrane lipids (see Fig. 16).

The extracted phospholipids from sweet potato and rat liver mitochondria gave similar Arrhenius data, with the change in activation energy taking place at the same temperature (Fig. 17). A plot of τ_c for 12NS in stearic acid is also shown to illustrate the same phenomenon (Fig. 10); therefore, the observation is valid whether the nitroxide is localized in small domains of molecular dimensions or dispersed in a bulk phase.

Gotto *et al.* (1970) used the spin label maleimide to study high- and low-density serum lipoproteins. This spin label serves as an alkylating agent, where the alkyl group is the nitroxide and has relative specificity for amines and sulfhydryl groups. When the nitroxide was alkylated onto the free amines and sulfhydryls of the delipidated apoprotein of high-density serum lipoprotein, the ESR signal appeared relatively free. Upon adding the lipid moiety back to the apoprotein, the signal was largely immobilized. This gave some indication that the lipid elements of the high-density serum lipoprotein surround and place the surface of the apoprotein in hydrophobic zones. This would largely inhibit the free flopping motion that might be expected to occur on surfaces of proteins in aqueous media by moieties containing hydration spheres.

Chapman *et al.* (1969) had previously shown that the maleimide spin label (X) that alkylates sulfhydryls and amine groups gave a two-component signal when red blood cell ghosts were treated, illustrating the heterogeneity of the binding sites. Holmes and Piette (1970) used the maleimide spin label (X) and iodoacetate-containing spin label to show at least two and probably three different types of binding sites on the sulfhydryls and amines of red blood cell proteins. These authors used the effects of chlorpromazine and sodium dodecyl sulfate to refine their observations. Chlorpromazine added an immobilized component, and treatment with sodium dodecyl sulfate, in contrast, caused a two-component signal to become a free, one-component signal.

If the heterogeneity of a system, with respect to polarity or viscosity, is discontinuous, then it may be possible to resolve three or more components. If the system is linearly heterogeneous, then the lines will broaden and resolution is not possible. A partially resolvable heterogeneous system is shown in Fig. 18, where 2-nitroxide tetradecane (2NT, XII) is dissolved in human high-density serum lipoprotein. Ascorbic acid has been shown to destroy nitroxide signals where molecular collisions are possible (the im-

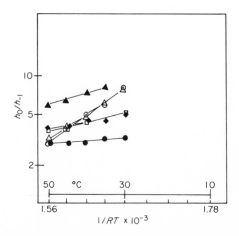

Fig. 18. Whole *Neurospora* hyphae of wild type and the mutant *cel* were labeled with the spin label fatty acid 12NS. ▲, Unsupplemented *cel*; ●, unsupplemented wild type; ○ and △, *cel* and wild type supplemented with 17:0; and □ and ◇, *cel* and wild type that have been supplemented with a mixture of oleic and elaidic acids. The motion parameter shown on the ordinant is graphed in the standard manner of an Arrhenius plot.

mediate proximity), probably by chemical reduction. In Fig. 18 addition of ascorbate removed part of what appeared to be a two-component signal at 0°C and still left a two-component signal showing that 2NT was clustered in three domains having three different viscosities. Since ascorbate is water-soluble, its effect is very reduced in hydrophobic zones and makes possible the selective destruction of signals originating from regions that contain water or have aqueous channels. The untreated sample at 30°C shows somewhat broadened lines and two third lines. After treatment with 0.3 equivalent of ascorbate, the lines become more narrow and homogeneous in appearance.

VI. Spin-Labeled Yeast

It was reported in 1968 by Keith *et al.* that 12NS was incorporated into phospholipids, neutral lipids, and free fatty acids of *Neurospora crassa* during growth and was found in these forms in the isolated mitochondria. The distribution of 12NS, at that time, was determined by spin content. Since the signal was destroyed by *Neurospora* and isolated mitochondria at easily measurable rates, there was some uncertainty as to the real distribution of the 12NS molecule. Table XVI shows the distribution of 12NS-^{3}H into a respiratory-competent desaturase mutant (KD115), a petite desaturase mutant (KD46), and wild type yeast. Both KD115 and wild type vary in the distribution of 12NS with fatty acid supplementation in the same way, while KD46 has a different pattern. The similarity between KD115 and S288C does not extend to the distribution of 18:1Δ^9 *cis*. Here the two desaturase mutants are more similar.

Most of the phospholipids in yeast can be accounted for by phosphatidyl choline, phosphatidyl ethanolamine, and phosphatidyl inositol. The distribution of 12NS from yeast grown on several fatty acid supplements into the major phospholipid classes is shown in Table XVII. Here again wild type and KD115 show a strong resemblance and KD46 shows its own unique distribution.

Tables XVI and XVII have shown that 12NS is incorporated into phospholipids and neutral lipids in a respiratory-competent mutant, in petite, and in wild type. While the respiratory-competent mutant and wild type, in general, show the same distribution of 12NS, the data indicate that 12NS may be treated as a unique fatty acid and therefore may not mimic the distribution of any other fatty acid. The different distributions in KD46 will simply be treated as phenomena for the present.

A mutant of *Neurospora crassa* deficient in the synthesis of long-chain fatty acids has been probed with spin labels (Henry and Keith, 1971).

TABLE XVI

DISTRIBUTION OF ^{3}H FROM ^{3}H-12NS[a] AND 18:1Δ^9 cis-1-^{14}C[b] INTO YEAST LIPIDS

Fatty Acid Supplement	Tracer	Yeast Strain[c]	Lipid Class[d]	Disinte- grations/Min	% Distribution
16:1Δ^9 $trans$	^{3}H-12NS	KD46	PL	19,253	19.8
			FFA	35,044	36.0
			NL	43,111	44.3
18:1^{9Ξ}	^{3}H-12NS	KD46	PL	16,098	20.5
			FFA	12,322	15.7
			NL	50,069	63.8
18:1Δ^9 cis	^{3}H-12NS	KD46	PL	88,055	52.3
			FFA	12,611	7.5
			NL	67,721	40.2
16:1Δ^9 cis	^{3}H-12NS	KD115	PL	2,293	14.9
			FFA	3.003	19.5
			NL	10,086	65.6
16:1Δ^9 $trans$	^{3}H-12NS	KD115	PL	2,150	11.1
			FFA	2,931	15.1
			NL	14,364	73.9
18:1^{9Ξ}	^{3}H-12NS	KD115	PL	2,281	10.4
			FFA	3,035	13.9
			NL	16,541	75.7
18:1Δ^9 cis	^{3}H-12NS	KD115	PL	5,605	25.5
			FFA	2,143	10.3
			NL	14,095	64.2
16:1Δ^9 cis	^{3}H-12NS	S288C	PL	3,183	16.0
			FFA	2,393	12.0
			NL	14,340	72.0
16:1Δ^9 $trans$	^{3}H-12NS	S288C	PL	2,968	14,8
			FFA	2,884	14.4
			NL	14,178	70.8
18:1^{9Ξ}	^{3}H-12NS	S288C	PL	1,549	13.6
			FFA	1,706	14.9
			NL	8,164	71.5
18:1Δ^9 cis	^{3}H-12NS	S288C	PL	4,476	22.7
			FFA	2,610	13.2
			NL	12,618	64.0
18:1Δ^9 cis	^{14}C-18:1Δ^9 cis	KD115	PL	31,695	69.4
			FFA	2,398	5.3
			NL	11,581	25.4
18:1Δ^9 cis	^{14}C-18:1Δ^9 cis	KD46	PL	473,096	70.6
			FFA	6,644	1.0
			NL	190,592	28.4
18:1Δ^9 cis	^{14}C-18:1Δ^9 cis	S288C	PL	28,503	44.6
			FFA	3,478	5.4
			NL	31,932	50.0

This mutant (*cel*, for fatty acid chain elongation) grows well in the presence of such saturates as palmitate (16:0) or odd-chain components such as pentadecanoate (15:0) and desaturates these to the corresponding Δ^9 *cis* unsaturates. This mutant is slightly "leaky" and will grow slowly without supplementation. Under nonsupplemented conditions the addition of 12NS during growth followed by ESR analysis of the hyphae results in an immobilized signal compared to wild type under the same growth conditions. Supplementation of *cel* and wild type in the same way followed by treatment with 12NS and ESR analysis results in almost identical spectra over a temperature range, as evidenced by h_0/h_{-1} (Fig. 19). The ratio h_0/h_{-1} is a motion parameter that is proportionate to τ_c, in the reason of interest. Under unsupplemented and very slow growth conditions, the immobilized signals coming from 12NS in *cel* reflect a high viscosity of lipid zones. We suppose that this rigid viscosity of lipid zones is characteristic of poor physiological conditions for organismic function. The more fluid lipid zones of wild type and supplemented *cel* probably are indicative of a more functional system.

From the standpoint of this organism's membrane physiology, the most important observation is that a saturated fatty acid is a requirement for growth. The organism is not deficient in desaturase activity and therefore can produce unsaturates when an appropriate saturate is available. The observation that growth cannot occur on pure Δ^9 *cis* unsaturated fatty acids, we believe, demonstrates the requirement for a heterogeneous fatty acid composition where part of this composition must be supplied by saturated fatty acids.

Several palmitate requirers in yeast have recently been isolated in this laboratory (Henry, 1971). These mutants fall into several complementation groups and probably represent chain elongation mutants for different sites in the fatty acid chain elongation enzyme complex. For present purposes, two aspects are of interest. First, these mutants will not grow on a highly

---

[a] ³H-12NS from J. C. Williams *et al.* (1972).

[b] ¹⁴C-18:1Δ^9 *cis* from New England Nuclear Corp.

[c] Wild type S288C and respiratory-competent KD115 were grown at 30°C with agitation in flasks of 40 ml of minimal medium (YNB + 2% glucose + 1% Tergitol) supplemented with $2 \times 10^{-4}\ M$ fatty acid. Petite KD46 was grown on YEPD (YEP + 2% glucose + 1% Tergitol) supplemented with fatty acid. At the time of inoculation, 1.5 μc of ³H-12NS was added to each flask and 1.5 μc of ¹⁴C-18:1Δ^9 *cis* was added to each flask containing 18:1Δ^9 *cis*. Stationary-phase pellets were harvested for tracer analysis. 18:1$^{9\varXi}$ was obtained from Lachat Chemicals, Inc., and repurified, and the other fatty acids are from the Hormel Institute.

[d] PL, phospholipid; FFA, free fatty acid; and NL, neutral lipid.

TABLE XVII

DISTRIBUTION OF ^{3}H-12NS^a AND 18:1Δ^9 cis-1-^{14}C^b INTO PHOSPHOLIPIDS

Fatty Acid Supplement	Tracer	Yeast Strain	Lipid Class[c]	Disinte-grations/Min	% Distribution
16:1Δ^9 cis	^{3}H-12NS	KD115	PC	1,173	58
			PE + PI	841	42
16:1Δ^9 trans	^{3}H-12NS	KD115	PC	1,080	58
			PE + PI	794	42
18:1$^{\Xi}$	^{3}H-12NS	KD115	PC	703	48
			PE + PI	757	52
18:1Δ^9 cis	^{3}H-12NS	KD115	PC	1,303	37
			PE + PI	2,193	63
16:1Δ^9 cis	^{3}H-12NS	KD46	PC	4,670	83
			PE + PI	963	17
16:1Δ^9 trans	^{3}H-12NS	KD46	PC	1,737	70
			PE + PI	761	30
18:1$^{\Xi}$	^{3}H-12NS	KD46	PC	2,197	82
			PE + PI	499	18
18:1Δ^9 cis	^{3}H-12NS	KD46	PC	8,083	93
			PE + PI	602	7
16:1Δ^9 cis	^{3}H-12NS	S288C	PC	2,339	57
			PE + PI	1,770	43
16:1Δ^9 trans	^{3}H-12NS	S288C	PC	1,330	68
			PE + PI	613	32
18:1$^{\Xi}$	^{3}H-12NS	S288C	PC	1,230	49
			PE + PI	1,296	51
18:1Δ^9 cis	^{3}H-12NS	S288C	PC	1,269	46
			PE + PI	1,513	54
18:1Δ^9 cis	18:1Δ^9 cis-1-^{14}C	KD115	PC	5,272	49
			PE + PI	5,568	51
18:1Δ^9 cis	18:1Δ^9 cis-1-^{14}C	KD46	PC	71,106	99
			PE + PI	385	1
18:1Δ^9 cis	18:1Δ^9 cis-1-^{14}C	S288C	PC	1,799	50
			PE + PI	1,819	50

[a] ^{3}H-12NS from J. C. Williams et al. (1972).

[b] 18:1Δ^9 cis-1-^{14}C from New England Nuclear Corp.

[c] PC, phosphatidyl choline; PE, phosphatidyl ethanolamine; and PI, phosphatidyl inositol. The growth conditions are identical to those presented in Table XVI.

purified Δ^9 cis unsaturated fatty acid; consequently, a saturated fatty acid is a growth requirement. These yeast mutants and the Neurospora cel mutant represent the only reported cases of such observations. The biological aspect of these mutants was dealt with earlier in this chapter. The second aspect is the ESR analyses of these mutants. Since these yeast mutants are all saturated fatty acid requirers, it is possible to control the fatty acid composition to some extent. Supplementation with 15:0, 16:0, and 17:0 yields cells that show phase transitions at different temperatures (Fig. 19). One of the yeast saturated fatty acid requirers (mutant 1) has been examined by ESR using 12NS and several other spin labels. Figure 20 shows the changes in the temperature of phase transition as a function of fatty acid supplementation, and Fig. 21 shows an Arrhenius plot of τ_c for the yeast (mutant 1 grown on 16:0) and its lipid extracts. The phospholipid extracts of untreated yeast, yeast cells subjected to heat treatment, and yeast cells subjected to heat treatment and freeze-thaw are represented. These all result in a phase transition at the same temperature. In addition, phospholipase-C, pronase, sodium azide, and microbial lipase-treated cells all show a phase transition at $13° \pm 1.5°C$. The slopes and absolute τ_c change somewhat among the different preparations but probably not much more than among several different preparations with the same treatment. Since the temperature of phase transition is unchanged with different treatments and since τ_c at various temperatures is not drastically changed among the different preparations, it appears that these properties are contained in the lipids and not in the proteins. Furthermore, it is difficult to state what the influences of membrane proteins are on the behavior and function of membrane lipids.

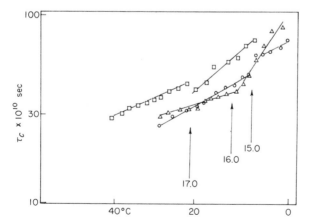

Fig. 19. An Arrhenius plot is shown where yeast mutant 1 has been grown on different fatty acid supplementations and then *in vitro* spin-labeled with 12NS. □, Growth on 17:0; △, growth on 16:0; and ○, growth on 15:0.

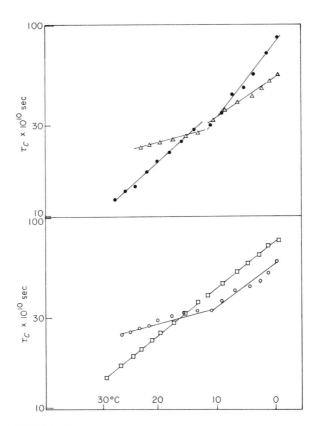

Fig. 20. ●, ESR signals taken from 12NS in extracted phospholipids after yeast mutant 1 had grown on a supplementation of 16:0; △, spectra from heat-treated yeast; ○, spectra from heat-treated and freeze-thawed yeast; and □, spectra taken from untreated mutant 1 grown on 16:0 that was spin-labeled with 12NS-C.

Figure 20 also shows the Arrhenius plot of 12-nitroxide stearate-cholesterol (12NS-C, IX) added *in vitro* to mutant 1, which was grown on palmitate. This line is straight and shows no indication of a phase change in this temperature range. We believe that this indicates that the nitroxide is in close proximity to the sterol skeleton.

Intact yeast or yeast phospholipids bind several nitroxides in the same general way as reflected by the ESR spectra. A spin-labeled sterol (6NC, VIII) is highly immobilized in yeast, indicating that most of it goes into phospholipid domains. This nitroxide is considerably more free in octadecane or triglycerides than in phospholipids.

While the temperature of phase transition is the same for all preparations of yeast having the same fatty acid composition, the spectra are not identical. The spin label fatty acid 4NS (III) illustrates this point. Figure 21 shows spectra of 4NS in Asolectin and in mutant 1. Seelig (1970) has shown that the definition of two high-field lines is proportionate to the degree of anchoring of the polar head group. The spectra in Fig. 21 show that Asolectin and the extracted lipids from mutant 1 have approximately the same degree of definition, although the Asolectin spectra are more immobilized. Both are considerably better defined than in the intact yeast. We postulate that yeast treated with a spin label fatty acid under *in vitro* conditions has only its cytoplasmic membrane labeled due to the difficulty of hydrophobic molecules such as fatty acids in passing com-

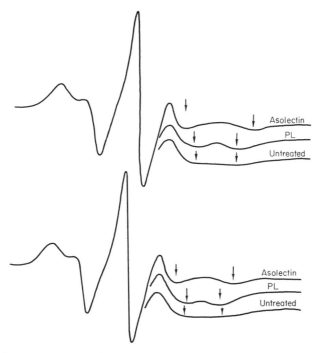

Fig. 21. The spectra were taken from the spin label fatty acid 4NS in Asolectin at 20° and 30°C. The partial spectra were taken from the phospholipids of yeast mutant 1, which was grown on the fatty acid supplementation 16:0. The other partial lined spectra show how the third line definition changed when 4NS was added to yeast that had grown on 16:0 as a fatty acid supplementation but had been untreated insofar as heat or any type of denaturation was concerned.

pletely through the membrane. For example, several hours are required to get "good" uptake of radioactive fatty acids into ester form, but a sizable amount is immediately bound by the cell and can be reisolated as free fatty acids. Therefore 4NS (III) acid added to intact yeast probably labels the membrane. We observe that 4NS in yeast is not anchored so well as in isolated phospholipids. We imagine that the reduction in anchoring is caused by extensive ionic interactions between membrane lipids and membrane proteins.

One can imagine that phase transitions are observable in isolated phospholipids, yeast preparations, or other lipid preparations for more than one reason. For example, if the spin label acting as an impurity is randomly dispersed in the more liquid phase but is clustered into impurity pools in the more solid phase, then we would expect to observe a change in line width (W) at the temperature of phase transition. In yeast systems, using 12NS, no such broadening occurs, as is shown by the plot of log W_0 versus $°K^{-1}$ (Fig. 22). Instead, there is a change in the way h_0/h_{-1} responds to temperature changes. Therefore we explain these phase transitions simply as an observable phenomenon and offer the following postulation as to the mechanism: The nitroxide is an impurity and resists ordering in its local domain. Therefore the boundaries of the nitroxides' influence are the important zones for consideration. The boundaries of the impurity pool are the smallest and offer the greatest restriction to motion at the lowest temperature.

As the temperature increases, the pool expands and the motion becomes more free. At the temperature of phase transition, the matrix molecules become less organized and the energy barriers to motion at the impurity boundaries change. This necessarily results in a change in E_a of motion parameters. We imagine that these energies will almost always be different but that some situation (matrix) can exist where they are equalized. For the case where 12NS-C yields no phase transition and we attribute this to a cholesterol effect, we imagine that the impurity pool is considerably expanded. The increase in pool size would necessarily make it more difficult for the nitroxide to sense changes in boundary size and matrix influences on the boundaries.

In general terms, we postulate that membrane lipids comprise the matrix for the deposition of membrane proteins. The dependency of membrane function on physical parameters must reside largely in the lipids. The lipids impose these physical limitations on the membrane proteins, which may be expressed as restrictions in enzyme activity. For example, in the sweet potato and rat liver mitochondrial preparations discussed earlier, enzyme-dependent oxygen uptake shows activity changes at the same temperatures

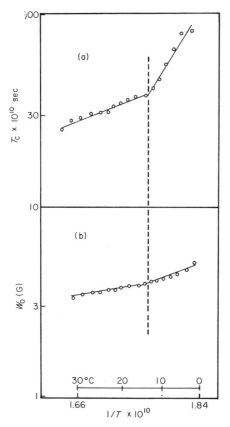

Fig. 22. (a) Arrhenius plot of the motion of 12NS in yeast mutant 1, which had been supplemented with 16:0. (b) How the midline width parameter of the same spectra varies with temperature.

where phase transitions were observed to occur in their mitochondrial phospholipids and other mitochondrial lipids.

In this laboratory, we feel that the lipid elements of membranes form the basis for membrane function, structure, and temperature dependency. Still, we recognize that they contain proteins ranging from 30 to 70% of membrane dry weight. Membranes form selective barriers between two environments and in addition to this generally recognized function must contain proteins arranged in ordered mosaics to enhance the interaction of enzymes. This quarternary ordering of elements must give specific membranes unique function. The literature dealing with membranes, proteins, and the postulated modes of interaction between membrane proteins and lipids tells us the current level that exists in understanding membranes as complex structures and will be dealt with briefly.

VII. Membrane Proteins

Since the finding of proteinaceous material in membrane preparations, many possible functions have been proposed for the protein moiety. Among those most favored are (1) transport, (2) maintenance of membrane structure, (3) recognition sites, and (4) membrane-bound enzymic syntheses. Much work has been directed toward an understanding of protein-lipid interaction in order to formulate a model for membrane structure and toward the role of proteins in transport systems. Recently, the importance of the topological arrangement of proteins in receptor sites for hormones, drugs, transport material, and cell contact systems has been stressed. Coincident with this emphasis is the belief that ordered arrays of membrane proteins may be of considerable advantage for complex multienzyme syntheses.

Fundamental to an understanding of membrane protein function is a knowledge of the structure and spatial relationships of membrane-associated proteins. From theoretical and experimentally derived considerations, the following structural models are tenable: (1) the Davson-Danielli unit membrane as modified by Robertson, (2) the subunit model, and (3) the mosaic scheme. Sjöstrand and Lucy independently proposed a membrane consisting of juxtaposed, protein-coated lipid micelles (Sjöstrand, 1963; Green and Perdue, 1966; Glauert and Lucy, 1968); their models are based upon thermodynamic considerations and membrane solubilization studies. The unit membrane model (Davson and Danielli, 1935; Robertson, 1960) consists of a phospholipid bilayer with the polar head groups facing the hydrophilic environment. Proteins are bound to the surfaces by hydrophilic interactions. This model was thought to satisfy both the requirements for thermodynamic stability and the staining pattern observed by electron microscopy. With the advent of transport studies and the concept of *trans* membrane pores, this model has been modified to allow for protein zones penetrating the membrane. Green's recent proposition that hydrophobic regions of the proteins extend into the methylene chain regions of the phospholipids (Vanderkooi and Green, 1970) resembles earlier theories by Benson (1966). According to Robertson (1960), the unit membrane surface proteins are postulated to be in an extended β conformation resembling the pleated sheet structure. This allows interaction of the peptide bond and polar side chains with the phospholipid polar head groups and the aqueous environment. Also, the lipophilic side chains would be able to achieve some penetration of the hydrophobic bilayer interior or interact with similar side chains on neighboring proteins. Although various degrees of protein folding that deviate from the β conformation and the presence of globular proteins are allowed, considerable β structure should be present. This has not been observed by optical methods (Stein and Fleisher, 1967; Urry and Ji, 1968). Choules has pointed out the difficulties in interpreting ORD-CD and argued for the presence of 30 to 45% β sheet in *Mycoplasma* membranes (Choules and Bjorklund, 1970).

Although the requirement of a lipid bilayer with the lipid chains perpendicular to the plane of the membrane is generally accepted, considerable controversy as to the protein disposition exists. Singer has proposed a major modification of the Davson-Danielli model in which the proteins, as in the original scheme, are considered to be mainly globular (Glaser *et al.*, 1970), but instead of forming a uniform layer on the lipid surface, the proteins are randomly scattered on, embedded in, and penetrating through the membrane. Thus, as in Green's postulations, hydrophobic as well as hydrophilic or ionic interactions are important. As more data are obtained, newer models are constantly being proposed that differ from each other in fine nuances. From the original, distinctly bipartite, unit membrane-subunit controversy, most authors have modified their competitors' theories and assumed a more equitable position. Differentiation between models depends on an accurate determination of the importance of hydrophobic interactions of proteins and lipids. Magnetic resonance spectroscopy appears to be the most useful technique for this. Optical data indicating considerable α-helical content and freeze-fracture electron photomicrographs lend support to the Singer formulation (Branton, 1969). Singer has also proposed that a spectrum of possible membrane structures ranging from the previously described format to protein membranes with only a little lipid intercalated may exist. The protein envelopes of insect viruses may be an example of the latter.

The Davson-Danielli model indicates that a structural protein may be a necessary constituent of all membranes. Although the subunit and mosaic models do not require any special protein to maintain structural integrity, the unit membrane surface protein of extended β conformation would argue strongly for its existence. Criddle *et al.* (1962) isolated a mitochondrial structural protein (MSP) from bovine heart mitochondria. The mitochondrial enzyme system with the ordered arrangement of the electron transport chain reasons for a structural protein to maintain alignment of the complexes. MSP was found to make up approximately 50% of the total mitochondrial protein, to bind respiratory chain components and phospholipids, and to be insoluble. The amino acid composition and physical properties of structural protein isolated from liver mitochondria, microsomes, erythrocytes, and chloroplasts were similar in a variety of species (Richardson et al., 1963; Hulton and Richardson, 1964; Edwards and Criddle, 1966; Woodward and Mûnkres, 1966). Recent investigations have shown that a major portion of the material isolated as structural protein from mitochondria is denatured ATPase, and the concept of membrane structural proteins is questioned (Senior and MacLennon, 1970). The Fraction I protein isolated from chloroplasts was once thought to be a structural protein, and now it also appears to be a denatured enzyme, RUDP carboxylase (Kawashima and Wildman, 1970). Laico et al. (1970) have recently reported the finding of a group of small (molecular weight about 5000) glycoproteins in bovine mitochondria, bovine and human erythrocytes, and bovine rods. The function of these self-aggregating proteins is unknown at present.

The nature of the units of electron transfer function and their relationship to membrane structure have been perceived through the efforts of many investigators using biochemical and electron microscopic techniques. The integrated enzyme assemblies of oxidative phosphorylation are well-characterized examples of enzymic systems functioning in a membrane matrix where it has been demonstrated that phospholipids play an essential role in several of the reactions. The role of membranes is also of paramount importance in other synthetic processes. Membrane-bound ATP production provides energy for protein synthesis. Hendler has reviewed the relationship between protein biosynthesis and membranes (1968). Bound ribosomes are a general cytological fact and tRNA, mRNA, and nascent peptides are found associated with membrane fragments. The activity of bound ribosomes is greater than for the free. Free ribosomes may be an experimental artifact, but their lesser capacity for protein synthesis could also be a function of the system investigated, as the theory that free ribosomes synthesize intracellular protein and bound ribosomes make protein for export has been advanced.

A brief mention of transport proteins should be made even though space does not allow a complete review of this ever-growing field. Little knowledge concerning the molecular basis of transport in eukaryotic single-cell organisms has been gained compared to bacterial systems. Biochemical genetics of yeast has indicated the presence of specific permeases for amino acids and sugars. The galactose system is well characterized genetically (Hartwell, 1970). A knowledge of the proteins involved might be inferred from the results in bacterial systems using osmotic shock techniques in which bacterial cells lose the ability to accumulate a variety of compounds, and protein factors that bind galactose, sulfate, and amino acids are found in the shock fluid. Kundig *et al.* (1970) have studied a phosphotransferase system in *E. coli* that is located on the cell surface and functions in the uptake of carbohydrates. An enzyme catalyzes the phosphorylation of a protein, which transfers the phosphoryl group to the glycoside. Specificity as to which sugar is phosphorylated is governed by the enzyme catalyzing the transfer to the carbohydrate. Osmotic shock disrupts the system and releases the intermediate transfer protein into the medium, and mutations in it or the enzymes correlate with loss of transport capacity (Tanaka and Lin, 1967; Kundig *et al.*, 1970).

Fox and Kennedy (1965) isolated and partially purified a protein from *E. coli* that functions in the lactose transport system. Genetic evidence indicates that the protein is coded for by the permease gene of the *lac* operon (Fox *et al.*, 1967). Further work concerning the turnover of the transport protein and incorporation into the membrane as a function of lipid composition has been published and is being continued (Wilson *et al.*, 1970; Wilson and Fox, 1972).

Although evidence of receptor sites for hormones, small molecules, and cell-cell contact is prevalent in multicellular organisms, few examples are well documented among the eukaryotic protists. Crandall and Brock (1968) have investigated the molecular basis of mating in the yeast *Hansenula wingei*. Sexual agglutination between haploid mating types was found to be due to glycopeptides on the cell surface. These proteins have also been isolated from the cytoplasm and may be incorporated in the cell membrane. Each glycoprotein is specifically synthesized by only one mating type and combines only with the glycoprotein of the opposite type. The factor from one is a multivalent agglutinin of high molecular weight. The complementary mating factor was a smaller, univalent glycopeptide that neutralized the biological activity of the agglutinin. *Saccharomyces cerevisiae* was found to release an oligopeptide hormone that causes the opposite mating type to undergo morphological alterations prior to conjugation and diploid zygote formation (Duntze *et al.*, 1970). Presumably, a complementary receptor site for the hormone exists that, when activated, causes the initiation of enzymic synthesis.

Many diverse functions have been elucidated for proteins incorporated into biological membranes. As yet there is no definite evidence for a protein component serving a purely structural role. Although the Robertson version of the Davson-Danielli model cannot be invalidated until there is a greater knowledge of the relative importance of hydrophobic and hydrophilic lipid-protein interactions, the random mosaic protein distribution as described by Singer allows for stabilization of membrane structure by catalytically active proteins.

VIII. Summary

The degree of specificity of the fatty acid requirement of a number of fatty acid biosynthetic mutants is described in this chapter. The general conclusion is that a certain amount of flexibility exists with regard to which fatty acids result in functional membranes. The exact basis of this specificity is unknown.

We have shown by physical considerations how various properties of fatty acids may serve to restrict or enhance their suitability as membrane elements at a particular growth temperature. These include *cis* versus *trans* double-bond configuration, double-bond position with respect to terminal groups, degree of unsaturation, and presence of other chain interrupter groups. However, other criteria may be imposed upon these considerations by the existence of requirements for proper membrane structure, function, and biosynthesis.

Membrane structure as related to fatty acid composition has been studied in *Mycoplasma* and *E. coli* where certain fatty acid supplementations resulted in abnormal cell morphology. Studies linking the function of various cytochromes to phospholipid effectors have been performed (Green and Fleisher, 1963). However, the question of whether fatty acids are needed for limited membrane synthesis is not clear. On the one hand, Mindich (1970), using glycerol mutants of *Bacillus subtilis*, has shown that membrane protein synthesis continues in the absence of fatty acid synthesis. Fox (1969), however, employing the *E. coli* desaturase mutants, has found that a functional lactose transport system depends on the presence of an unsaturated fatty acid at the time of induction of the *lac* operon. He postulates that the fatty acid or a derivative may be required to activate the transport site on the membrane.

Another possibility determining the basis of the fatty acid specificity is that it is merely a reflection of an acylase or fatty acid permease specificity. Those fatty acids that get into the cell are metabolized; those that can be esterified to glycerol phosphate are incorporated into membranes.

To examine the relationship between fatty acid composition and the formation of functional mitochondria, Proudlock *et al.* (1969) looked at the mitochondria of unsaturated fatty-acid-depleted yeast. The mitochondria were found to contain cytochromes and to respire at normal rates. However, the depletion resulted in loss of mitochondrial oxidative phosphorylation.

Other investigators have used lipid requirers to elucidate the enzymic steps of the fatty acid biosynthetic pathway. Such studies have been reported by Bloomfield and Bloch with anaerobic yeast and by Silbert and

Vagelos with the *E. coli* desaturase mutants. Later studies with the *E. coli* mutants have demonstrated the existence of at least two genes (cistrons) responsible for unsaturated fatty acids (Cronan *et al.*, 1969). The metabolism of various fatty acids can be examined by supplementing the growth medium of auxotrophs with radioactive forms of specific fatty acids.

Luck (1963) performed the first density labeling of membranes by using a choline auxotroph of *Neurospora*. By measuring the density of mitochondria as a function of time, he was able to present convincing evidence that mitochondria arise from preexisting mitochondria by division. A similar approach has been described by Fox *et al.* (1970) and employs *E. coli* desaturase auxotrophs grown on bromostearic acid. Membranes isolated from cells grown on this fatty acid exhibited a greater density ($\Delta \rho = 0.06$ g/cm^3) than membranes derived from cells grown on $18{:}1\Delta^9$ *cis* or $18{:}1\Delta^9$ *trans*. This difference is enough to allow a complete separation of the two kinds of membranes.

In our ESR treatment we concentrate mainly on the information obtainable from a derived parameter of spin label motion, Arrhenius activation energy (E_a). This parameter has been used to illustrate that membranes and membrane lipids undergo "hydrocarbon melts" at the same temperature. Furthermore, heat treatment, freeze-thaw treatment, freeze-thaw and heat treatment, incubation with pronase, and treatment with lipase or phospholipase enzymes do not alter the temperature of phase transition (E_a^l, E_a^s intercept). The intercept of E_a^s with E_a^l (E_a in solid and liquid phases) seems to be a function of the fatty acid composition; however, these treatments do change the absolute values of E_a^s, E_a^l, and τ_c so that the physical state of the hydrocarbon zones is influenced by the phospholipid polar groups and by interaction with membrane proteins.

These interactions represent relatively long-range effects and do not alter the temperature at which the hydrocarbon melt occurs. In addition to information obtained from activation energy considerations, there are several other ESR parameters that give information. The hyperfine coupling constant (A_N) is proportionate to the polarity of the N $\rightarrow$ 0 group's local environment. The A_N for 12NS varies from about 16.1 G in water to 15.2 G in octadecane. At 60°C, the A_N for 12NS in *Neurospora crassa* mitochondria was measured at 15.2 to 15.3 G and similar values (15.2 to 15.4) were obtained for 12NS in yeast, *Mycoplasma*, rat liver mitochondria, fish liver mitochondria, sweet potato mitochondria, and potato mitochondria. Consequently, the spin label, 12NS, probes hydrophobic zones of membranes.

Vitamin C in aqueous environments destroys the signal originating from nitroxides. Since vitamin C is water-soluble and does not partition into hydrocarbon solvents, it is expected that we would not destroy the ESR signals originating from biological membranes where the nitroxide probe is dissolved in hydrophobic zones. However, several equivalents of vitamin C destroy the 12NS signal in all membranes that we have analyzed. This indicates that the membranes have a high density of aqueous channels, which come in close proximity to the nitroxides' local environment, or that the nitroxide spends some time in the domain of vitamin C caused by motion of the spin label or some type of membrane dynamics.

Acknowledgments

Research supported by USPHS Grant AM-12939, an institutional grant from the American Cancer Society (520), and AEC Contract Project Agreement 194.

References

Andreasen, A. A., and Steir, T. J. B. (1953). *J. Cell. Comp. Physiol.* **41**, 23.
Bard, M. (1971). Unpublished data.
Benson, A. A. (1966). *J. Amer. Chem. Soc.* **43**, 265.
Bloch, K., Baronowsky, P., Goldfine, H., Lennarz, W. J., Light, R., Norris, A. T., and Scheuerbrandt, G. (1961). *Fed. Proc., Fed. Amer. Soc. Exp. Biol.* **21**, 921.
Bloomfield, D. K., and Bloch, K. (1960). *J. Biol. Chem.* **235**, 337.
Branton, D. (1969). *Annu. Rev. Plant Physiol.* **20**, 209.
Chapman, D., Barratt, M. D., and Kamat, V. B. (1969). *Biochim. Biophys. Acta* **173**, 154.
Choules, G. L., and Bjorklund, R. F. (1970). *Biochemistry* **9**, 4759.
Crandall, M. A., and Brock, T. D. (1968). *Bacteriol. Rev.* **32**, 139.
Criddle, R. S., Bock, R. M., Green, D. E., and Tisdale, H. (1962). *Biochemistry* **1**, 827.
Cronan, J. E., Jr., Birge, C. H., and Vagelos, P. R. (1969). *J. Bacteriol.* **100**, 601.
Davson, H. A., and Danielli, J. F. (1935). *J. Cell. Comp. Physiol.* **5**, 495.
Deierkauf, F. A., and Booij, H. L. (1968). *Biochim. Biophys. Acta* **150**, 214.
Demain, A. L., Hendlin, D., and Newkirk, J. A. (1959). *J. Bacteriol.* **78**, 839.
Duntze, W., Mackey, V., and Manney, T. R. (1970). *Science* **168**, 1473.
Edwards, D. L., and Criddle, R. S. (1966). *Biochemistry* **5**, 588.
Engelman, D. (1970). *J. Mol. Biol.* **47**, 115.
Esfahani, M., Barnes, E. M., Jr., and Wakil, S. J. (1969). *Proc. Nat. Acad. Sci. U.S.* **64**, 1057.
Forrester, A. R., Hay, J. M., and Thompson, R. H. (1968). "Organic Chemistry of Stable Free Radicals." Academic Press, New York.
Fox, C. F. (1969). *Proc. Nat. Acad. Sci. U.S.* **63**, 850.
Fox, C. F., and Kennedy, E. P. (1965). *Proc. Nat. Acad. Sci. U.S.* **54**, 891.
Fox, C. F., Carter, J. R., and Kennedy, E. P. (1967). *Proc. Nat. Acad. Sci. U.S.* **57**, 698.
Fox, C. F., Law, J. H., Tsukagoshi, N., and Wilson, G. (1970). *Proc. Nat. Acad. Sci. U.S.* **67**, 598.
Glaser, M., Simpkins, H., Singer, S. J., Sheetz, M., and Chan, S. I. (1970). *Proc. Nat. Acad. Sci. U.S.* **65**, 721.

Glauert, A. M., and Lucy, J. A. (1968). *In* "The Membranes" (A. J. Dalton and F. Haguenau, eds.), pp. 1–30. Academic Press, New York.

Gorter, E., and Grendel, F. (1925). *J. Exp. Med.* **41**, 439.

Gotto, A. M., Kon, H., and Birnbaumer, M. E. (1970). *Proc. Nat. Acad. Sci. U.S.* **65**, 145.

Green, D. E., and Fleisher, S. (1963). *Biochem. Probl. Lipids, Proc. Int. Conf., 7th, 1962* Vol. 1, p. 325.

Green, D. E., and Perdue, J. F. (1966). *Proc. Nat. Acad. Sci. U.S.* **55**, 1295.

Griffith, A. H., and Waggoner, A. S. (1969). *Accounts Chem. Res.* **2**, 17.

Gulik-Krzywicki, Branton, T. D., and Keith, A. D. (1971). Unpublished data.

Gunstone, F. D. (1967). "Introduction to the Chemistry and Biochemistry of Fatty Acids and their Glycerides." Chapman & Hall, London.

Hamilton, C. L., and McConnell, H. M. (1968). *In* "Spin Labels in Structural Chemistry and Molecular Biology," (A. Rich and N. Davidson, eds.), p. 115. W. Freeman, San Francisco, California.

Hartwell, L. H. (1970). *Annu. Rev. Genet.* **4**, 373.

Hendler, R. W. (1968). "Protein Biosynthesis and Membrane Biochemistry." Wiley, New York.

Henry, S. (1971). Unpublished data.

Henry, S., and Keith, A. (1971). *J. Bacteriol.* **106**, 174.

Hofmann, K., O'Leary, W. M., Yoho, C. W., and Liu, T. (1959). *J. Biol. Chem.* **234**, 1672.

Holman, R. T. (1969). *Progr. Chem. Fats Other Lipids* **7**, 275.

Holmes, D. E., and Piette, L. H. (1970). *J. Pharmacol. Exp. Ther.* **173**, 78.

Horecker, B. L. (1967). *In* "Aspects of Yeast Metabolism" (A. K. Mills and H. Krebs, eds.), p. 321. Davis, Philadelphia, Pennsylvania.

Hsia, J., Schneider, H., and Smith, I. C. P. (1970). *Biochim. Biophys. Acta* **202**, 399.

Hubbell, W. L., and McConnell, H. M. (1968). *Proc. Nat. Acad. Sci. U.S.* **61**, 12.

Hubbell, W. L., and McConnell, H. M. (1969a). *Proc. Nat. Acad. Sci. U.S.* **63**, 16.

Hubbell, W. L., and McConnell, H. M. (1969b). *Proc. Nat. Acad. Sci. U.S.* **64**, 20.

Hulton, H. O., and Richardson, S. H. (1964). *Arch. Biochem. Biophys.* **105**, 288.

Hutchings, B. L., and Boggiana, E. (1947). *J. Biol. Chem.* **169**, 229.

Kawashima, N., and Wildman, S. G. (1970). *Annu. Rev. Plant Physiol.* **21**, 325.

Keana, J. F. W., Keana, S. B., and Beetham, D. (1967). *J. Amer. Chem. Soc.* **89**, 3055.

Keith, A. D., Waggoner, A. S., and Griffith, D. H. (1968). *Proc. Nat. Acad. Sci. U.S.* **61**, 819.

Keith, A. D., Resnick, M. R., and Haley, H. B. (1969). *J. Bacteriol.* **98**, 415.

Keith, A. D., Bulfield, G., and Snipes, W. (1970). *Biophys. J.* **10**, 618.

Kitay, E., and Snell, E. E. (1950). *J. Bacteriol.* **60**, 49.

Kivelson, D. (1960). *J. Chem. Phys.* **33**, 1094.

Klein, H. P. (1957). *J. Bacteriol.* **73**, 530.

Kundig, W., Ghosh, S., and Roseman, J. (1970). *Proc. Nat. Acad. Sci. U.S.* **52**, 1067.

Ladbrooke, B. D., and Chapman, D. (1969). *J. Chem. Phys. Lipids* **3**, 304.

Laico, M. T., Ruoslahti, E. I., Papermaster, D. S., and Dreyer, W. J. (1970). *Proc. Nat. Acad. Sci. U.S.* **67**, 120.

Lampen, J. O. (1966). *Symp. Soc. Gen. Microbiol.* **16**, 111.

Lands, W. E. M. (1965). *Annu. Rev. Biochem.* **34**, 331.

Lein, J., and Lein, P. S. (1949). *J. Bacteriol.* **58**, 595.

Letters, R. (1967). *In* "Aspects of Yeast Metabolism" (A. K. Mills and H. Krebs, eds.), p. 303. Davis, Philadelphia, Pennsylvania.

Libertini, L. J., Waggoner, A. S., Jost, P. C., and Griffith, O. H. (1969). *Proc. Nat. Acad. Sci. U.S.* **64**, 13.

Light, R. J., Lennarz, W. J., and Bloch, K. (1962). *J. Biol. Chem.* **237**, 1793.

Luck, D. J. L. (1963). *Proc. Nat. Acad. Sci. U.S.* **49**, 233.

Lyons, J. M., and Asmundson, C. M. (1965a). *J. Amer. Oil Chem. Soc.* **42**, 40.

Lyons, J. M., and Asmundson, C. M. (1965b). *J. Amer. Oil Chem. Soc.* **42**, 1056.

Lyons, J. M., and Raison, J. K. (1969). *Plant Physiol.* **45**, 386.

Lyons, J. M., and Raison, J. K. (1970). *Comp. Biochem. Physiol.* **24**, 1538.

McConnell, H. M., and McFarland, B. G. (1970). *Quart. Rev. Biophys.* **3**, 91.

Metcalfe, J. C. (1970). *In* "Permeability and Function of Biological Membranes," p. 222. North-Holland Publ., Amsterdam.

Meyer, F., and Bloch, K. (1963). *J. Biol. Chem.* **238**, 2654.

Meyer, F., Light, R. J., and Bloch, K. (1963). *Biochim. Biophys. Acta* **1**, 415.

Mindich, L. (1970). *J. Mol. Biol.* **49**, 415.

Overton, E. (1899). *Vierteljahresschr. Naturforsch. Ges. Zuerich* **44**, 88.

Perkins, D. D., Glassey, M., and Bloom, B. A. (1962). *Can. J. Genet. Cytol.* **4**, 187.

Poneleit, C. G., and Alexander, D. E. (1965). *Science* **147**, 1585.

Proudlock, J. W., Haslam, J. M., and Linnane, A. W. (1969). *Biochem. Biophys. Res. Commun.* **37**, 847.

Raison, J. K., Lyons, J. M., Mehlhorn, R. J., and Keith, A. D. (1971). *J. Biol. Chem.* **246**, 4036.

Resnick, M. A., and Mortimer, R. K. (1966). *J. Bacteriol.* **92**, 547.

Richardson, S. H., Hulton, H. O., and Green, D. E. (1963). *Proc. Nat. Acad. Sci. U.S.* **50**, 821.

Robertson, J. D. (1960). *Progr. Biophys. Biophys. Chem.* **10**, 343.

Rodwell, A. (1968). *Science* **160**, 1350.

Rozantsev, E. G. (1970). "Free Nitroxyl Radicals." Plenum, New York.

Schweizer, E., and Bolling, H. (1970). *Proc. Nat. Acad. Sci. U.S.* **67**, 660.

Seelig, J. (1970). *J. Amer. Chem. Soc.* **72**, 3881.

Selinger, Z., and Holman, R. T. (1965). *Biochim. Biophys. Acta* **106**, 56.

Senior, A. E., and MacLennon, D. H. (1970). *J. Biol. Chem.* **245**, 506.

Silbert, D. F., and Vagelos, P. R. (1967). *Proc. Nat. Acad. Sci. U.S.* **58**, 1579.

Silbert, D. F., Ruch, F., and Vagelos, P. R. (1968). *J. Bacteriol.* **95**, 1658.

Sjöstrand, F. S. (1963). *J. Ultrastruct. Res.* **9**, 561.

Smith, I. C. P. (1970). *In* "Biological Applications of Electron Spin Resonance Spectroscopy" (J. R. Bolton, D. Borg, and H. Schwartz, eds.), p. 000. Wiley, New York.

Snipes, W., and Keith, A. (1970). *Res. Develop.* **21**, 22.

Stein, J. M., and Fleisher, S. (1967). *Proc. Nat. Acad. Sci. U.S.* **58**, 1392.

Stein, J. M., Tourtellotte, M. E., Reinert, J. C., McElhaney, R. N., and Rader, R. L. (1969). *Proc. Nat. Acad. Sci. U.S.* **63**, 104.

Suomalainen, H., and Keränen, A. J. A. (1963). *Biochim. Biophys. Acta* **70**, 493.

Suomalainen, H., and Keränen, A. J. A. (1968). *Chem. Phys. Lipids* **2**, 296.

Suomalainen, H., and Nurminen, T. (1970). *Chem. Phys. Lipids* **4**, 247.

Tanaka, S., and Lin, E. C. C. (1967). *Proc. Nat. Acad. Sci. U.S.* **57**, 913.

Tourtellotte, M., Branton, D., and Keith, A. (1970). *Proc. Nat. Acad. Sci. U.S.* **66**, 909.

Urry, D. W., and Ji, T. H. (1968). *Arch. Biochem. Biophys.* **128**, 802.

Van Deenen, L. L. M. (1966). *Progr. Chem. Fats Other Lipids* **8**, 1.

Vanderkooi, G., and Green, D. E. (1970). *Proc. Nat. Acad. Sci. U.S.* **66**, 615.

Waggoner, A. S., Kingzett, T. J., Rottschaefer, S., and Griffith, O. H. (1969). *Chem. Phys. Lipids* **3**, 245–253.

Williams, J. C., Mehlhorn, R. J., and Keith, A. D. (1972). *Chem. Phys. Lipids* **7,** 260.

Williams, W. L., Broquist, H. P., and Snell, E. E. (1947). *J. Biol. Chem.* **170,** 619.

Wilson, G., and Fox, C. F. (1971). *J. Mol. Biol.* **55,** 49.

Wilson, G., Rose, S. P., and Fox, C. F. (1970). *Biochem. Biophys. Res. Commun.* **38,** 617.

Wisnieski, B. J., and Kiyomoto, R. (1972). *J. Bacteriol.* **109,** 186.

Wisnieski, B. J., Keith, A. D., and Resnick, M. R. (1970). *J. Bacteriol.* **101,** 160.

Woodward, D. O., and Munkres, K. O. (1966). *Proc. Nat. Acad. Sci. U.S.* **55,** 872.

Author Index

Numbers in italics refer to the pages on which the complete references are listed.

A

Aaronson, S., 6, 12, *36*, 63, 95, 109, *136*, *137*, *138*, *139*, 226, *229*, *231*
Abbott, B., 69, *136*
Abdulla, Y. H., 184, *190*
Abelson, P. H., 213, 227, *231*
Abraham, A., 206, *229*, 247, *255*
Ach, L. L., 72, *143*
Ackman, R. G., 82, 86, 90, 95, 97, 102, 135, *136*
Adam, H. K., *36*
Adams, G. A., 227, *230*
Adelberg, E. A., 105, *143*
Aees-Jorgensen, E., 74, 134, *136*
Aftergood, L., 74, 134, *136*
Agranoff, B. W., 167, *190*, *192*
Ailhaud, G. P., 46, *136*, 239, *256*
Akagi, S., 7, 9, *40*
Akamatsu, Y., 53, *136*
Akhtar, M., 32, *37*
Alberts, A. W., 46, 48, *138*, *140*
Alcaide, A., 7, 8, *37*
Alexander, D. E., 266, *320*
Alexander, G. J., 32, *40*
Alfin-Slater, R. B., 74, 134, *136*
Allen, C. F., 131, *136*, 202, 213, *229*, 238, 244, 250, *255*
Allen, M. B., 124, *136*, *137*
Alt, G. H., 18, *37*
Anand, S. R., 112, *143*

Anding, C., 10, 11, *37*
Andreasen, A. A., 54, *136*, 264, 265, *318*
Angus, W. W., 171, 174, *190*
Ansell, G. B., 147, 149, *190*
Appleby, R. S., 71, 76, 87, 95, 131, 133, *141*
Appelman, D., 212, *229*
Arigoni, D., 19, *38*
Arnon, D. I., 246, *255*
Arsenault, G. P., 15, *37*
Arstila, A. V., 147, *195*
Asmundson, C. M., 285, 286, *320*
Audette, R. C. S., 112, *136*
Avivi, L., 8, 12, *37*, *38*

B

Baccarini-Melandri, A., 241, 243, *256*
Bachhawat, B. K., 206, *229*, 247, *255*
Baer, E., 199, *229*
Bailey, J. M., 76, *136*
Bailey, R. W., 202, 204, 209, *232*
Baisted, D. J., 23, *37*
Baker, H., 6, 12, *36*, 226, *229*
Bakerman, H. A., 48, *140*
Balamuth, N., 42, 115, *139*
Ballauf, A., *40*
Band, R. N., 188, *191*
Bandurski, R. S., 206, *232*
Bangham, A. D., 182, *190*
Baraud, J., 52, 117, *137*, *141*

323

SUBJECT INDEX

A

Absidia glauca, sterols, 15
Acanthamoeba castellanii, phospholipids, 154, 163, 180
Acanthamoeba palestinensis, phospholipids, 188
Acanthamoeba sp., fatty acids, 66, 71, 76, 119
Acanthopeltis japonica, sterols, 7
Acetylenic acids, as inhibitors of unsaturated fatty acid biosynthesis, 62
Acetylenic acids, properties, 288
Achlya bisexualis, sterols, 15, 36
Achlya caroliniana, sterols, 15
Acyl carrier protein
  function, 46
  occurrence, 46
  structure, 48
AEP phosphonolipids, *see also* Phosphonolipids, occurrence in cilia, 153
Aerobacter cloacae, sterols, 13
Agaricus campestris
  fatty acids, 111
  sterols, 19
Agmenellum quadruplicatum, fatty acids, 84
Ahnfeltia stellata, sterols, 7
Alaria crassifolia, sterols, 9
Amantia muscaria, fatty acids, 111
2-Aminoethyl-phosphonic acid, 153, 174
2-Aminoethyl phosphonolipids, *see also* Phospholipids, occurrence of, 152, 154–155
2-Aminoethyl phosphonolipids, *see also*

Phospholipids, structure of, 151
Amoeba proteus, phospholipids, 188
Amphidinium carterii, fatty acids, 89
Anabaena cylindrica
  chloroplast lipid, 235
  fatty acids, 84, 129
  phospholipids, 235
Anabaena flos-aquae
  chloroplast lipid, 235
  fatty acids, 84, 129
  phospholipids, 235
Anabaena variabilis, fatty acids, 65, 84, 129
Anacystis cyanea, fatty acids, 84
Anacystis marinus, fatty acids, 84
Anacystis montanus, fatty acids, 84
Anacystis nidulans
  fatty acids, 65, 79, 84, 129
  glycolipids, 244, 246
  sterols, 12
Ankistrodesmus braunii, fatty acids, 66, 71, 101
Antheridiol, function in reproduction of fungi, 36
Antithamniou plumula, sulfolipids, 210
Apistonema carteri, sterols, 12
Aplanopsis terrestris, sterols, 15
Apodachlya brachynema, sterols, 15
Apodachlya minima, sterols, 15
Apodachlyella completa, sterols, 15
Arachidonic acid, *see* Fatty acids, polyunsaturated
Arthrobacter, fatty acids, 46–48
Ascophyllum nodosum, sterols, 9
Ascosterol, occurrence of, fungi, 17